Sensing Techniques for Food Safety and Quality Control

Food Chemistry, Function and Analysis

Series Editors:
Gary Williamson, *University of Leeds, UK*
Alejandro G. Marangoni, *University of Guelph, Canada*
Juliet A. Gerrard, *University of Auckland, New Zealand*

Titles in the Series:
1: Food Biosensors
2: Sensing Techniques for Food Safety and Quality Control

How to obtain future titles on publication:
A standing order plan is available for this series. A standing order will bring delivery of each new volume immediately on publication.

For further information please contact:
Book Sales Department, Royal Society of Chemistry, Thomas Graham House, Science Park, Milton Road, Cambridge, CB4 0WF, UK
Telephone: +44 (0)1223 420066, Fax: +44 (0)1223 420247
Email: booksales@rsc.org

Visit our website at www.rsc.org/books

Sensing Techniques for Food Safety and Quality Control

Edited by

Xiaonan Lu
University of British Columbia, Vancouver, Canada
Email: xiaonan.lu@ubc.ca

ROYAL SOCIETY OF CHEMISTRY

THE QUEEN'S AWARDS FOR ENTERPRISE: INTERNATIONAL TRADE 2013

Food Chemistry, Function and Analysis No. 2

Print ISBN: 978-1-78262-664-0
PDF eISBN: 978-1-78801-052-8
EPUB eISBN: 978-1-78801-197-6
ISSN: 2398-0656

A catalogue record for this book is available from the British Library

The Royal Society of Chemistry is a charity, registered in England and Wales, Number 207890, and a company incorporated in England by Royal Charter (Registered No. RC000524), registered office: Burlington House, Piccadilly, London W1J 0BA, UK, Telephone: +44 (0) 207 4378 6556.

Visit our website at www.rsc.org/books

Printed in the United Kingdom by CPI Group (UK) Ltd, Croydon, CR0 4YY, UK

Preface

Ensuring the safety and quality of food products is critical to consumers and it is also one of the most important research topics in food science and technology of the past several decades. There are numerous factors that can affect food quality and safety. These factors are sometimes complicated, inter-correlated and difficult to track and control. For example, the pollutants in irrigation water, soil and the environment can be readily migrated and/or adsorbed into plant-based food products, such as fresh produce. As another example, the quality of harvested food products will be either positively or negatively impacted by post-harvest storage and delivery, such as the modified atmospheric packaging technique and cold chain delivery. That is also why food scientists have to pay attention to every single step "from the farm to the fork".

Sometimes, food quality and food safety are two factors that need to be balanced. For example, a thermal processing technique needs to be applied for inactivation of the potential pathogenic bacteria in food products, but this will negatively impact the quality of the foods, such as the texture, colour and flavour. In this case, part of the quality of the food products has to be sacrificed to ensure the safety of the foods. As another example, pasteurized milk is a popular dairy product that can be produced by high temperature short time (HTST) processing. However, HTST is not able to fully inactivate spoilage bacteria, which is why pasteurized milk can only be stored in a refrigerator for about three weeks. Theoretically, sterilized milk has zero microbes due to a complete inactivation of the entire pathogenic and spoilage microorganisms. Although the shelf life of sterilized milk products is significantly extended, the taste and flavour has been significantly jeopardized. That is why consumers still prefer pasteurized milk products instead of sterilized milk products. In short, these two representative examples tell us that food safety and food quality are the two

Food Chemistry, Function and Analysis No. 2
Sensing Techniques for Food Safety and Quality Control
Edited by Xiaonan Lu

Published by the Royal Society of Chemistry, www.rsc.org

parameters we need to consider at all times. Food safety is the first priority. Once the safety of the food products is guaranteed, we should maximize the quality of the food products.

There are numerous ways to control the safety and quality of food products. Food law and regulation is the guideline for all food manufacturers to follow. For example, Hazard Analysis and Critical Control Points (HACCP) has been validated to be a very effective systematic preventive approach to food safety from biological, chemical and physical hazards in the production process. Unfortunately, HACCP may not be fully applied and followed for every single food processing event. In this case, detection will be a very important strategy to ensure food safety and quality by the following two approaches: (1) detection of food quality and safety during food processing and (2) detection of food quality and safety after food processing. By applying the first approach, the processing efficiency can be monitored. This will maximally ensure the processing techniques that can control the safety and quality of food products. The second approach is regarded as the "last defence" to ensure that only safe and high-quality food products will be delivered to the consumers.

During the past several decades, a variety of methods and techniques have been developed for the determination and control of food safety and quality. Instrumental analytical methods play a key role. Nowadays, chromatography-based instruments are widely available in every corner of the world. High performance liquid chromatography and liquid chromatography-mass spectrometry have been applied to determine various types of food chemical contaminations and food compositions. On the other side, conventional plating methods and various molecular techniques have been applied for accurate determination of pathogenic and spoilage bacteria in agri-food products. For all of the aforementioned determinations, there are three major limitations. First, food is a very complicated matrix. Therefore, an effective separation technique is extremely crucial to remove the interference of other food components that may impact on the determination by an instrumental detector, such as mass spectrometry. Second, the level of the target analyte (*e.g.*, pesticide residue, *Salmonella* and norovirus) in the food products may be extremely low. Microbial enrichment is inevitable in most cases once the concentration of the targeted microorganism could not be directly detected by the current techniques. Third, sample preparation is time consuming and labour intensive. Essentially, this is due to the first two limitations (*i.e.*, food sample matrix and low level of target analyte). Besides those aforementioned technical limitations, instrumental analysis is expensive and requires lab space and experienced personnel. Therefore, instrumental analysis could not achieve high-throughput screening of a large number of food product samples in a relatively short time period.

The concept of "sensing technique" or "sensor" is an object with the purpose to detect events and send the information to a PC that informs the output device for generation of the corresponding output. A sensing technique is a very general concept that includes chemical sensing techniques, biosensing techniques, electrochemical sensing techniques and others. Compared to instrumental analysis for food safety and quality control, sensing techniques have unique advantages. They are rapid, sensitive, portable and can provide high-throughput screening. In addition, some types of sensing techniques can partially reduce or even avoid sample preparation procedures as well as the enrichment of the target analyte(s). Therefore, we have drafted this book entitled "*Sensing Techniques for Food Safety and Quality Control*". The most advanced sensing techniques and their applications in food safety and quality will be systematically introduced. Our major aim for this book is to let the readers become more familiar with the technical advancement of sensors in agri-food products.

In Chapter 1, the Raman spectroscopic sensing technique will be introduced. Raman spectroscopy has been recently applied in food analysis over the past two decades. A derivative of the Raman spectroscopic technique, called surface enhanced Raman spectroscopy (SERS), integrates nanotechnology and laser technology to create a new generation of sensing platform that can detect food contamination and quality parameters in an ultra-sensitive and high-throughput manner. The author, Dr Zhong Zhang at the University of Nebraska-Lincoln, is one of the leading experts in the United States in developing and applying Raman and SERS sensing techniques to study food chemical and microbiological hazards as well as food quality.

In Chapter 2, the quantum dot sensing technique will be introduced. Quantum dots are extremely small semiconductor particles with unique electronic and optical properties. Quantum dots are widely applied in solar cells, quantum computing, transistors and recently sensing techniques. As the authors of this chapter, Dr Russ Algar's group at the University of British Columbia is one of the leading teams in Canada applying quantum dots for biological and sensing applications.

Microfluidic "lab-on-a-chip" is a device that can integrate several laboratory functions onto a single device or chip of a few square millimetres to achieve a fully automated and high-throughput screening. The application of the "lab-on-a-chip" concept in food analysis has boomed during the past five years. The polymer-based microfluidic "lab-on-a-chip" device was created first, followed by the development of the paper-based microfluidic device, also known as "microfluidic v2.0". Therefore, Chapter 3 and Chapter 4 are individually drafted to introduce the two generations of microfluidic techniques and their applications in food safety and quality control. Chapter 3 was written by Dr Xian Huang's group at Tianjin University. Dr Huang is one of the leading experts in China in developing microfluidic techniques for the study of biological systems. Chapter 4 was drafted by Dr Jie Xu's group at the University of Illinois, Chicago. Dr Xu has a great deal of experience in developing both polymer-based and paper-based "lab-on-a-chip" devices to study food safety and quality.

In Chapter 5, the colorimetric sensing technique will be introduced. This technique will facilitate a visual detection of food chemical and microbiological hazards by the naked eye. Therefore, a separate detector is not required, which will be suitable for in-field monitoring of food safety and quality. Dr Xiaonan Lu's group at the University of British Columbia drafted this chapter. His lab has a great deal of experience in applying colorimetric sensors to detect food chemical contaminants.

The next three chapters are related to developing very effective recognition elements that can separate and enrich the target analytes from the complicated food matrices. These recognition elements include antibody (Chapter 6), molecularly imprinted polymers (Chapter 7) and aptamer (Chapter 8) and all can be readily integrated into a sensing platform. In Chapter 9, the production of antibody and its application in immunoassay is systematically introduced by Dr Shuo Wang's group at Tianjin University of Science and Technology. Dr Wang is the authority in developing immunoassay in Asia. Currently, enzyme-linked immunosorbent assay (ELISA) is still the most successful knowledge translation for commercialization in food analysis. In Chapter 7, as the "artificial antibody", the technique of molecularly imprinted polymers for sensing food safety and quality is introduced by Dr Yiwei Tang's group at Bohai University. Molecularly imprinted polymers are one of the most exciting techniques developed in analytical chemistry for food analysis during the past several decades. This type of "artificial antibody" has the potential to replace some natural antibodies in the future due to its easy synthesis and extreme stability in the environment. In Chapter 8, the aptamer-based sensing technique is introduced. Aptamers are oligonucleotides or peptides that can bind to a specific target molecule. Aptamers have been gradually applied to replace antibody that can recognize and capture target analyte in a complicated matrix, such as foods. As the authors of this chapter, Dr Maria DeRosa's group at Carleton University is one of the leading groups in Canada studying aptamer-based sensing technology and its bio-application.

In Chapter 9 and Chapter 10, two novel carbon materials (*i.e.*, carbon nanotubes and graphene) are introduced. These two materials have many unusual properties. For example, graphene is about 200 times stronger than the strongest steel. In addition, the bonding between the atoms in carbon nanotubes is very strong and the tubes have extreme aspect ratios. These unusual properties have been smartly applied as the read-out components in a sensor. The application of carbon nanotubes as sensing materials is introduced in Chapter 9 by Dr Wei Xue's group at Rowan University while the application of graphene as a sensing material is introduced in Chapter 10 by Dr Xian Zhang at the University at Buffalo. Both Dr Xue and Dr Zhang have extensive research experience in developing and applying advanced carbon nanomaterials.

Finally, the smartphone-based sensing technique is introduced in Chapter 11 by Dr Jane Ru Choi at the University of Malaya. Rapid development of smartphones with various embedded sensors has enabled applications in biomedical diagnosis. Recently, their application in food analysis has started to emerge. Dr Choi is an experienced researcher of applying smartphones to detect target analyte in various matrices, including clinical samples and agri-food samples.

Taken together, this book will comprehensively introduce the advancement of sensing techniques for food safety and quality control. By applying various sensors for food analysis, a more rapid, high-throughput, convenient and reliable determination of agri-food products will be achieved.

I am grateful for the support from all members in Lu Food Safety Engineering Laboratory at UBC as well as the company of beloved Vivian and TUTU.

Xiaonan Lu
University of British Columbia, Vancouver, Canada

Contents

Food Chemistry, Function and Analysis No. 2
Sensing Techniques for Food Safety and Quality Control
Edited by Xiaonan Lu

Published by the Royal Society of Chemistry, www.rsc.org

CHAPTER 1

Raman Spectroscopic Sensing in Food Safety and Quality Analysis

ZHONG ZHANG

Department of Food Science and Technology, University of Nebraska-Lincoln, 1901 21st Street, Lincoln, NE 68588, USA
Email: zzhang64@unl.edu

1.1 Raman spectroscopy

1.1.1 Basics of Raman Spectroscopy

Photons, which make up light, may interact with a molecule when the incident light is directed onto the molecule. The photons could be adsorbed by the molecule, if the photon energy matches the energy gap between the ground state and the excited state of the molecule. The molecule is promoted to a higher energy excited state after the adsorption of photon energy.[1,2]

On the other hand, if the photon energy does not match the differences between two energy levels of a molecule, the photon can also interact with the molecule and be scattered from it.[3] For example, when a single frequency of light is used to irradiate samples, the photons interact with the molecules and polarize the electron clouds near the nuclei to form a virtual state for this molecule. The lifetime of this virtual state is very short, leading to the re-radiation of the photons. When the energy of a re-radiated photon is equal to the energy of an incident photon, the scattering is called

Food Chemistry, Function and Analysis No. 2
Sensing Techniques for Food Safety and Quality Control
Edited by Xiaonan Lu

Published by the Royal Society of Chemistry, www.rsc.org

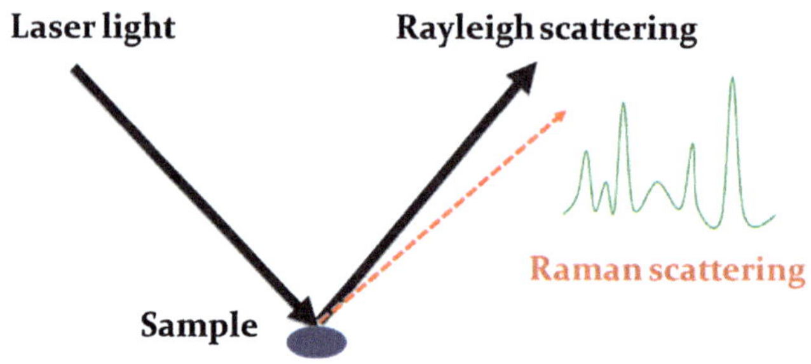

Figure 1.1 Generation of Raman scattering and a typical Raman spectrum.

"Rayleigh scattering", which is the dominant scattering and does not involve any energy transfer. When the molecule in the ground state is promoted to the virtual state and then returns to the excited state, the scattering is called "Stokes scattering". This process involves energy transfer from photons to the molecule.[1] Likewise, when a molecule in the excited state is promoted to the virtual state and then returns to the ground state, the scattering is called "anti-Stokes scattering". This scattering process transfers the energy from the molecule to photons. Raman scattering is the inelastic scattering of a photon, including both Stokes and anti-Stokes scattering. However, Raman spectroscopy usually records Stokes scattering because anti-Stokes scattering is very weak compared to Stokes scattering. For example, Coherent Anti-Stokes Raman Spectroscopy (CARS) collects anti-Stokes Raman scattering. Due to the weak scattering intensity, it has not been widely used for routine analysis (Figure 1.1).

By recording the Raman scattering and calculating the energy loss of incident photons, Raman spectroscopy provides detailed vibrational information for the molecules in a sample.[4] The Raman spectra are like "fingerprints" to the molecules, which are very useful for the identification of molecules. However, as stated above, the dominant scattering of the photon–molecule interaction is Rayleigh scattering. Raman scattering is a very weak effect. Therefore, many methods have been used to enhance the signals of Raman spectroscopy, resulting in a new term called "surface enhanced Raman spectroscopy" (SERS).

1.1.2 The Raman Spectrometer

The Raman spectrometer consists of three major components: an excitation source, a sampling apparatus and a detector. Modern Raman spectrometers use a laser as the excitation source, a spectrometer for the detector and either a microscope or a fibre-optic probe for the sampling apparatus. The laser is very important in Raman spectrometry because it directly influences the Raman response from the targeted molecules.

A laser is a device that emits light through a process of light amplification due to the stimulated emission of electromagnetic radiation. The word "laser" is an acronym for Light Amplification by Stimulated Emission of Radiation. A laser device consists of a gain medium, an energy supply and an optical cavity. The gain medium is used to amplify light by stimulated emission. The energy is usually supplied by electrical current. The optical cavity is used to amplify the light. It is made by a pair of mirrors. The light could bounce between the two mirrors and pass the gain medium, where the light is amplified each time. One of the mirrors has a small transparent area, which allows the laser to spread out and form a narrow beam. This light beam could be used for Raman spectroscopy.

Laser sources for Raman spectroscopy are divided into three categories, including ultra-violet laser, visible laser and near-infrared laser. The ultra-violet laser has a wavelength of 244 nm, 257 nm, 325 nm and 364 nm. The visible laser has a wavelength of 457 nm, 473 nm, 488 nm, 514 nm, 532 nm, 633 nm and 660 nm. The near-infrared laser has a wavelength of 785 nm, 830 nm, 980 nm and 1064 nm. For Raman spectroscopy, the most widely used lasers are the visible lasers with a wavelength of 532 and 633 nm, and the near-infrared laser with a wavelength of 785 nm.

A high-performance laser is required to obtain high-quality Raman spectra. It is also important to select a suitable excitation wavelength in order to get the best Raman response. The Raman intensity is positively proportional to λ^{-4}, where λ represents the laser wavelength. Thus, the infrared laser (785 nm) will result in a lower Raman intensity compared to the visible laser (514 nm and 633 nm). Second, the spatial resolutions are different between different lasers. The laser spot can be calculated according to the equation that diameter $= 1.22\ \lambda/\mathrm{NA}$. NA is the numerical aperture of the microscope. For example, when a 0.75/50 microscope lens is used, the laser spot has a diameter of 1.28 μm and 0.73 μm for a 785-nm laser and a 532-nm laser, respectively. Thus, the 514-nm laser has a smaller laser spot. Third, the laser performance also depends upon the target chemicals. The infrared laser (785 nm) is useful in suppressing the fluorescence for organic chemicals. The 514-nm and 633-nm lasers are useful for the resonance Raman experiments because most of the excitation wavelength of fluorescence is close to this region. Ultra-violet lasers are useful for the analysis of macromolecules, such as proteins, because the ultra-violet laser can be used to achieve the resonance for those biomolecules. Finally, when a laser is used for surface enhanced Raman spectroscopy (SERS), the laser performance depends upon the SERS substrate as well. The maximum SERS enhancement is achieved when the laser wavelength is close to the localized surface plasmon resonance (LSPR) of the substrate.

1.1.3 Surface Enhanced Raman Spectroscopy (SERS)

Surface enhanced Raman spectroscopy is a technique that can greatly enhance the Raman signals from molecules or bacteria adsorbed onto the surface of a substrate (Figure 1.2).[5,6] The enhancement factor can be as high as 10^{10}. Traditional analytical methods, such as High Performance Liquid

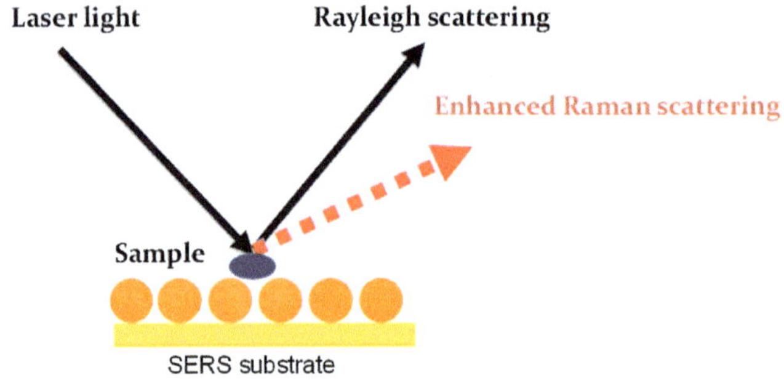

Figure 1.2 Scheme of surface enhanced Raman scattering.

Chromatography (HPLC), Gas Chromatography (GC) and Enzyme-Linked Immunosorbent Assay (ELISA), are time consuming and labour intensive, and often require complicated sample pre-treatment.[7] The running and maintenance costs for HPLC and GC are also very high. In contrast, SERS does not require complex sample pre-treatment and expensive solvent/column for operation. Furthermore, SERS only requires a few seconds for a single measurement, which is much faster than HPLC and GC. Due to its high sensitivity, unique spectroscopic fingerprinting features and non-destructive data acquisition, the SERS technique has extensive potential to be used for the detection and identification of chemical and microbiological contaminants in foods as well as food quality analysis.[8]

1.1.4 Statistical Analysis for SERS Methods

Unlike the HPLC profile, the SERS spectrum is highly dependent upon the SERS substrates and the equipment used. The SERS spectrum often has relatively big variations compared to other analytical methods. It is highly necessary to use statistical methods to analyze the SERS spectra and interpret them accurately. The most commonly used methods are Principal Component Analysis (PCA) and Partial Least Squares (PLS).

PCA is a statistical procedure that converts a set of correlated variables into a set of linearly uncorrelated variables called "principal components".[9] The number of principal components is smaller than the number of variables. By using an appropriate orthogonal transformation, a series of principal components can be identified. The first principal component must have the largest possible variance and the second principal component has the second largest possible variance. This analogy is applied to other principal components. In other words, PCA is a transformation method. This orthogonal transformation could transform the data to a new coordinate system. In this new coordinate system, the largest variance is projected onto the first coordinate and the second largest variance is projected onto the second coordinate. The PCA technique is used to find those coordinate systems and evaluate the contribution of each component, which means that it can be used to find the major contributing peaks for a specific molecule from the complex Raman spectra.

For example, in order to comprehensively study an object, we have to include a variety of factors that have influence on the object. However, not all of the factors contribute equally to the behaviour of this object. Some factors contribute more than other factors. The factors may also be related to each other. We hope we can find the internal structure of the data that can best explain the variance of these data. Let's assume there are a few factors named X_1, X_2, $X_3 \ldots X_p$. We hope we can find one equation as follows.

$$C_j = a_{j1}x_1 + a_{j2}x_2 + a_{j3}x_3 + \cdots + a_{jp}x_p \qquad j = 1,2,3,\ldots p$$

The equation will lead to the following equations ($n \le p$)

$$C_1 = a_{11}x_1 + a_{12}x_2 + a_{13}x_3 + \cdots + a_{1p}x_p \qquad j = 1,2,3,\ldots p$$
$$C_2 = a_{21}x_1 + a_{22}x_2 + a_{23}x_3 + \cdots + a_{2p}x_p \qquad j = 1,2,3,\ldots p$$
$$C_3 = a_{31}x_1 + a_{32}x_2 + a_{33}x_3 + \cdots + a_{3p}x_p \qquad j = 1,2,3,\ldots p$$
$$C_n = a_{n1}x_1 + a_{n2}x_2 + a_{n3}x_3 + \cdots + a_{np}x_p \qquad j = 1,2,3,\ldots p$$

We hope that the variance of C_1 is the largest, the variance of C_2 is the second largest, and so on. The eigenvectors $(a_{j1}, a_{j2}, a_{j3}, \ldots a_{jp})$ are orthogonal. In addition, any two of the variances $(C_1, C_1, \ldots, C_n)$ are not correlated. The sum of the variance of C equals the variance of X as in the following equation.

$$\mathrm{Vari}(C_1) + \mathrm{Vari}(C_2) + \cdots \mathrm{Vari}(C_p) = \mathrm{Vari}(x_1) + \mathrm{Vari}(x_2) + \cdots + \mathrm{Vari}(x_p)$$

Then, what we have to do is to identify the matrix of a_{ij}. That is the correlated eigenvector matrix. We can solve the matrix $A(a_{ij})$ using the singular value decomposition method or other methods. After solving the matrix, we could calculate the variance of each component and the contribution of each component to the total variances. In this way, we can obtain the most important components and understand the data structure more effectively. The number of principal components is usually much smaller than the number of factors. PCA is a very useful technique in analyzing Raman data from different chemicals or microorganisms. For example, Raman spectroscopy was used to distinguish three types of pesticides in apples.[12] These three pesticides have extensive similar peaks that are derived from the matrix. In order to differentiate the pesticides, PCA could be used to eliminate the matrix effect and find the most important peaks that contribute to the separation of these pesticides.

PLS is a technique that combines the features of PCA and multiple linear regressions. It is particularly useful to predict a set of dependent variables from a large set of independent variables. For example, it can be used in SERS when we test a series of concentrations of pesticides (dependent variables). Normally, a pesticide has a lot of characteristic peaks (independent variables), which are correlated to each other (proportional). By using PLS, we can build a model to effectively predict the pesticide concentrations by correlating with the intensities of the most important characteristic peaks.

In general, PLS is used to find the fundamental relations between the **Y** matrix (Dependent variables) and the **X** matrix (Predictor variables) when the number of predictor variables is higher than the number of dependent variables.

To be specific, let's assume we have **I** observations for **J** predictors. Therefore, we will be able to have **I** observations of dependent variables **K** ($K<J$). The observations will be stored in the matrix of $\mathbf{I}\times\mathbf{K}$ (**Y**) and the predictors will be stored in the matrix of $\mathbf{I}\times\mathbf{J}$ (**X**), such as in the following two matrices.

$$\begin{vmatrix} Y_{11} & Y_{21} & Y_{31} & Y_{K1} \\ Y_{12} & Y_{22} & Y_{32} & Y_{K2} \\ Y_{13} & Y_{23} & Y_{33} & Y_{K3} \\ \cdots & \cdots & \cdots & \cdots \\ Y_{1I} & Y_{2I} & Y_{3I} & Y_{KI} \end{vmatrix} \begin{vmatrix} X_{11} & X_{21} & X_{31} & X_{J1} \\ X_{12} & X_{22} & X_{32} & X_{J2} \\ X_{13} & X_{23} & X_{33} & X_{J3} \\ \cdots & \cdots & \cdots & \cdots \\ X_{1I} & X_{2I} & X_{3I} & X_{JI} \end{vmatrix}$$

The purpose of PLS is to predict **Y** by **X** and to find their common structures ($1<\mathbf{K}<\mathbf{J}$). PLS can find the components from **X** that are also relevant to **Y**. PLS searches a set of components (latent vectors) that can perform a simultaneous decomposition of **X** and **Y** under the condition that these components explain the maximum covariance between **X** and **Y**. Then, we perform the regression step where the decomposition of **X** is used to predict **Y**.

In the first step, the independent variables are decomposed as $\mathbf{X}=\mathbf{TP}^{\mathrm{T}}+\mathbf{E}$ with $\mathbf{TT}^{\mathrm{T}}=\mathbf{I}$, where **T** is the score matrix, **P** is the loading matrix, **I** is the identity matrix and **E** is the error matrix. In a similar way, **Y** is estimated as $\hat{\mathbf{Y}}=\mathbf{TBC}^{\mathrm{T}}+\mathbf{F}$, where **B** is the diagonal matrix with regression weights and **F** is the error matrix. The column of the **T** matrix is the latent vector. Then, the following step is regression. We have to find two sets of weights **w** and **c** in order to create a linear combination of the columns of **X** and **Y** for the purpose of maximizing the covariance of **X** and **Y**. That is to say that we have to find the first pair of vectors as the following equation shows:

$$\mathbf{t}=\mathbf{Xw} \text{ and } \mathbf{u}=\mathbf{Yc}\ (\mathbf{w}^{\mathrm{T}}\mathbf{w}=1,\ \mathbf{t}^{\mathrm{T}}\mathbf{t}=1,\ \mathbf{t}^{\mathrm{T}}\mathbf{u}=\text{maximum, covariance of } \mathbf{tu} \text{ is maximum})$$

After the first latent vector is identified, it is subtracted from **X** and **Y**. All the latent vectors are identified until X becomes an empty matrix.

The dependent variables are then predicted by multivariate regression using the equation of $\hat{\mathbf{Y}}=\mathbf{TBC}^{\mathrm{T}}=\mathbf{XB}_{\mathrm{PLS}}$ where $\mathbf{B}_{\mathrm{PLS}}=\mathbf{P}^{\mathrm{T}}+\mathbf{BC}$.[10] Finally, through cross-validation, we can identify the latent variables for the best generalization to predict new dependent variables.

1.2 Sensing of Food Contaminations by SERS

The general detection scheme of SERS is illustrated in Figure 1.3. Food contaminants will be extracted by water/organic solvent and directly placed

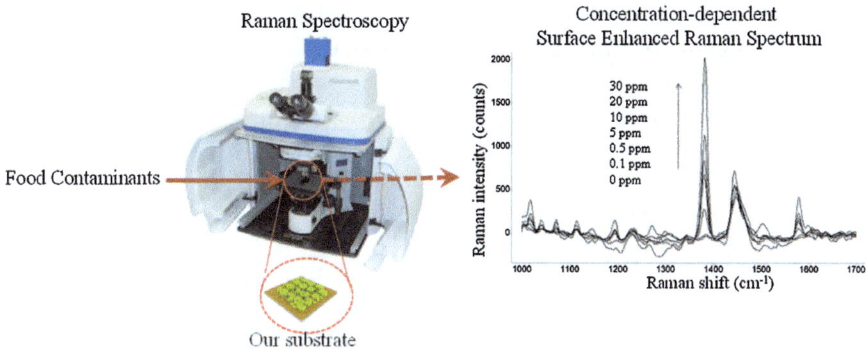

Figure 1.3 Schematic illustration of the detection of food contaminants by SERS. Raman spectrum reproduced with permission from Journal of Food Science; Z. Zhang, Q. Yu, H. Li, A. Mustapha and M. Lin; 2015, 80, N450–N458. Copyright © 2015 WILEY-VCH Verlag GmbH & Co. KGaA, Weinheim.

onto the commercial or custom-made SERS substrate. The Raman laser will focus on the substrate, which can significantly enhance the Raman signals. A variety of food contaminants, including pesticides, antibiotic residues and bacterial toxins, have been measured using a Raman spectrometer coupled with highly reproducible SERS substrates. SERS can provide a fast and cost-effective method for the identification of food contaminants while providing similar accuracy compared to HPLC and GC methods. The rapid SERS methods with the emergance of high-performance substrate will make *in situ* food safety monitoring possible in the food processing chain.

1.2.1 SERS Detection of Chemical Contaminations in Foods

Chemical contaminants, such as illegal chemical additives, growth hormones and natural toxins, represent a major group of contaminants in foods. There are many routes for chemical contamination to enter the food supply chain. Some chemical contaminants are intentionally added into foods to boost the nutritional values or alter the organoleptic properties of foods. Natural toxins could occur because of inappropriate food storage conditions, for example, food mycotoxins. Growth hormones are added into the feeding material of animals to increase the growth rate. Illegal pesticides are used to control pests in food crops and increase the crop yield. For example, in 2006, Greenpeace tested some vegetables sold in two grocery stores in Hong Kong and discovered that 70% of tested samples contained pesticide residues. Thirty percent of tested vegetable samples exceeded safety levels and several samples were tested to be positive for illegal pesticides. The disappearance of illegal pesticides could also impose potential risks for drinking water. Chemical contamination has been posing serious risks to human health in recent years. Constant testing is one of the most effective ways to prevent contaminated foods entering consumer markets. The most employed testing methods are chromatographic-based detection methods, such as HPLC, HPLC-MS and GC-MS. However, with the fast development of Raman equipment and SERS substrates, the SERS technique presents a new opportunity for rapid food analysis.

1.2.1.1 Label-free Detection of Chemical Contaminations by SERS

Chemical contaminants, such as pesticides and melamine, are often small molecules. The small molecules could be extracted by solvents and directly placed onto SERS substrates for analysis. Depending upon the contaminants and the performance of SERS substrates, the detection limit of those chemical contaminants often falls into the range of parts per billion (ppb) or parts per million (ppm).

Since the outbreak of illness from melamine in 2007, a number of papers have been published on the detection of melamine by SERS methods. Melamine, a notorious chemical contaminant, was added to dairy products to boost their protein content. The detection of melamine by SERS has been proposed by many researchers. In early studies, commercial substrates were used to study the feasibility of the SERS technique. Q-SERS, a commercial SERS substrate, was used to detect melamine by Liu and coauthors.[11] Q-SERS is made of aggregated gold nanoparticles on a silicon wafer. The gaps of the gold nanoparticles provide "hot spots" for SERS detection. Melamine is first extracted by an organic solvent/water solution from dairy products. It was reported that the detection limit for SERS using Q-SERS substrate was 2 ppm of melamine in liquid milk.

Another commercial substrate, Klarite, is a pattern of inverted gold pyramids developed by the Renishaw Company in the UK using a lithography method. These substrates have also been extensively used in the SERS measurement of melamine in dairy products. It was reported that SERS with Klarite substrates could rapidly detect 0.1% (w/w) melamine in wheat gluten, 0.05% in chicken feed, 0.05% in cakes and 0.07% in noodles, respectively.[12–14] PLS was used in SERS spectral analysis. Other types of substrates, such as silver dendrites, were also used in melamine analysis.[15] The authors used molecularly imprinted polymers rather than an organic solvent to extract the melamine and deposited the melamine onto silver nano-dendrites for SERS analysis. It was suggested that the limit of detection and limit of quantification were 0.012 mmol L^{-1} and 0.039 mmol L^{-1} of melamine in whole milk, respectively. Fractal-like gold nanostructures were also proposed to detect melamine in agri-food products.[16] It was reported that the enhancement factor of the fractal-like gold nanostructures could reach an impressive enhancement level of $\sim 4 \times 10^7$. They could be used to detect melamine and other chemical contaminants, such as dye, and the lowest detectable concentration for the dye molecules was at the ~0.2 ppb level.

The SERS technique was also used to detect pesticides and other chemical contaminations. Various high-performance substrates have been developed for the analysis of pesticides. Dithiol-functionalized metal nanoparticles were used to create plasmonic "hot spots" and enhance Raman spectroscopic response to the pesticides. It was reported that the dithiol aids in the nanoparticle linkage and creates inter-particle junctions where sensitive "hot spots" required for SERS enhancement are present, creating a specific environment in the nanogaps between silver and gold nanoparticles. They showed a high sensitivity of SERS for the detection of organochlorine pesticides with a limit of detection reaching 10^{-8} M. Zhang and others also present a new way to assemble vertically aligned gold nanorod arrays for the detection of pesticides.[17,18] According to the report, the standing nanorod arrays were closely packed onto the gold film, which generated a strong electromagnetic field and uniformly distributed SERS "hot spots" on the surface of the array. Their results demonstrated that SERS could detect as low as 0.1 ppm of carbaryl and a good prediction was made by the multiple linear regression models ($R > 0.97$). In addition, SERS has also been used to detect atrazine, mycotoxins and antibiotics in foods using similar procedures.[19–21]

The extraction of pesticides from fresh produce is required for the aforementioned SERS methods. However, the detection would be largely simplified if the SERS measurement could be performed directly on the fresh produce. For example, shell-isolated SERS was proposed to measure pesticides on the surface of fresh produce.[22] Shell-isolated SERS signals amplification was provided by gold nanoparticles with an ultrathin silica or alumina shell. It was proposed that shell-isolated SERS could be directly used to detect the pesticides on the surface of citrus fruits without solvent extraction.

1.2.1.2 *Indirect Detection of Chemical Contaminations by SERS*

Not all the chemical contaminants can generate enough Raman spectroscopic responses for an acceptable detection limit by the SERS method. Therefore, a labelling chemical, or so-called Raman dye, was used to amplify the signals derived from the chemical contaminants. These Raman dyes, such as rhodamine B and mercaptobenzoic acid, can generate intensive SERS signals. The detection scheme is usually based upon the changes of Raman signals of the dye that is induced by the addition of the target analyte. The use of Raman dye in SERS detection can sometimes achieve greater sensitivity compared to the direct detection method. For example, a novel "turn-off" biosensing strategy for the detection of thrombin was reported. It was based upon SERS and the mediation of spacing between 4-mercaptobenzoic acid (4-MBA) labelled gold nanoparticles (AuNPs). The detection limit was determined to be 160 fM.[23,24] A method for the detection of mercury ions was also designed using droplet-based microfluidics combined with SERS.[25] Quantitative analysis of mercury(II) ions was achieved by calculating the spectral peak area of rhodamine B with a detection limit between 100 and 500 ppt. A similar strategy was used to detect As ions coupled with the microfluidic sampling technique.[26]

1.2.2 SERS Detection of Microbiological Contaminations in Foods

Multiple outbreaks of illness have been associated with foodborne pathogens, such as *E. coli* O157:H7 in ground beef in the US recently.[27] *E. coli* O157:H7 can cause severe illness such as diarrhoea and acute kidney failure. It is an urgent requirement to develop rapid and convenient methods to monitor foodborne pathogens, such as *E. coli* O157:H7, *Salmonella* and *Listeria*, during food harvesting and food processing. The rapid identification of foodborne pathogens will enable the food supply chain to take faster action on the contaminated products and improve overall food safety.

1.2.2.1 Label-free Detection of Bacteria by SERS

Raman signals are enhanced by the strong electromagnetic field between the junctions or small gaps among the gold/silver nanomaterials. A small distance, such as 1–2 nm, is necessary for the generation of a strong electromagnetic field. However, the size of bacteria means it is difficult to fit into a small nanogap. Bacteria are usually identified through Raman signals from the bacteria cell membrane and cell wall. It was reported that SERS coupled with *in situ* coating of bacteria with silver nanoparticles could be utilized to discriminate between three strains of *E. coli* and one strain of *Staphylococcus epidermidis* by hierarchy cluster analysis. Using SERS mapping, a detection limit of 2.5×10^2 cells mL^{-1} can be achieved.[28] SERS, along with novel silver nanorod array substrates and PCA analysis, has also been used for the direct detection of pathogenic bacteria, such as *E. coli* O157:H7, *E. coli* DH 5α, *Staphylococcus aureus*, *S. epidermidis* and *S.* Typhimurium.[29] In addition, it has been shown that SERS was able to detect single bacterial cells adsorbed onto the silver dendrites with a limit of detection as low as 10^4 CFU mL^{-1}.[30,31]

SERS can also be used to detect bacteria spores. Gold SERS-active substrates were successfully applied to detect and discriminate among five *Bacillus* spores (*B. cereus* ATCC 13061, *B. cereus* ATCC 10876, *B. cereus* sp., *B. subtilis* sp. *and B. stearothermophilus* sp.). PCA results indicate that the Raman shift range between 900 and 1200 cm^{-1} contributed significantly to the total data variance. It was also pointed out that a dipicolinic acid band at 998 cm^{-1} could serve as a biomarker for differentiation among bacterial spores.[32]

1.2.2.2 Indirect Detection of Pathogens by SERS

Label-free SERS methods have shown promising results in discriminating between different bacteria. However, in the scenario of foodborne pathogens, it often requires a low detection limit to ensure food microbiological safety. Such a low detection limit is very challenging to achieve by label-free SERS methods. Therefore, many researchers have developed so-called indirect detection methods using immuno-magnetic enrichment and SERS nanoprobes. The SERS nanoprobes are labelled with Raman dye for signal amplification. For example, Wang and coauthors have used both immune-enrichment and Raman nanoprobes to detect both *S. Typhimurium* and *S. aureus* (Figure 1.4). It was validated in a blind test that the limit of detection is 10^3 CFU mL^{-1} with high specificity.[33,34] Similar strategies have been used in the detection of *E. coli*. The authors used gold-coated magnetic spherical nanoparticles, which were prepared by immobilizing biotin-labelled anti-*E. coli* antibodies onto avidin-coated magnetic nanoparticles, in the separation and concentration of *E. coli* cells. The limit of detection was reported to be 8 and 24 CFU mL^{-1}, respectively.[35]

DNA-based amplification coupled with SERS has also been used to detect foodborne pathogens. DNA-based methods have many advantages compared to antibody-based methods. The extraction and amplification of DNA fragments from the target bacteria make the detection more specific. For example, *S. Enteritidis* has been quantified by a loop-mediated isothermal amplification SERS assay. According to the study, the target DNA was amplified by LAMP and then labelled with Au-nanoprobes composed of gold nanoparticle-modified with specific cy5/DNA probes, which allows detection by SERS. The sensitivity of the developed assay is much higher compared to traditional PCR methods.[36] A novel approach of using SERS for the

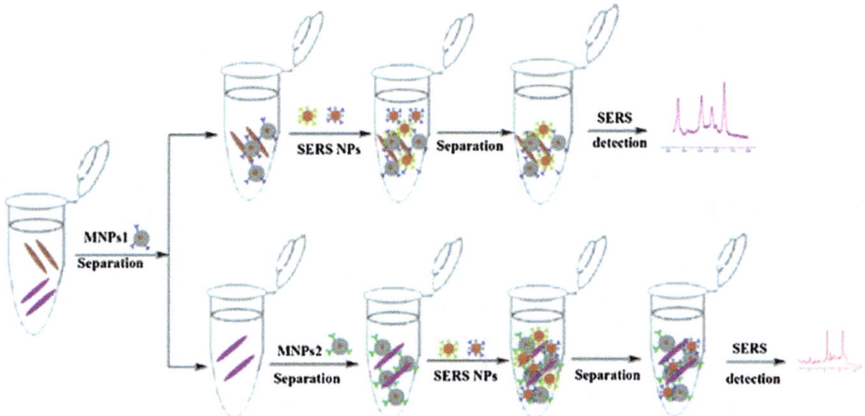

Figure 1.4 A rapid and sensitive method was developed for separation and detection of multiple pathogens in a food matrix by magnetic-SERS nanoprobes. In this scheme, pathogens were first immuno-magnetically captured with MNPs@SiO_2 and pathogen-specific SERS probes (gold nanoparticles integrated with a Raman reporter) were functionalized with corresponding antibodies to allow the formation of a sandwich assay to complete the sensor module for the detection of multiple pathogens in selected food matrices.
Reproduced from *Analytical and Bioanalytical Chemistry*, Separation and detection of multiple pathogens in a food matrix by magnetic SERS nanoprobes, 399, 2011, 1271–1278, Y. Wang, S. Ravindranath and J. Irudayaraj. With permission of Springer.

sequence-specific detection of DNA was proposed. The study utilized magnetic nanoparticles (MNPs) for the enrichment of the target molecules and the detection of the target DNA by labelled Raman dye.[37]

1.3 Determination of Food Components and Food Quality by SERS

In recent years, there has been increasing interest in the use of novel vibrational spectroscopic methods, such as SERS, as a rapid analytical technique to analyze food composition and determine food quality.[12,13] Raman signals are generated by the inelastic scattering of the incident light from a sample and the shift of frequency or wavelength of the scattered light results from characteristic molecular vibrations.[38] Characteristic information of different food components could be obtained by the fingerprinting Raman spectra without complicated sample pre-treatment, such as liquid–liquid extraction, HPLC and GC.[14] The SERS technique has been widely applied for the analysis of lipids, proteins, oligosaccharides and other food components.[39]

1.3.1 Analysis of Food Proteins by SERS

Proteins are bio-functional macromolecules consisting of several polypeptides, which typically folded into a globular or fibrous form.[40] Food protein can also interact with lipids, polysaccharides and minerals. The content of food proteins is a very important nutritional index of food products. The interaction of food protein with other components also influences food quality and food properties. However, current analytical methods for food proteins, such as HPLC, infrared spectroscopy and gel electrophoresis, are complex and often need cumbersome purification. As a new approach to analyze food proteins, SERS could rapidly reflect the fingerprinting information of various proteins in the food product without destroying the food and complex pre-treatments. Unlike the conventional analytical methods, SERS not only differentiates bands representing vibrational modes of the peptide backbone and its side chains, but can also provide information about spectral positions, intensities of protein secondary, tertiary and quaternary structures, side chain orientations and the local environments.[41] In addition, SERS has a great advantage compared with fluorescence-based methods. The broad emission spectra from molecular fluorophores in the proteins make simultaneous detection of multiple proteins impossible. Furthermore, susceptibility to photo bleaching of fluorescence-based methods also makes the detection limit very high.[42] In comparison, these problems do not exist in the SERS technique. An increasing number of studies validated that SERS has a great potential to be applied in the analysis of food proteins and deciphering the interactions of protein–protein, protein–polysaccharides and protein–lipids.[43]

1.3.2 Analysis of Food Lipids by SERS

Most of the lipids identified in foods are in the form of triglycerides, cholesterol and phospholipids.[44] Some dietary fats are necessary to facilitate absorption of fat-soluble vitamins (A, D, E and K) and carotenoids. Therefore, the quality of oil is important to the health of consumers. In recent years, Raman spectroscopy has been used to test the quality of oil, such as sunflower and olive oil. Muik and others investigated the chemical changes taking place during lipid oxidation in several edible oils by Raman spectroscopy.[45] The authors detected the formation of saturated and unsaturated aldehydes using Raman spectra of pure chemicals. The Raman shifts are very close among different cooking oils. However, the ratio of peak intensity is different among the oils. As a result, the oil composition could be differentiated by Raman spectroscopy.

The SERS technique has also been used to differentiate cooking oil and adulterated illegal cooking oil in China. The illegal cooking oil is made from swill and cooked oil with a high peroxide value.[46] The 1660 cm^{-1} shift of the C=C stretching mode of *cis* unsaturated fatty acid will vanish and the intensity of the C=O vibration will increase. For conventional Raman scattering, this change could be detected only if the oil has been adulterated with a large portion of illegal oil. To detect the low concentration of illegal oil, specific noble metal nanoparticles, such as silver nanoparticles, have to be supplemented into the oil to enhance Raman signal intensity. In brief, SERS is a rapid and useful method for the detection of oil quality and safety.

1.3.3 Analysis of Polysaccharides by SERS

The conventional methods to measure the concentration of polysaccharides and detect the structure of polysaccharides are HPLC, mass spectroscopy (MS) and nuclear magnetic resonance (NMR).[47] Both MS and NMR methods require high-purity samples of polysaccharides. For NMR methods, radioactive elements have to be used and this can pose health risks for the operators. However, these disadvantages could easily be avoided by using SERS techniques. Mrozek and colleagues successfully applied SERS in the analysis of polysaccharides in aqueous solution.[48] Different types of monosaccharides were differentiated by SERS when the concentrations were 0.01 M (Figure 1.5). They concluded that the SERS approach is a valuable method for sensitive detection and characterization of carbohydrates. The identification of oligosaccharide and quantification of a mixture could be fulfilled by SERS. They demonstrated the feasibility of using Raman spectroscopy for the analysis of small quantities of chemically similar oligosaccharides and their mixtures at a concentration of 1 mM. The conventional HPLC methods sometimes could not separate similar oligosaccharides and often lead to the overlap of peaks. In comparison, by using the SERS technique, not only could the characterization of oligosaccharides be accomplished quickly, but quantification of oligosaccharides could also

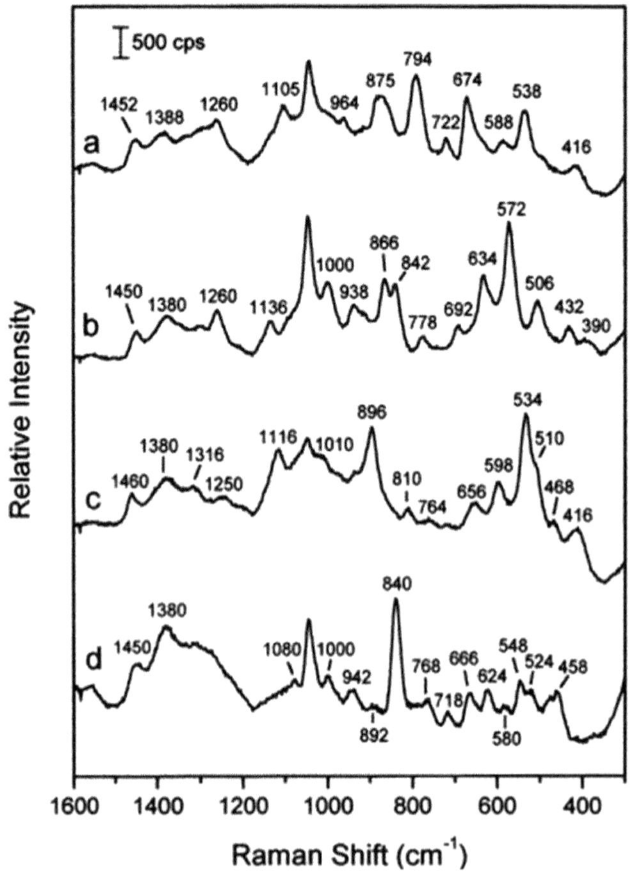

Figure 1.5 SERS spectra for 5 μL aliquots of 10^{-2} M solutions of (a) D-ribose, (b) D-arabinose, (c) D-xylose and (d) D-lyxose.
Mrozek, M. F., Weaver, M. J., Detection and identification of aqueous saccharides by using surface-enhanced Raman spectroscopy, *Analytical Chemistry*, 2002, **74**, 4069–4075. Copyright 2002 American Chemical Society.

be fulfilled at the same time without sample pre-treatment. They also identified that Raman spectra of millimolar concentrations of aqueous mono- and oligosaccharides can be obtained by drop coating deposition onto gold or silver foil substrates in a nano-size range. Their study also indicated that spectra of individual sugars were identified with 100% accuracy and mixtures of the two sugars were quantified with an average error of 2.7% in the relative maltotetraose/stachyose. The SERS technique provides new ways for rapid and accurate detection of polysaccharides in the food industry. As a result, SERS could be applied to monitor the production process of sugar-containing products. In addition, rapid detection of saccharide, such as glucose, has opened a new door for controlling the product quality where the glucose is used as a quality indicator.

1.4 Summary

With the development of high-performance Raman-active substrates in recent years, SERS coupled with various extraction/enrichment methods has been validated to be a rapid and promising tool for the detection of chemical and bacterial contaminations in agri-foods. SERS has also shown its potential to be used in the analysis of food composition and food quality due to its fingerprinting spectral features and fast measurement performance. SERS is expected to be used to replace chromatography-based methods in the near future for the analysis and control of standard food quality and safety.

References

1. E. Smith and G. Dent, *Modern Raman Spectroscopy: A Practical Approach*, John Wiley & Sons, 2005.
2. M. Kerker, D.-S. Wang and H. Chew, *Appl. Opt.*, 1980, **19**, 3373–3388.
3. N. B. Colthup, L. H. Daly and S. E. Wiberley, *Introduction to Infrared and Raman Spectroscopy*, Elsevier, 1990.
4. J. R. Ferraro, *Introductory Raman Spectroscopy*, Academic Press, 2003.
5. G. McNay, D. Eustace, W. E. Smith, K. Faulds and D. Graham, *Appl. Spectrosc.*, 2011, **65**, 825–837.
6. W. Premasiri, D. Moir, M. Klempner, N. Krieger, G. Jones and L. Ziegler, *J. Phys. Chem. B*, 2005, **109**, 312–320.
7. J. Zheng and L. He, *Compr. Rev. Food Sci. Food Saf.*, 2014, **13**, 317–328.
8. M. Moskovits, *J. Raman Spectrosc.*, 2005, **36**, 485–496.
9. M. Linting, J. J. Meulman, P. J. Groenen and A. J. van der Koojj, *Psychol. Methods*, 2007, **12**, 336.
10. G. Sanchez, PLS Path Modeling with R, Trowchez Editions, Berkeley, 2013. http://www.gastonsanchez.com/PLS_Path_Modeling_with_R.pdf.
11. B. Liu, M. Lin and H. Li, *Sens. Instrum. Food Qual. Saf.*, 2010, **4**, 13–19.
12. M. Lin, L. He, J. Awika, L. Yang, D. Ledoux, H. Li and A. Mustapha, *J. Food Sci.*, 2008, **73**, T129–T134.
13. M. Lin, *Front. Chem. Eng. China*, 2009, **3**, 427–435.
14. M. Lin, The Application of Surface-Enhanced Raman Spectroscopy to Identify and Quantify Chemical Adulterants or Contaminants in Foods, in *Handbook of Vibrational Spectroscopy*, ed. J. M. Chalmers and P. R. Griffiths, Wiley-Blackwell, 2001.
15. Y. Hu, S. Feng, F. Gao, E. C. Li-Chan, E. Grant and X. Lu, *Food Chem.*, 2015, **176**, 123–129.
16. L. He, N.-J. Kim, H. Li, Z. Hu and M. Lin, *J. Agric. Food Chem.*, 2008, **56**, 9843–9847.
17. Z. Zhang and M. Lin, *J. Mater. Chem. C*, 2014, **2**, 4545–4551.
18. Z. Zhang, Q. Yu, H. Li, A. Mustapha and M. Lin, *J. Food Sci.*, 2015, **80**, N450–N458.
19. S. Bonora, E. Benassi, A. Maris, V. Tugnoli, S. Ottani and M. Di Foggia, *J. Mol. Struct.*, 2013, **1040**, 139–148.

20. D. K. Singh, E.-O. Ganbold, E.-M. Cho, K.-H. Cho, D. Kim, J. Choo, S. Kim, C. M. Lee, S. I. Yang and S.-W. Joo, *J. Hazard. Mater.*, 2014, **265**, 89–95.
21. L. He, M. Lin, H. Li and N. J. Kim, *J. Raman Spectrosc.*, 2010, **41**, 739–744.
22. J. F. Li, Y. F. Huang, Y. Ding, Z. L. Yang, S. B. Li, X. S. Zhou, F. R. Fan, W. Zhang, Z. Y. Zhou and B. Ren, *Nature*, 2010, **464**, 392–395.
23. Y. Wu and R. Y. Lai, *Biotechnol. J.*, 2016, **11**, 788–796.
24. Z. Wu, Y. Liu, X. Zhou, A. Shen and J. Hu, *Biosens. Bioelectron.*, 2013, **44**, 10–15.
25. C.-W. Wang, Z.-H. Lin, P. Roy and H.-T. Chang, *Front. Chem.*, 2013, **1**, 20.
26. W.-S. Kim, J.-H. Shin, H.-K. Park and S. Choi, *Sens, Actuators, B*, 2016, **222**, 1112–1118.
27. L. Benjamin, M. Jay-Russell, E. Atwill, M. Cooley, D. Carychao, R. Larsen and R. Mandrell, *Epidemiol. Infect.*, 2015, **143**, 81–93.
28. H. Zhou, D. Yang, N. E. Mircescu, N. P. Ivleva, K. Schwarzmeier, A. Wieser, S. Schubert, R. Niessner and C. Haisch, *Microchim. Acta*, 2015, **182**, 2259–2266.
29. H. Chu, Y. Huang and Y. Zhao, *Appl. Spectrosc.*, 2008, **62**, 922–931.
30. P. Wang, S. Pang, J. Chen, L. McLandsborough, S. R. Nugen, M. Fan and L. He, *Analyst*, 2016, **141**, 1356–1362.
31. C. Fan, Z. Hu, A. Mustapha and M. Lin, *Appl. Microbiol. Biotechnol.*, 2011, **92**, 1053–1061.
32. L. He, Y. Liu, M. Lin, A. Mustapha and Y. Wang, *Sens. Instrum. Food Qual. Saf.*, 2008, **2**, 247–253.
33. Y. Wang, S. Ravindranath and J. Irudayaraj, *Anal. Bioanal. Chem.*, 2011, **399**, 1271–1278.
34. M. L. Lesaicherre, T. L. Paxon, F. J. Mondello, M. C. Burrell and A. Linsebigler, SPIE Defense, Security, and Sensing, 2009.
35. B. Guven, N. Basaran-Akgul, E. Temur, U. Tamer and İ. H. Boyacı, *Analyst*, 2011, **136**, 740–748.
36. M. S. Draz and X. Lu, *Theranostics*, 2016, **6**, 522.
37. K. K. Strelau, A. Brinker, C. Schnee, K. Weber, R. Möller and J. Popp, *J. Raman Spectrosc.*, 2011, **42**, 243–250.
38. G. Keiser, *Optical Fiber Communications*, Wiley Online Library, 2000.
39. D. Yang and Y. Ying, *Appl. Spectrosc. Rev.*, 2011, **46**, 539–560.
40. F. Chiti and C. M. Dobson, *Annu. Rev. Biochem.*, 2006, **75**, 333–366.
41. E. A. Carter and H. G. M. Edwards, *Infrared Raman Spectrosc. Biol. Mater.*, 2001, **24**, 421.
42. C. L. Zavaleta, B. R. Smith, I. Walton, W. Doering, G. Davis, B. Shojaei, M. J. Natan and S. S. Gambhir, *Proc. Natl. Acad. Sci.*, 2009, **106**, 13511–13516.
43. X. X. Han, B. Zhao and Y. Ozaki, *Anal. Bioanal. Chem.*, 2009, **394**, 1719–1727.
44. S. Park and P. Addis, *J. Food Sci.*, 1985, **50**, 1437–1441.
45. B. Muik, B. Lendl and A. Molina-Díaz, *Chem. Phys. Lipids*, 2005, **134**, 173–182.
46. Y. Wang, W. C. Ma and G. Y. Chen, *Adv. Mater. Res.*, 2012, **347**, 2684–2687.
47. D. Martens and W. Frankenberger, *Chromatographia*, 1990, **29**, 7–12.
48. M. F. Mrozek and M. J. Weaver, *Anal. Chem.*, 2002, **74**, 4069–4075.

CHAPTER 2

Quantum Dots in the Analysis of Food Safety and Quality

K. DAVID WEGNER,[a,b] MICHAEL V. TRAN,[a] MELISSA MASSEY[a] AND W. RUSS ALGAR*[a]

[a] University of British Columbia, Department of Chemistry, 2036 Main Mall, Vancouver, BC V6T 1Z1, Canada; [b] CEA Grenoble, INAC/SyMMES/LEMOH, UMR5819 SPrAM (CEA-CNRS-Université Grenoble Alpes), 17 rue des Martyrs, 38054 Grenoble cedex 9, France
*Email: algar@chem.ubc.ca

2.1 Introduction

In the 1960s, it was feared that the large rate of increase in the human population would cause problems with the global food supply, leading to mass starvation, especially in low-income countries.[1] In reality, global food production increased faster than the human population in the subsequent decades.[1] The Green Revolution, which peaked in the late 1960s, dramatically changed agriculture and led to higher food production and crop yields.[2,3] Some of the means of this revolution were the development of genetically improved crops, new agricultural technologies and the increased utilization of fertilizers, antibiotics and pesticides.[1,3,4] Unfortunately, the extensive use of chemical products to prevent loss of crops or to strengthen immunity in livestock breeding can be problematic. For example, humans are not immune to the toxic effects of pesticides, and contamination of groundwater and chemical accumulation in the food chain have been shown to have toxicological risks and adverse health effects.[4,5] Moreover, the availability of food does not guarantee its safety and quality.

Food Chemistry, Function and Analysis No. 2
Sensing Techniques for Food Safety and Quality Control
Edited by Xiaonan Lu

Published by the Royal Society of Chemistry, www.rsc.org

Apart from anthropogenic contamination with chemicals, food safety can be compromised by contamination with pathogens. These pathogens and their related toxins can pose a serious health hazard at levels as low as 10–100 colony forming units (cfu). Many safety standards require less than one pathogen cell per 25 g of food,[6] where one cell in 1 mL of food matrix would be roughly equivalent to one part per trillion.[7] Some of the most common foodborne pathogens are *Escherichia coli*, *Staphylococcus aureus*, *Salmonella* and *Listeria monocytogenes*.[4] When consumed, these bacteria multiply in the gastrointestinal system and produce toxins. Common symptoms of infection with these pathogens are typically localized to the gastrointestinal tract and include diarrhoea, abdominal pain, nausea and vomiting.[8] Symptoms can appear as early as 12 hours after consumption of contaminated food.[7] In addition to bacteria, many moulds (a class of fungi) produce toxic secondary metabolites called mycotoxins. Some of the foregoing pathogenic organisms present the additional danger that the produced toxins can persist in food for a long time after the death of the pathogen.[8]

The contamination of food with pathogens and toxins is a global challenge. Although many people recover from foodborne illnesses, deaths do occur, even in high-income countries with widespread access to modern health care and other resources. In Canada, a 2008 listeriosis outbreak associated with contaminated cold cuts (luncheon meat) resulted in 22 deaths,[9] and, in 2000, a water supply contaminated with pathogenic *E. coli* made thousands of people ill and resulted in seven deaths.[10] In 2011, a listeriosis outbreak in the United States of America resulted in 33 deaths.[11] That same year, an outbreak of pathogenic *E. coli* in Germany made thousands ill and resulted in 53 deaths.[12] In 2016, a pathogenic *E. coli* outbreak in the United Kingdom was associated with contaminated salad leaves and made over 160 people ill and resulted in two deaths.[13] In low-income countries that lack the many resources of high-income countries, contaminated drinking water is one of the leading sources of illness.[14]

Beyond pathogens and toxins, testing of food and production facilities for cross-contamination with allergens is critically important to consumer health. In 2012, the leading cause of food recalls in the United States was reported to be undeclared allergens.[15] The misrepresentation and adulteration of food is also a growing concern, with many instances thought to be motivated by financial gain or other commercial benefits. For example, toxic melamine contamination of milk and infant formula in China in 2008, and in pet foods in North America and Europe in 2007, have been speculated to be deliberate.[16,17] The standard chemical test for food protein content will detect melamine as protein, such that melamine could be used to boost the apparent protein content of the product. The melamine contamination in China resulted in the hospitalization of more than 50 000 babies and six deaths among the 290 000 poisoned.[17] Misrepresentation of food is a health concern for reasons such as the potential for severe allergic reactions to mislabelled food, and harm to the unborn children of pregnant mothers.

Disclosure that food has been prepared with ingredients from genetically modified organisms (GMOs) is also an increasing requirement, albeit that concerns about the use of GMOs remain under debate.[18]

The detection of chemical residues, toxins, pathogens and allergens contaminating food and water is clearly of utmost importance to society. Numerous strategies have been developed in order to detect, isolate and identify potential threats in food. The oldest and still most often used method for detection of foodborne pathogens is growing colonies on culture plates and monitoring size, colour and shape.[19,20] The grown bacteria can be further investigated using fluorescent staining and imaging with different microscopy techniques. Biochemical tests are another approach and detect metabolic or enzymatic products of a pathogenic agent. These tests mainly target alcohols, acids or gases to identify the threat.[20] Other common and sensitive techniques exploit antibody–antigen recognition and include enzyme-linked immunosorbent assays (ELISA) and western blotting.[20,21] The antibodies are selected to bind to specific epitopes on the target pathogen or to bind to specific toxins, enabling both identification and quantification of a food contaminant. Another useful group of techniques are based on nucleic acid amplification, especially polymer chain reaction (PCR) methods, and analyze genes that are unique to specific pathogens or pathogen families.[21] For the most part, the foregoing techniques have been presented in order of increasing sensitivity, specificity and sophistication. Apart from these techniques, other methods currently in use or in development include gas and liquid chromatography, flow cytometry, electrochemical methods, infrared and Raman spectroscopy, fluorescence spectroscopy and chemiluminescence or bioluminescence.[8,19,21]

Although the main techniques noted above are often sensitive and selective, they do have drawbacks. Culturing bacteria typically requires between 24 and 72 h to obtain a result, which can be too long to prevent wider outbreak of a pathogen or initiate treatment of infected persons.[21] The variety in pathogen sizes can also be problematic as, for example, viruses with a size of ~100 nm are only observable with electron microscopy and special growing techniques.[21] Although powerful, molecular diagnostic techniques like ELISA and PCR require specialized equipment, trained personnel and sometimes tedious and technical laboratory procedures. Many newer or emerging detection methods emphasize multiplexing (*i.e.*, the simultaneous measurement of several targets within one sample) to help address limitations of throughput. Examples include multiplexed PCR, DNA microarrays and array-based immunoassays.[22] Despite the success of these laboratory techniques, there remains great demand for rapid, sensitive and selective detection that is simple, portable, low-cost and thus suitable for on-site use outside of the laboratory environment. Non-laboratory environments may include the field, food production or packaging lines, and resource-limited settings such as rural/remote communities and low-income countries. The developing world has some of the most serious problems with contaminated food and water, and the World Health Organization has proposed the ASSURED criteria for an ideal assay, which should be affordable, sensitive, specific, user-friendly, rapid and robust, equipment-free and deliverable for the end user.[23] (Equipment-free does not necessarily preclude portable devices, but rather larger bench-top equipment that requires an electrical outlet.) In all cases, the ideal strategy is proactive screening during production, processing and packaging to ensure that food is safe before it is delivered to the consumer, instead of responding to outbreaks and recalling food that is already available to the consumer. Improved technologies, particularly in terms of speed of analysis and lower cost per analysis, would greatly facilitate implementation of both proactive and globally applicable strategies for food safety.

Nanotechnology has great potential to help meet the demands of modern food safety diagnostics.[24] Many materials, when shrunk from the bulk to the nanometre size scale ($10^{-9}-10^{-7}$ m), exhibit new effects or new properties; for example, enhanced surface reactivity, quantum confinement, plasmonic behaviour, enhanced electrical conductivity and enhanced magnetic properties.[8] Nanomaterials are typically defined as materials that have at least one side or internal structure on the scale of nanometres. A nanoparticle is a type of nanomaterial that is typically defined as a discrete particle that has all three dimensions on the order of 100 nm or less.[25] Nanoparticles (NPs) can be tailored in shape, size and composition – the combination of which generally determines their special properties. NPs also have large surface area-to-volume ratios and efficient diffusion, which is often advantageous *versus* bulk interfaces and quite suitable for microfluidic devices where mechanical mixing is not possible.[8] Finally, modification of NP surfaces with different functional groups, chemistries and cargos allows engineering and adaptation of nanoparticles to optimally match their specific purpose.

NPs are of interest in many areas of food and agriculture, including but not limited to packaging materials, delivery systems for agrochemicals and probes for pathogen and toxin detection.[26] In primary agricultural production, nano-formulated agrochemicals can enhance the efficacy of conventional formulations of fertilizers and pesticides, and nanoparticles are useful for environmental remediation (*e.g.*, water, soil).[25] Nano-sieves can be used for bacterial filtration in food processing.[25] In food preservation and packaging, the antimicrobial activity of certain nanoparticles can help lengthen storage time and shelf-life and keep products fresher.[25] Most importantly for this chapter, many nanoparticles also have optical, electronic or magnetic properties that make them useful and advantageous as probes or signal transduction agents for the detection of foodborne pathogens, chemical products and toxins.[19,24] Some of the most popular types of nanoparticles for assays and diagnostics are gold NPs and nanorods, carbon nanomaterials (*e.g.*, nanotubes, graphene), magnetic NPs and semiconductor quantum dots.[21] Whereas other chapters in this book address several of the foregoing nanomaterials, this chapter focuses on research and development (R&D) toward the use of quantum dots for the analysis of food safety and quality. Quantum dots have attracted tremendous interest in the areas of chemical and biological analysis and imaging because of their unique and highly advantageous fluorescence properties, as well as several other useful characteristics. The following sections first summarize the key properties of quantum dots, then review selected examples of their application in the context of food-related assays for pathogens, pesticides, antibiotics and GMOs. The chapter concludes with a summary and perspective on future directions for R&D in this area.

2.2 Quantum Dots

2.2.1 Overview

Quantum dots (QDs) are fluorescent semiconductor nanocrystals that are approximately spherical in shape and 1–10 nm in diameter.[27–30] These colloidal particles are composed of hundreds to thousands of atoms arranged in a crystalline lattice (Figure 2.1). A wide variety of semiconductor materials are used to synthesize QDs, including II–VI materials (*e.g.*, CdSe, CdTe), III–V materials (*e.g.*, InP, InAs), IV–VI materials (*e.g.*, PbS, PbSe) and group IV materials (*e.g.*, Si).[31,32] The nanometre size of QDs leads to unique optical properties that are remarkably different from the corresponding bulk semiconducting material. In particular, semiconductor QDs are brightly fluorescent whereas their bulk analogues are not. When a semiconductor material absorbs a photon of light, an electron is excited across the band gap to form an electron–hole pair called an exciton. In the bulk size regime, the energy of the band gap is independent of the size of the semiconductor crystal; however, QDs are smaller than the so-called Bohr exciton radius of the bulk semiconductor, leading to a phenomenon called *quantum confinement*.[33] With quantum confinement, the band gap energy becomes dependent on the size of the nanocrystal, increasing as the size of the QD decreases (Figure 2.2). Other consequences of quantum confinement include much stronger absorption of light and a significant fluorescence quantum yield, where the wavelengths (*i.e.*, energies) of absorption and

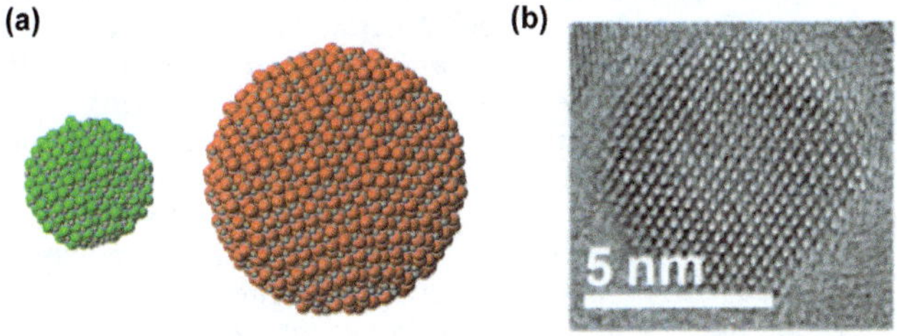

Figure 2.1 (a) Atomistic illustration of two sizes of QD nanocrystal. (b) High-resolution TEM image of a QD. Panel (b) reprinted with permission from W. R. Algar, K. Susumu, J. B. Delehanty, I. L. Medintz, Semiconductor Quantum Dots in Bioanalysis: Crossing the Valley of Death, *Anal. Chem.*, 2011, **83**, 8826–8837. Copyright 2011 American Chemical Society.

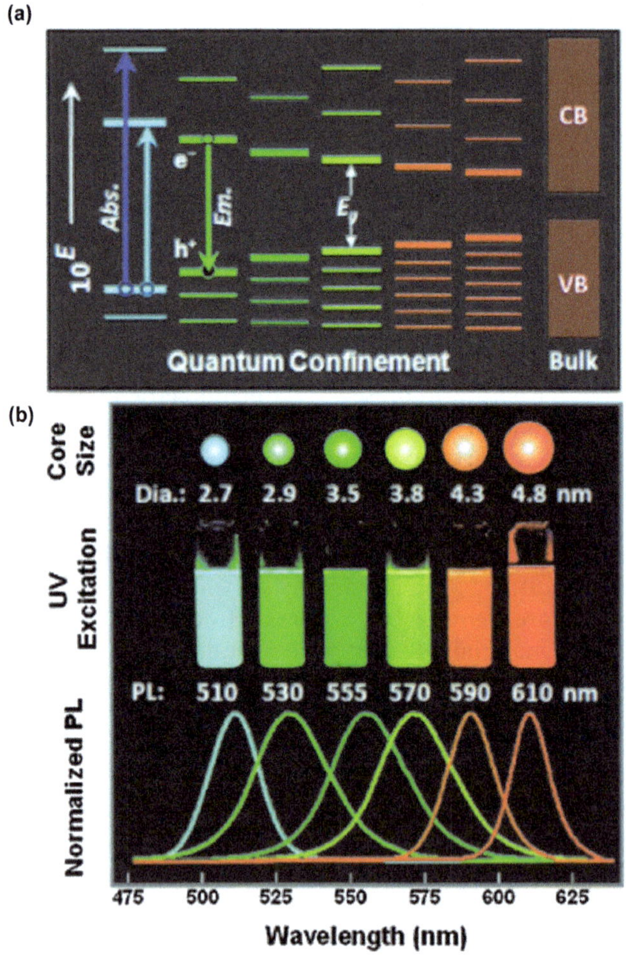

Figure 2.2 (a) Qualitative energy diagram for QD nanocrystals with increasing size from left to right. The conduction band (CB) and valence band (VB) of the bulk semiconducting material are shown for reference. The band gap energy is E_g and absorption (Abs.) and emission (Em.) transitions are shown, where h^+ is a hole and e^- is an electron. (b) Illustration of different sizes of core QDs, photograph of solutions of the illustrated QDs under UV illumination and the corresponding photoluminescence (PL) spectra. Since the emission from QDs is not strictly fluorescence, "photoluminescence" is a more accurate term for their emission; however, we use "fluorescence" in this chapter for simplicity.
Reprinted with permission from W. R. Algar, K. Susumu, J. B. Delehanty, I. L. Medintz, Semiconductor Quantum Dots in Bioanalysis: Crossing the Valley of Death, *Anal. Chem.*, 2011, **83**, 8826–8837. Copyright 2011 American Chemical Society.

fluorescence emission can be tuned by the size of the QD. To date, the most widely utilized QDs are CdSe/ZnS core/shell nanocrystals, where the CdSe core determines the optical properties and the ZnS shell protects and enhances those properties. CdSe/ZnS QDs have been widely utilized because of their well-established methods of synthesis and outstanding optical properties. Driven in part by the desire to develop materials without Cd, other QD materials are in development (*e.g.*, InP, Si).[34–36] The composition of the QD material is important not just because of the benefits or liabilities of the constituent elements, but also because the composition of the QD determines the range over which its fluorescence can be size-tuned (Figure 2.3). It is also possible to tune the fluorescence of alloyed QDs (*e.g.*, $CdSe_xS_{1-x}$) by changing the composition without a change in nanocrystal size. Despite significant progress, the properties of most non-Cd QD materials do not yet match those of Cd-based QDs. The following text is largely written in the context of CdSe and related materials (*e.g.*, CdTe, CdSeS) as benchmark QDs. These materials remain popular because of their highest-quality optical properties, established methods for synthesis, their fluorescence across the visible spectrum and their commercial availability.[37] Some alternative QD materials are now also available commercially (*e.g.*, InP).

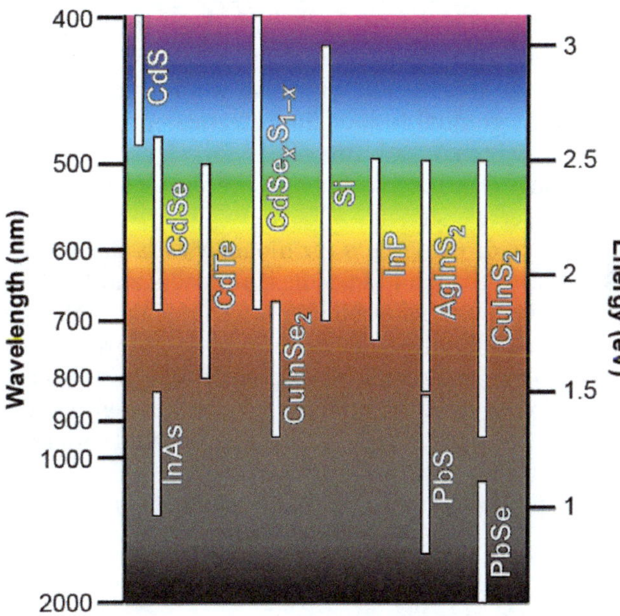

Figure 2.3 Approximate wavelength ranges over which the fluorescence of various QD materials can be tuned through control of nanocrystal size. The visible spectrum is between *ca.* 400 and 650 nm.
Reprinted with permission from N. Hildebrandt, C. M. Spillmann, W. R. Algar, T. Pons, M. H. Stewart, E. Oh, K. Susumu, S. A. Díaz, J. B. Delehanty, I. L. Medintz, *Chemical Reviews*, 2016, DOI: 10.1021/acs.chemrev.6b00030 (http://pubs.acs.org/doi/abs/10.1021/acs.chemrev.6b00030). Copyright 2016 American Chemical Society.

2.2.2 Advantages in Bioanalysis

QDs exhibit both optical and physical properties that are highly advantageous for bioanalysis.[37–39] Optically, QDs have broad absorption spectra that begin at the band gap energy and extend into the ultra-violet region, such that visible QD fluorescence can be excited over a wide range of wavelengths. In addition to being size-tunable, the fluorescence emission spectra of QDs are symmetric and narrow with full-width-at-half-maxima of 25–35 nm for relatively monodisperse samples. Taken together, these properties allow (i) the wavelength of QD fluorescence to be optimized and matched to the needs of an application, (ii) the fluorescence from multiple colours of QDs to be excited simultaneously and (iii) each colour of fluorescence to be measured with minimal interference from other colours (*i.e.*, minimal crosstalk). This multiplexing capability is very attractive for bioanalysis related to food safety and quality because many samples may be at risk of contamination with more than one type of toxin or pathogen, and it is possible to detect as many toxins as there are colours of QDs.[40] Importantly, QDs have quantum yields that are comparable to those of conventional organic fluorescent dyes, and also have molar extinction coefficients (10^4–10^7 $M^{-1}\,cm^{-1}$) that are typically 10–100-fold larger than those for organic dyes (10^4–10^5 $M^{-1}\,cm^{-1}$).[27] These features make QDs very brightly fluorescent and thus potentially ideal for detection of small amounts of toxins or pathogens. QDs also exhibit superior resistance to photobleaching (*i.e.*, gradual loss of fluorescence emission under continuous illumination), which is advantageous for signal integration, kinetic measurements, long-term monitoring and the use of high-power excitation to maximize fluorescence signals.

Another advantage of using QDs for bioanalysis is the ability to functionalize their surfaces with small molecules, polymers and biomolecules such as antibodies, enzymes, other proteins, nucleic acids (DNA, RNA, ssDNA, aptamers), lipids and peptides.[41] In many cases, the surface area of a QD is sufficiently large to conjugate multiple copies of these biomolecules, the role of which is most often targeted binding to specific analytes of interest. As will be discussed further below, QDs are typically coated with small molecule ligands or polymers that determine their physical properties as a colloid, including stability with respect to pH and ionic strength, as well as biocompatibility.

2.2.3 Synthesis and Functionalization Strategies

The optical properties of QDs are tailored through both the size of the nanocrystal and the type of semiconductor material. To maximize their advantages in bioanalysis, QDs should be synthesized in a controlled manner to minimize nanocrystal defects, obtain high brightness and achieve monodispersity. The first documented synthesis of colloidal QDs was reported by Brus and colleagues in 1984.[42] They synthesized various sizes of core-only CdS QDs stabilized with a styrene/maleic anhydride copolymer in water. Since then, the standard method for synthesizing QDs has become significantly different. The synthesis of high-quality, monodisperse and brightly fluorescent QDs became routine when researchers began injecting organometallic precursors of Cd and a chalcogen (*i.e.*, S, Se or Te) into hot coordinating solvents such as tri-octylphosphine oxide (TOPO) under an inert atmosphere.[43] Other key advances were the synthesis of core/shell QDs, where the core is epitaxially coated with a shell of a structurally compatible and higher band gap energy semiconductor (*e.g.*, CdSe/ZnS),[44] and also the use of safer and less volatile precursors.[45] Another important advance was the realization that impurities in technical grade solvents were active in the synthesis of QDs but difficult to reproduce.[46] Synthesis methods have thus been adapted to use solvents (*e.g.*, octadecene) free of such impurities with controlled addition of coordinating ligands (*e.g.*, long-chain alkyl acids; long-chain primary, secondary or tertiary alkyl amines) for growth of QDs. Several reviews provide more details on the intricacies of QD synthesis.[47–51]

QDs can also be synthesized in water, which may be an attractive strategy given that most applications in food analysis will require QDs with aqueous compatibility; however, the overall quality of these QDs is significantly lower than QDs synthesized *via* high-temperature organometallic methods. In particular, these aqueous-synthesized QDs often have poor quantum yields and large size polydispersity, resulting in lower brightness and spectrally broader PL.[52]

High-quality QDs synthesized by high-temperature organometallic methods are stabilized by hydrophobic ligands and are not suitable for most applications in food analysis until rendered hydrophilic. The two most prevalent methods for generating hydrophilic QDs are (i) exchange of the native hydrophobic ligands with hydrophilic ligands *via* mass action, and (ii) encapsulation with an amphiphilic polymer or phospholipids (Figure 2.4).[53–55] The hydrophilic ligands or polymers serve two purposes: maintaining a stable colloidal dispersion in an aqueous environment and providing functional groups for the conjugation of biomolecules to the QD.

2.2.4 Bioconjugation Strategies

Two general strategies are commonly used for the conjugation of biomolecules to QDs: direct covalent coupling to the ligand or polymer coating on the QD; and non-covalent adsorption or metal-affinity coordination. The covalent coupling strategy forms new chemical bonds between functional groups on a biomolecule (*e.g.*, amines, thiols, carboxyls) and functional groups on the QD coating. This strategy requires activating agents such as carbodiimides (*e.g.*, 1-ethyl-3-(3-dimethylaminopropyl)carbodiimide, EDC; couples amine and carboxyl groups) with *N*-hydroxysuccinimide (NHS; increases the efficiency of the reaction), reactive groups such as maleimides (reacts with thiols) and succinimidyl esters (reacts with amines), or

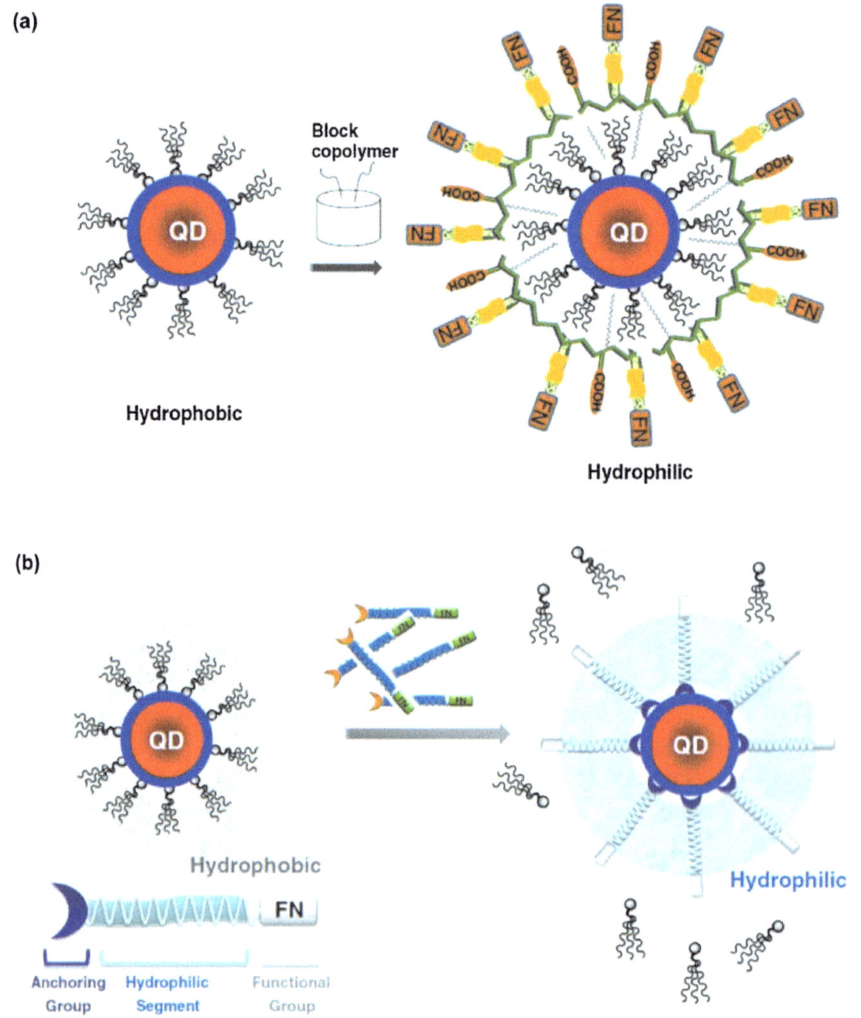

Figure 2.4 Strategies for rendering QDs hydrophilic and colloidally stable dispersions in aqueous media. (a) Encapsulation within an amphiphilic block copolymer. FN represents a functional group that is suitable for downstream chemical ligations. (b) Exchange of hydrophobic ligands with hydrophilic ligands.
Reproduced from *Adv. Drug Delivery Rev.*, **64**(2), H. Mattoussi, G. Palui, H. B. Na, Luminescent quantum dots as platforms for probing in vitro and in vivo biological processes, 138–166. Copyright 2012 with permission from Elsevier B. V.

heterobifunctional crosslinkers such as succinimidyl 4-(*N*-maleimidomethyl)-cyclohexane-1-carboxylate (SMCC; links amine and thiol groups). The conjugation of biomolecules through adsorption often relies on electrostatic interactions between the biomolecule and the QD coating, whereas metal-affinity coordination relies on strong dative interactions between metal atoms on the inorganic surface of the QD and thiol-terminated linkers or polyhistidine tags on a biomolecule. Another approach to bioconjugation is to use either of the two strategies above to conjugate a QD with (strept)avidin, which permits simple attachment of biotinylated biomolecules. Additional and more detailed information about the bioconjugation of QDs can be found in several published reviews.[28,41,56]

Many types of biomolecules may be conjugated to QDs; however, antibodies and aptamers remain the most common for toxin and pathogen analysis (Figure 2.5). The role of both of these classes of biomolecule is to bind to a toxin or pathogen of interest with high affinity and specificity. Aptamers are advantageous in comparison to antibodies because they can be selected *in vitro* and chemically synthesized, resulting in lower cost and less batch-to-batch variation. Aptamers are also more robust than antibodies.[57] Although many aptamers exhibit binding affinity and specificity that is comparable to an antibody, there are also many aptamers that have lower affinity for their targets than antibodies, and there is a multitude of targets for which aptamers have not yet been selected. The conjugation of aptamers to QDs relies on linkers attached to either the 5′ or 3′ terminus of the aptamer. These linkers tend to have terminal amine or thiol groups for crosslinking with functional groups (*e.g.*, amine, carboxyl) associated with the ligand or polymer coating on the QD, or sometimes thiol groups are able to coordinate to the inorganic shell of the QD. Antibodies have many amine and carboxyl groups available for crosslinking, as well as disulfide bridges that can be reduced to yield free thiols for crosslinking.[58] Alternatively, aptamers and antibodies can both be biotinylated to bind to QDs modified with (strept)avidin. Beyond crosslinking, a particularly effective and versatile approach to bioconjugation with QDs has been the high-affinity binding of polyhistidine-tagged proteins, antibodies, peptides and oligonucleotides to the inorganic surface of QDs. These and other bioconjugation strategies have been reviewed in detail.[28,56,59]

2.3 Applications of QDs in Food Safety and Quality Analysis

Most analytical applications of QDs, including those in food safety and quality analysis, are in the R&D phase. Ongoing areas of research with QDs that are anticipated to lead to future commercial products are photovoltaics, solid-state lighting and biomedical technologies.[39,60,61] One of the few current applications of QDs in consumer products is the newest generation of televisions.[62] With respect to biomedical applications, most interest lies in using QDs as fluorescent tags in either imaging or molecular diagnostic assays, the latter of which is most relevant to food safety and quality analysis. This section describes several representative examples of QD-based assays for foodborne or waterborne pathogens, pesticides and

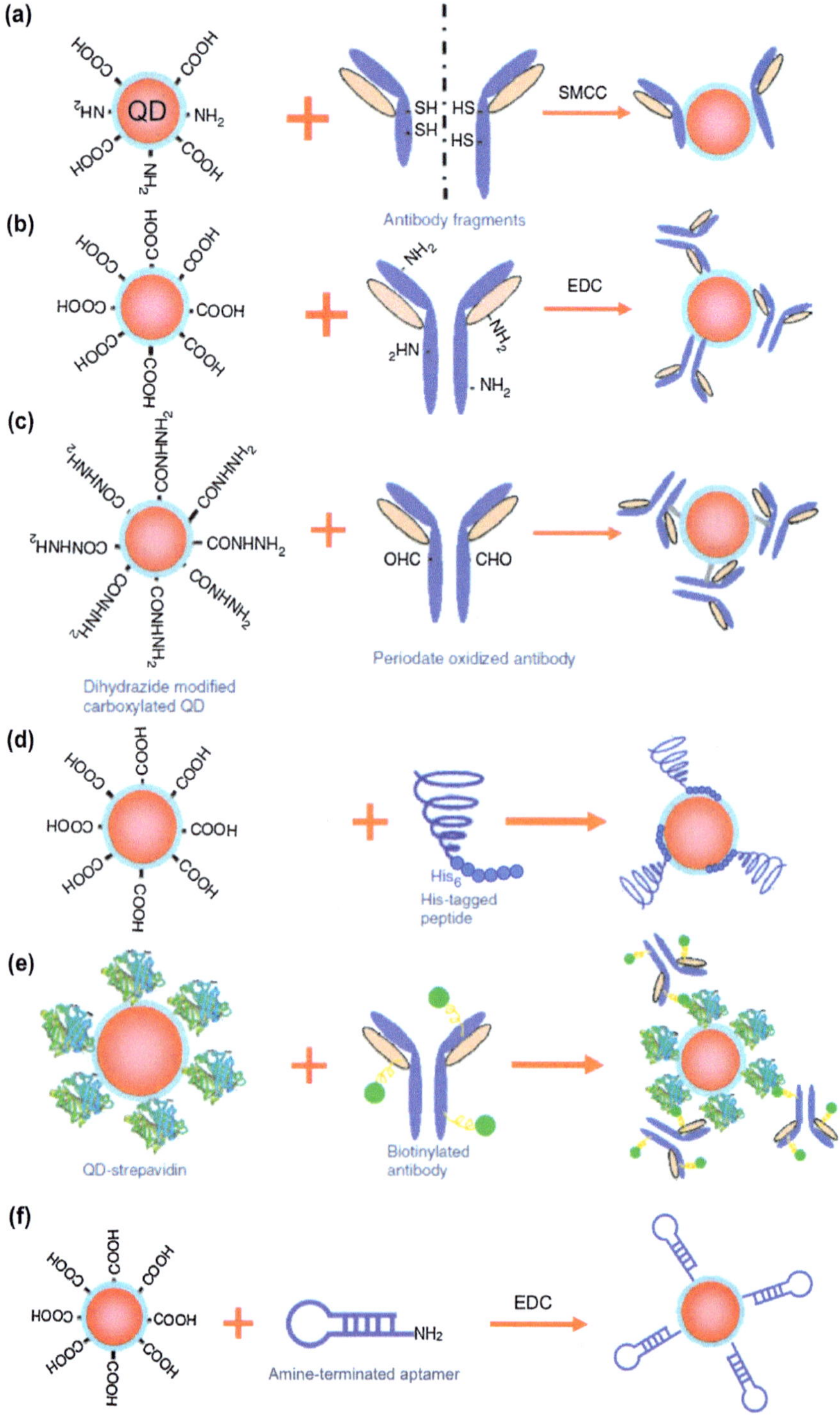
(a)
QD
COOH
NH2
SH
HS
SMCC
Antibody fragments
(b)
COOH
NH2
EDC
(c)
CONHNH2
OHC
CHO
Dihydrazide modified carboxylated QD
Periodate oxidized antibody
(d)
COOH
His6
His-tagged peptide
(e)
QD-strepavidin
Biotinylated antibody
(f)
COOH
NH2
EDC
Amine-terminated aptamer

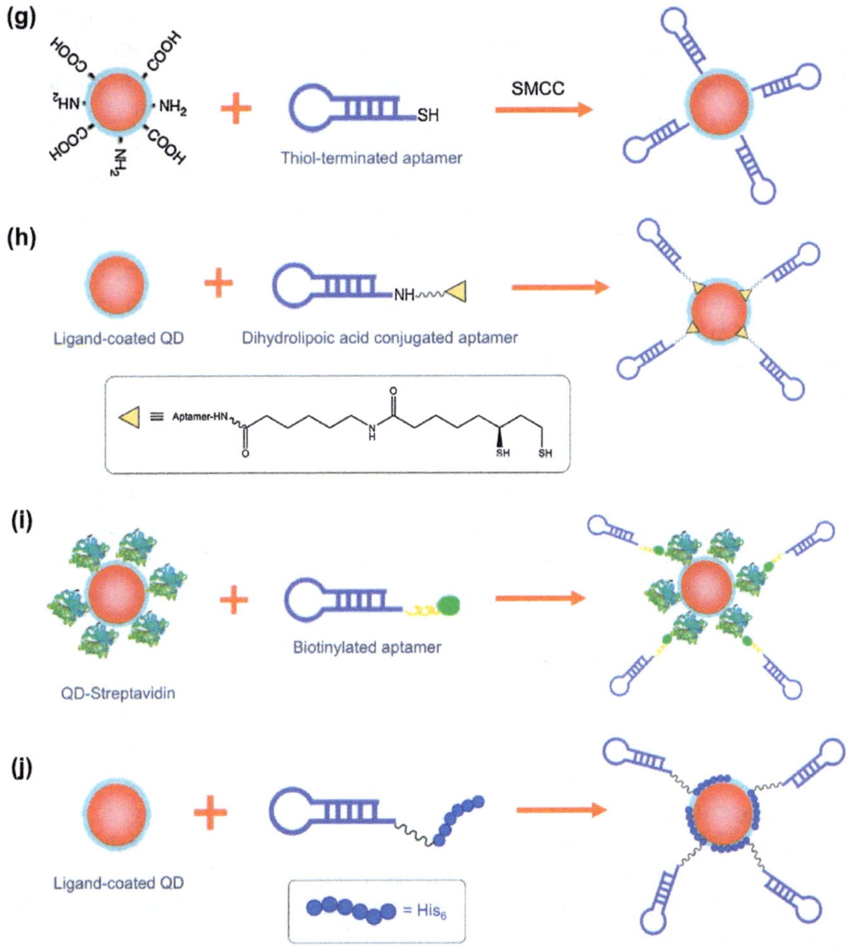

Figure 2.5 Common bioconjugation strategies with peptides, proteins and antibodies. (a) Coupling between thiol and amine groups using SMCC. (b) Coupling between carboxyl and amine groups using EDC activation (and often NHS, not shown). (c) Hydrazone coupling between hydrazide and aldehyde groups. (d) Binding of polyhistidine-tagged peptides (and proteins, not shown) to the inorganic shell of ligand-coated QDs. QDs can also be functionalized with coatings that display Ni^{2+}-nitrilotriacetic acid groups for binding polyhistidine-tagged peptides (not shown). (e) Binding between streptavidin-QD conjugates and biotinylated antibodies. (f) Coupling between carboxyl and amine groups using EDC activation (and often NHS, not shown). (g) Coupling between thiol and amine groups using SMCC. (h) Binding of a dithiol-terminated aptamer to the inorganic shell of ligand-coated QDs. (i) Binding of a biotinylated aptamer to streptavidin-QD conjugates. (j) Binding of polyhistidine-tagged aptamer to the inorganic shell of ligand-coated QDs.
Adapted by permission from Macmillan Publishers Ltd: Y. Xing, Q. Chaudry, C. Shen, K. Y. Kong, H. E. Zhau, L. W. Chung, J. A Petros, R. M. O'Regan, M. V. Yezhelyev, J. W. Simons, M. D. Wang, S. Nie, *Nat. Protocols*, 2007, **2**, 1152–1165. Copyright 2007 Nature Publishing Group.

antibiotic residues; however, interested readers can find a much greater body of published research on biomedical applications that share common concepts and technical approaches (see ref. 37–39, 61 for examples of reviews).

It should be noted that concern has been expressed about the heavy metal component of many popular QD materials (*e.g.*, CdSe/ZnS). Several publications address the complex toxicology of QDs in detail.[63–68] It is not advisable to formulate a general statement about the toxicity of QDs, as it depends on a multitude of factors, including the composition, functionalization, quality and other properties of the QD material, as well as experimental parameters and conditions, and the type of cell or organism. In the case of assays for food safety and quality, only very small amounts of QDs should be required and their use is *in vitro*, thus eliminating or minimizing concerns about the potential for toxicity in these applications. The current availability of consumer electronics with QDs is an indicator that regulatory approval will be possible. Another challenge may be bulk manufacturing of QDs because synthesis is often prone to batch-to-batch variation and difficulties with scale-up, although various solutions have been proposed.[69]

2.3.1 Foodborne Pathogens

The most problematic foodborne pathogenic bacteria are *Escherichia coli*, *Listeria monocytogenes*, *Salmonella*, *Staphylococcus aureus* and *Yersinia enterocolitica*.[70] One detection strategy for these and other bacteria is the use of fluorescent dye-labelled antibodies for pathogen binding and signal generation in fluorescence-based assays. QDs can replace these dyes and provide better analytical figures of merit. A pair of studies in which *E. coli* O157:H7 cells were labelled with biotinylated antibodies have shown that streptavidin-QDs can provide detection limits that are two orders of magnitude better than streptavidin-FITC (fluorescein isothiocyanate), which is a common fluorescent dye.[6,71] One of these studies used centrifugal isolation of the labelled *E. coli* cells,[6] whereas the other used immunomagnetic separation with capture antibodies coated on magnetic particles.[71] It was also possible to detect individual *E. coli* cells under a fluorescent microscope because of both the superior brightness of the QD labels and their ~500-fold slower rate of photobleaching (Figure 2.6).[6] This greater photostability is important when a detected pathogen cell has a low density of fluorophores on its surface, which, for example, can be caused by low affinity of the antibody toward the cell-surface antigen or steric considerations.[6]

Although the use of streptavidin-QDs with biotinylated antibodies is common, alternative methods have been used to conjugate antibodies to QDs. For example, QDs have been conjugated with antibodies using EDC, a zero-length crosslinker,[70] and a related method used EDC to conjugate QDs with protein A, which binds to the Fc region of an antibody.[72] This latter strategy can potentially control the orientation of the conjugated antibody

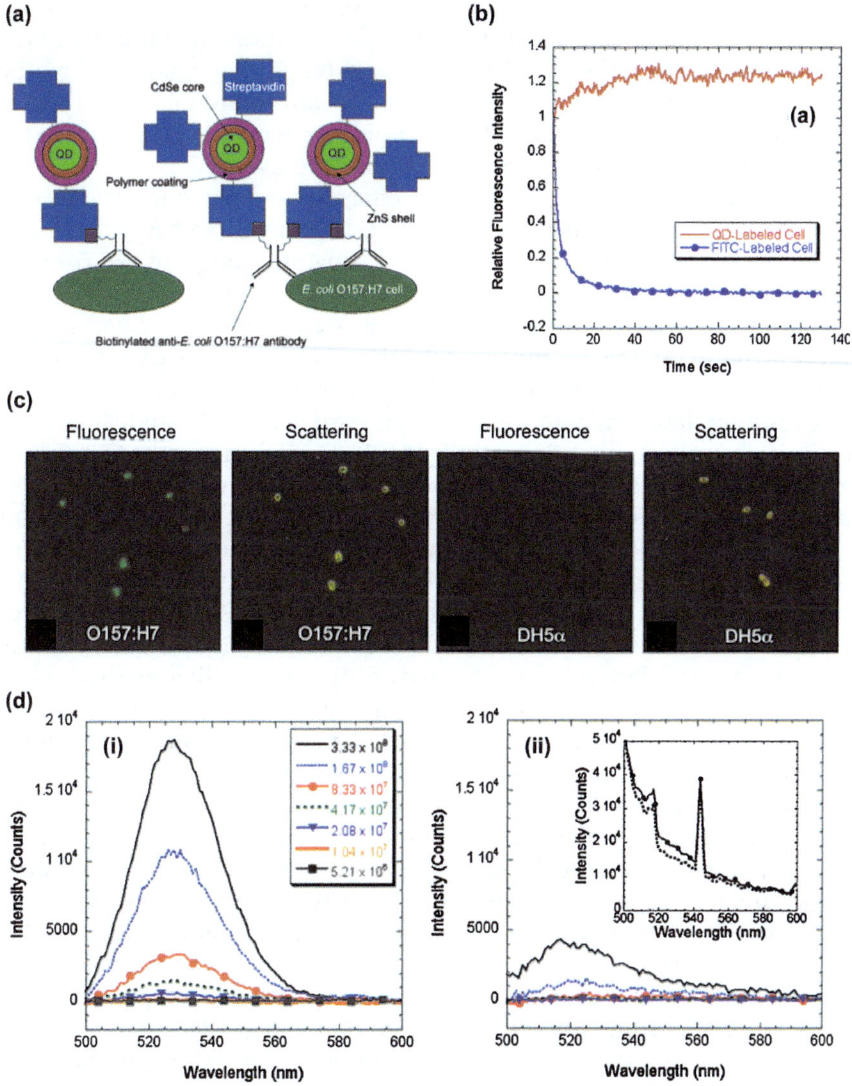

Figure 2.6 (a) Detection scheme for *E. coli* cells using biotinylated anti-*E. coli* antibodies and streptavidin-CdSe/ZnS QD. (b) Comparison of the photobleaching rates of a single *E. coli* cell decorated with FITC-labelled antibodies *versus* CdSe/ZnS QD-labelled antibodies. (c) Fluorescence microscopy images and darkfield scattering images for the detection of *E. coli* O157:H7 cells (left) and for *E. coli* DH5α cells (right) with streptavidin-QDs and biotinylated anti-*E. coli* O157:H7 antibodies. (d) Emission spectra in the presence of different concentrations of *E. coli* cells using (i) streptavidin-QDs and (ii) FITC-labelled streptavidin. Reproduced with permission from M. A. Hahn, J. S. Tabb, T. D. Krauss, Detection of Single Bacterial Pathogens with Semiconductor Quantum Dots, *Anal. Chem.*, 2005, 77, 4861–4869. Copyright 2005 American Chemical Society.

to a better degree than direct EDC conjugation or biotin-(strept)avidin strategies, and may result in better retention of antibody binding activity. Oriented immobilization is also possible through conjugation of amine-functionalized QDs with the thiol groups of dithiothreitol (DTT)-reduced antibodies using SMCC.[7] The SMCC crosslinking method outperformed two other biotin-streptavidin conjugation methods in assays.

Immunomagnetic separation is an efficient tool for enrichment of a target pathogen and thus increasing the sensitivity of an assay. Antibody-conjugated magnetic particles are used to separate a targeted cell from the surrounding environment using an external magnetic field and without damaging the cells.[73] Magnetic enrichment steps can help reduce analysis times by reducing the amount of culturing required prior to a diagnostic method.[74,75] One potential drawback of magnetic separation is that multiple binding, labelling and washing steps add complexity and time to the analysis, as well as more opportunity for loss of bound target.[76] As an example of one-step detection, *Salmonella typhimurium* was co-labelled with antibody-conjugated magnetic polymer nanoparticles (for capture and isolation) and QD-doped polymer nanoparticles (for detection).[76] After capture and washing, the bacteria cells were detected using fluorescence microscopy with a detection limit of 10 cfu mL^{-1}. Fluorescence signals decreased when the assay was tested with milk, serum and urine, but the assay remained selective. The sensitivity of the co-labelling method was 5-fold better than with the use of two separate magnetic and fluorescent labelling steps.[76] Another potential disadvantage of magnetic separation is that magnetic particles will sometimes interfere with fluorescence detection because they have significant background fluorescence and can reduce the excitation efficiency of fluorescent labels through competitive scatter and absorption of the excitation light.[72] The brightness and wavelength-tunable fluorescence emission of QDs make them less prone to this sort of interference than conventional fluorescent dyes, but QDs are not necessarily immune to this interference. An alternative to immunomagnetic separation is the utilization of a porous polycarbonate membrane.[77] The membrane is coated with antibodies that bind and capture the target pathogen while the sample flows through the membrane. These membranes can have low non-specific binding and negligible background fluorescence, but are prone to clogging with complex samples. QD-antibody conjugates have also been used as fluorescent labels to detect bacteria captured on these membranes. In one study, QD labels were found to be able to detect 100 cells mL^{-1}, which was 100-fold better than an analogous assay with fluorescein dye-labelled antibodies.[77]

Magnetic separation methods with QDs have also been demonstrated with aptamers. As an example, two different types of aptamer-NP conjugates were utilized for the detection of *S. aureus* and *S. typhimurium*.[78] One NP conjugate was a QD for detection of *S. aureus*, and the other was an upconversion nanoparticle (UCNP) for detection of *S. typhimurium*. Amine-terminated oligonucleotides were conjugated to carboxyl-functionalized QDs and magnetic nanoparticles (MNPs) were functionalized with short oligonucleotide capture probes that were complementary to a portion of the aptamer sequence. Hybridization linked the MNPs and QDs, and the fluorescence of the isolated MNPs was measured as a baseline. Upon incubation with the target pathogen, *S. aureus*, the aptamer-QD conjugates bound their target in preference to the oligonucleotide probes and were released from the MNP. The QD-aptamer conjugates that remained linked to the MNPs were collected using magnetic separation, and the measured fluorescence was inversely proportional to the amount of pathogen target. (The detection of *S. typhimurium* with UCNPs was analogous.) The limits of detection were 16 and 28 cfu mL^{-1} for *S. aureus* (QDs) and *S. typhimurium* (UCNPs), respectively, with a linear range between 50 and 10^6 cfu mL^{-1}.[78] Curiously, the signal chosen to be measured was the decrease in PL intensity in response to aptamer-target binding; however, measurement of fluorescence in the supernatant after magnetic separation would have offered a potentially preferable "turn on" assay format.

Cotton swabs have been used for sampling of bacteria and visual detection of *Salmonella* with QDs.[79] The swab was treated with a specially designed medium during preparation. Swabs were immersed in a sample and incubated for 18–24 h at 37 °C. The presence of *Salmonella* was observable as a black precipitate, which arose from the reduction of ammonium ferric citrate by hydrogen sulfide produced by *Salmonella*. *E. coli* strains, which are also likely to be present in samples, do not produce hydrogen sulfide. Swabs that were positive for the black precipitate were selected by eye and further incubated with antibody-conjugated QDs and then washed for confirmation of *Salmonella*. (Although not a characteristic of *E. coli*, the production of hydrogen sulfide is not unique to *Salmonella*.) Observation of QD fluorescence on the swab under a standard stereomicroscope equipped with a blue light-emitting diode (LED) excitation source confirmed a positive test for *Salmonella*. If no QD fluorescence was visible, a more sensitive laser scanning confocal fluorescence microscope was used to check for weak QD fluorescence. *Salmonella*-positive swabs were then used to colonize agar plates for culturing and identification according to international standards.[79] Although the initial incubation period is a disadvantage, the subsequent stages of visual detection (*i.e.*, black spot by eye, QD fluorescence by eye or on stereomicroscope) were advantageous in their simplicity and suitability for a non-specialized laboratory. In principle, positive swabs could be sent to a specialized lab for confirmation of results with conventional methods.

The above examples of assays for pathogen detection with QDs utilized either antibodies or aptamers, both of which are thought to target unique cell surface proteins expressed by a pathogen. An alternative approach is to detect genes that are unique to a pathogen. DNA microarrays have been widely used for the high-throughput fluorescent analysis of hundreds to thousands of genes, and are expected to provide results much faster than culture methods or immunological methods that require separate conditions for each individual pathogenic species. QDs have been used as fluorescent reporters to enhance the sensitivity of microarrays for the detection of different strains of 12 different species of foodborne pathogens in parallel.[80] A total of 78 different bacterial strains were isolated from 105 mock-contaminated food samples (pork, chicken, fish and milk) when tested for the following species: *Salmonella* spp., *Shigella* spp., *Escherichia coli* spp., *Proteus* spp., *Staphylococcus aureus*, *Yersinia enterocolitica*, *Listeria monocytogenes*, *Vibrio parahaemolyticus*, *Campylobacter jejuni*, *β-hemolytic Streptococcus*, *Enterococcus faecalis* and *Vibrio fluvialis*. Streptavidin-coated CdSe/ZnS QDs were bound to biotinylated PCR products that had been pre-hybridized to oligonucleotide probes immobilized on glass slides (*i.e.*, a sandwich hybridization assay). The fluorescence intensity was thus proportional to the quantity of PCR product. With QDs as the fluorescent reporter, the detection limit was 10 cfu mL^{-1}.[80]

Aside from brightness, one of the main advantages of QDs is their amenability to multiplexed fluorescence measurements. Among many other applications in bioanalysis, this advantage has been exploited for pathogen detection. In one study, two different QD-antibody conjugates with peak fluorescence wavelengths at 525 nm and 705 nm were used for the simultaneous detection of *E. Coli* O157:H7 and *Salmonella typhimurium* using a sandwich assay with initial magnetic separation of the bacteria. The detection limit was 10^4 cfu mL^{-1}, which, although not as sensitive as methods such as real-time PCR, offered a significantly shorter analysis time (2 h *versus* 16–48 h for the real-time PCR with an enrichment step).[75] Another study demonstrated the use of three different QD-antibody conjugates with peak fluorescence at 504 nm, 557 nm and 604 nm for the detection of *Salmonella enteritis*, *Staphylococcus aureus* and *E. coli*, respectively (Figure 2.7).[70] In this case, the immunolabelled cells were isolated by centrifugation prior to fluorescence measurements. The estimated detection limits were 10^2 cfu mL^{-1} with an analysis time of less than 2 h. The assay method was tested with pathogen-spiked samples of soda water, mineral water and apple juice, as well as dilutions of milk, tomato and chicken sauces, pudding, cheese and luncheon meat. Qualitative detection was possible, and quantitation improved with greater dilution (10^2–10^5-fold) or less complex sample matrices (*e.g.*, soda and mineral water).[70] A similar assay with immunomagnetic separation and QD-antibody conjugates with peak fluorescence at 525 nm, 565 nm and 625 nm was demonstrated for parallel detection of *Salmonella typhimurium*, *Shigella flexneri* and *E. coli* O157:H7 with detection limits of 10^3 cfu mL^{-1} with dilutions of apple juice and milk.[81]

The success of multiplexed assays often depends on the minimization of crosstalk, which can arise from cross-reactivity (*i.e.*, binding to more than one target) and non-specific binding of antibodies, as well as from optical crosstalk (*e.g.*, a fluorescent label yields signal in more than one detection channel, or excitation light contaminates the signal in a detection channel).

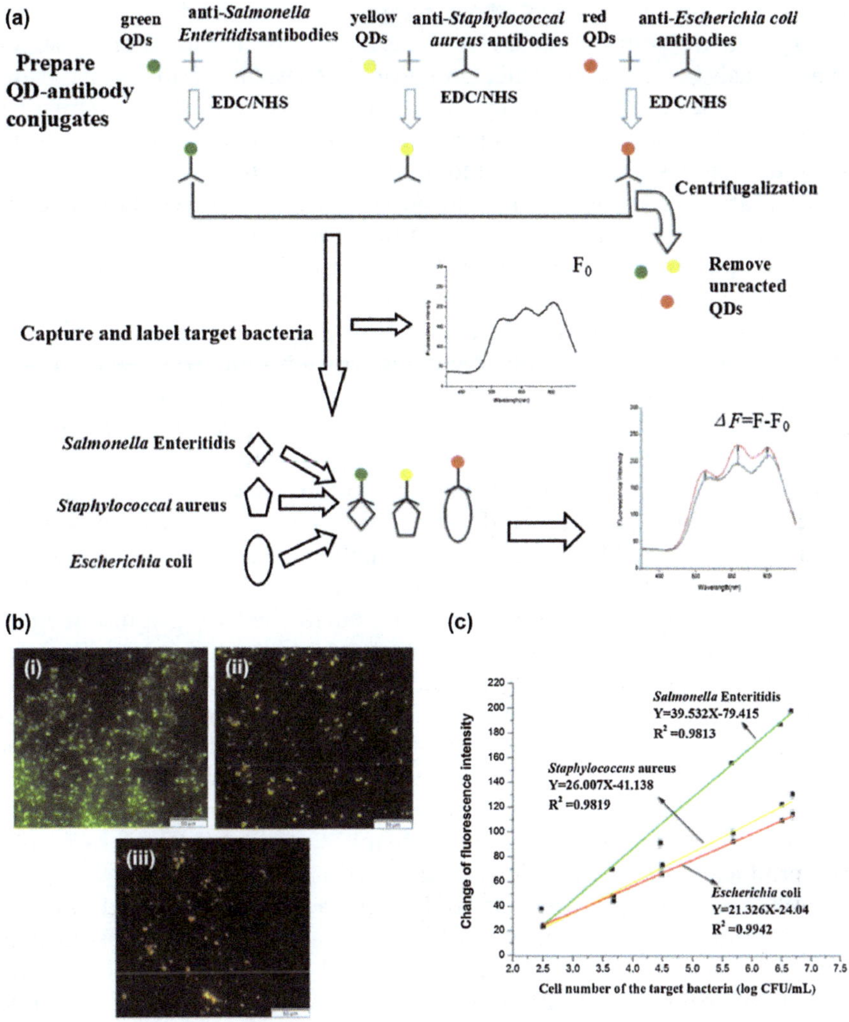

Figure 2.7 (a) Preparation scheme for the conjugation of three colours of QDs with antibodies for the detection of *Salmonella Enteritidis*, *Staphylococcal aureus* and *Escherichia coli*. (b) Fluorescence microscopy images of (i) green QD-labelled *Salmonella*, (ii) yellow QD-labelled *Staphylococcus* and (iii) red QD-labelled *E. coli*. (c) Change in the fluorescence intensity as a function of the log-transformed number of each cell type for the simultaneous detection of the three targeted bacteria.
Reprinted from *LWT - Food Sci. Technol.*, **61**(2), B. Wang, Q. Wang, Z. Cai, M. Ma, Simultaneous, rapid and sensitive detection of three food-borne pathogenic bacteria using multicolour quantum dot probes based on multiplex fluoroimmunoassay in food samples, 368–376. Copyright 2015 with permission from Elsevier Ltd.

The former is generally addressed through careful antibody selection, control experiments and optimization of assay chemistry; the latter is helped tremendously by the fluorescence properties of QDs. Limitations when using fluorescent dyes are the typically small spectral shift (*i.e.*, Stokes shift) between the absorption and fluorescence spectra, often requiring multiple excitation sources for different colours of dye, and the broad fluorescence spectra that can significantly overlap with one another and contaminate the detection channel of other dyes. In contrast, multiple colours of QD can be excited at one common wavelength far from the fluorescence wavelengths, and their symmetric and narrow fluorescence spectra can be much better separated or deconvoluted.[82] Spectrally multiplexed fluorescence detection is thus technically easier to implement and typically more effective with QDs.

In addition to sensitivity and multiplexing capability, one of the most sought-after characteristics of an assay for the detection of foodborne pathogens and contaminants is portability. Smartphones and cell phones with integrated cameras are examples of portable devices that are already ubiquitous throughout the world. The built-in digital cameras can be combined with peripheral attachments to enable microscopic imaging and fluorescence detection. In one such example, glass capillaries were coated with antibodies on their interior surface and flushed with liquid samples spiked with *E. coli*.[83] The bacteria were captured by the antibodies and, in a second and a third step, labelled with biotinylated antibodies and streptavidin-QDs to complete the sandwich immunoassay. The capillaries were then further used as light guides to deliver excitation light from blue LEDs, and the fluorescence intensity measured from camera images of the capillaries was used to quantify the bacteria number in the sample. This system had a detection limit of 5–10 cfu mL^{-1}. Assays of *E. coli* in fat-free milk produced similar detection limits.[83]

Another portable assay format are lateral flow test strips, which are particularly attractive because they provide the potential for on-site testing with portability, minimal reagent consumption and elimination of the need for complex instrumentation. Although lateral flow assays have typically used antibodies for target binding, a lateral flow test strip assay for *E. Coli*, *Listeria* and *Salmonella* has been developed using aptamers and QDs (Figure 2.8).[84] As noted above for other formats, the use of QDs with lateral flow devices has been shown to improve sensitivity.[85] Two aptamers were used as part of a sandwich assay for each type of bacterium: a biotinylated reporter aptamer was conjugated to a streptavidin-coated QD, and an amine-terminated capture aptamer was covalently immobilized on a nitrocellulose membrane. A limit of detection (LOD) of 300–600 live bacterial cells per lateral flow test was achieved. This was a significant improvement over the same assay performed with a colloidal gold conjugate, which yielded a detection limit of 3000–6000 cfu.

Microfluidics is yet another technology that is expected to help enable portable assays for food safety and quality. A key ambition of this technology is to create "lab-on-a-chip" devices that integrate multiple steps of sample

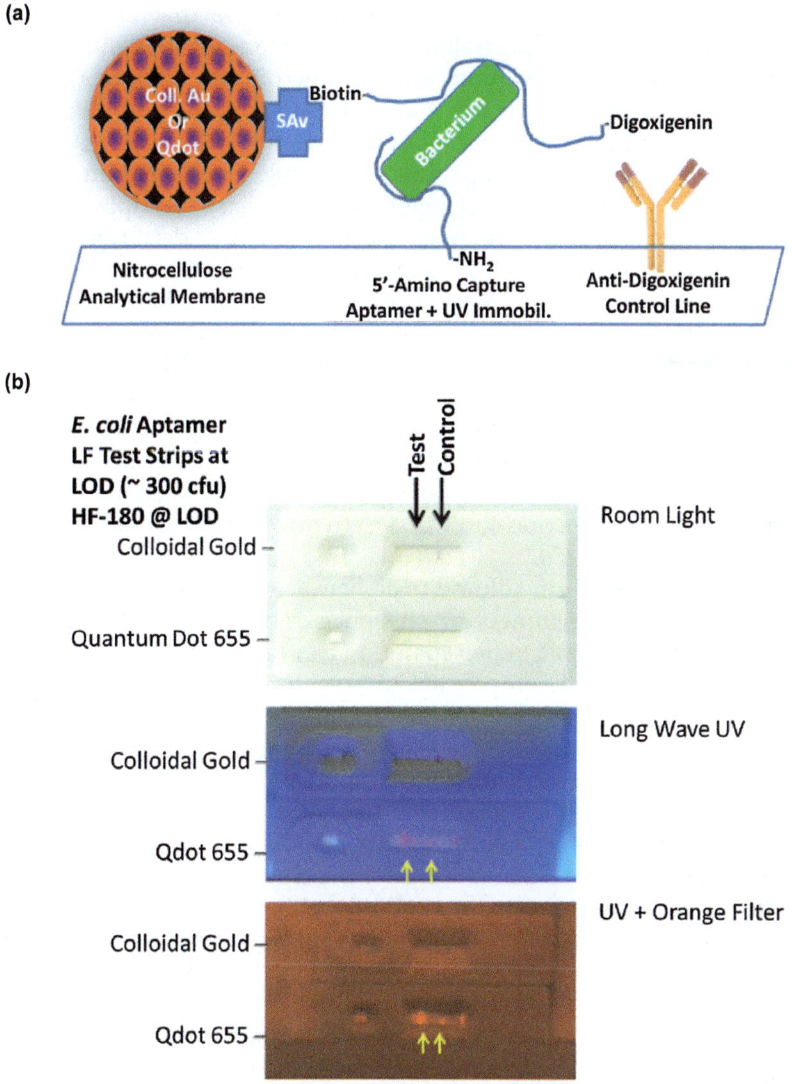

Figure 2.8 Lateral flow test strip assay for *E. Coli*, *Listeria* and *Salmonella* that uses QD-aptamer conjugates for detection. (a) Schematic showing the sandwich assay format for the detection of bacteria. A biotinylated aptamer for the bacterium of interest is conjugated to a streptavidin-coated QD (or Au NP, which was used as a comparison), and a second aptamer is immobilized onto the nitrocellulose membrane of the test strip. Binding of the target bacterium to the aptamers gives a fluorescent read-out on the test strip. The reporter aptamer is also functionalized with digoxigenin and acts as a control tracer by binding to immobilized anti-digoxigenin. (b) Comparison of the performance of a QD-based test strip with an Au NP-based test strip. The 10-fold improvement in sensitivity by using QDs as a reporter allows a visual read-out of the test results. Reprinted with permission from ref. 84, J. G. Bruno, *Pathogens*, 2014, **3**, 341–355 under the Creative Commons Attribution License. Copyright 2014 the authors.

preparation and analysis into a hand-held package that uses very small volumes of sample, and has straightforward operation and high throughput. The high surface area-to-volume ratio in microfluidic channels and more efficient mass transport can promote rapid binding in solid-phase affinity assays.[86] One example of a microfluidic chip for foodborne pathogen detection was designed with 12 inlet ports, 3 mixing zones and 6 immunoassay detection zones with immobilized capture antibodies.[87] Samples containing *Salmonella typhimurium* were first injected into the detection zones for capture of the bacteria. Next, primary antibodies and QDs conjugated with secondary antibodies were injected through separate inlets, bound to one another in the mixing zones, then delivered to the detection zone to label the captured bacteria. The mixing zones enhanced the efficiency of primary antibody–secondary antibody binding, and the flow velocity and shear stress were expected to help minimize non-specific adsorption. The detection zones were designed with integrated columns to increase its effective area and decelerate fluid flow to increase the interaction between the reactants.[87] The device had a detection limit of 37 cfu mL^{-1} and a dynamic range between 37 cfu mL^{-1} and 3.7×10^5 cfu mL^{-1}.

Given the mismatch in volume scale between real food samples and the microfluidic chip, enrichment or pre-concentration steps will often be necessary, and immunomagnetic separation can be a suitable strategy. In one example, a microfluidic chip made of poly(dimethylsiloxane) (PDMS) had magnets embedded beneath the microchannel to divide the microchannel into different zones for capture and detection of pathogens.[88] Magnetic particles were conjugated with capture antibodies for a sandwich immunoassay with QD-antibody conjugates. *E. coli* and *S. typhimurium* were detected in the range of 10^3–10^7 cfu mL^{-1}, and green-fluorescent QDs and red-fluorescent QDs were used in tandem for multiplexed detection.[88] In another example, a more sophisticated microfluidic chip was developed to eliminate the manual washing steps that are often required with immunomagnetic separation (Figure 2.9).[73] The chip design combined dynamic magnetophoretic separation with a stationary magnetic trap for detection of *S. typhimurium*. It was compatible with large sample volumes, complex sample matrices and high-throughput analyses. The principle was control of the flow direction of magnetic particles by manipulating the magnetic force against the hydrodynamic force. The first step in the assay was incubation of the sample with magnetic particles conjugated with capture antibodies. This solution was then injected into the microfluidic device where it entered a separation zone. This separation zone was a confluence of sample and buffer streams with laminar flow and had embedded nickel wires that induced a lateral magnetic force. The lateral magnetic force moved magnetic particle-tagged *S. typhimurium* into the buffer stream and detection outlet channel whereas untagged pathogens or other bacteria (*e.g.*, *E. coli*) continued in the sample stream through another outlet channel. The magnetic forces experienced by the magnetic particles in the separation zone were much larger than the hydrodynamic force that kept untagged cells and material in the

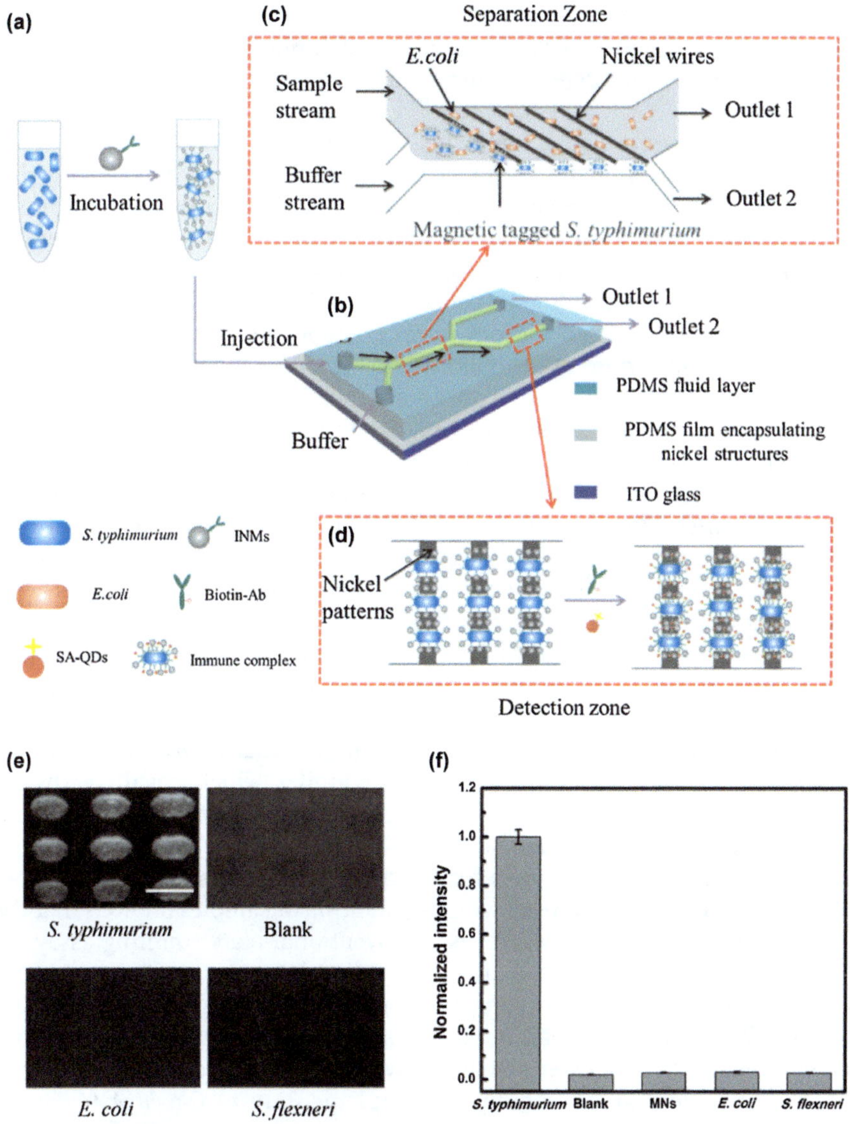

Figure 2.9 Working scheme of a microfluidic immunomagnetic assay device. (a) Labelling of the pathogen with antibody-conjugated magnetic particles. (b) Layout of the integrated chip. (c) Magnetophoretic separation of labelled pathogen cells from non-pathogen cells. (d) Magnetic trapping of pathogen cells and labelling with QDs for fluorescent detection. (e) Fluorescence images of the control experiments to test the specificity of the system using *S. typhimurium* (positive sample) and three negative samples: PBS (Blank), *E. coli* and *S. flexneri*. (f) Normalized PL intensities for the fluorescence images in (e).
Reprinted from *Biosens. Bioelectron.*, **74**, P.-L. Guo, M. Tang, S.-L. Hong, X. Yu, D.-W. Pang, Combination of dynamic magnetophoretic separation and stationary magnetic trap for highly sensitive and selective detection of Salmonella typhimurium in complex matrix, Z.-L. Zhang, 628–636. Copyright 2015 with permission from Elsevier B.V.

sample stream. In the detection zone, nickel patterns captured the magnetic particle-tagged *S. typhimurium* for labelling with biotinylated antibodies and streptavidin-QDs. The detection limit was 3.4×10^3 cfu mL^{-1} with a 50 min assay time.[73]

A common theme of the foregoing assay formats was that fluorescence intensity was measured and, as a result of washing steps to remove QDs that were not bound to pathogen cells, that fluorescence intensity was proportional to the number of QDs in the sample and, in turn, the number of pathogen cells. A fluorescence technique called Förster resonance energy transfer (FRET) allows measured fluorescence intensities to be correlated to the number of binding events without washing steps. FRET is non-radiative energy transfer from a donor fluorophore to an acceptor chromophore when the two are within a few nanometres of one another.[30] The result is quenching of the fluorescence of the donor and sensitization of fluorescence from the acceptor (if fluorescent). This technique is widely used in biology for binding assays, including immunoassays. A putative FRET assay for the detection of *Salmonella enteriditis* was reported with green- and orange-fluorescent QD-antibody conjugates.[89] In the absence of the *Salmonella*, the QD-antibody conjugates bound to each other because the orange-fluorescent QDs were conjugated with goat anti-rabbit IgG antibodies and the green-fluorescent QDs were conjugated with rabbit anti-*S. enteriditis* antibodies, with suspected FRET from the green-fluorescent QD to the orange-fluorescent QD. The presence of the target bacteria increased the signal from the green-fluorescent QD and decreased the signal from the orange-fluorescent QD. FRET was lost with dissociation of the antibody–antibody complex, which was driven by the affinity of the donor antibody for its target on the bacteria cells. The linear relationship between the intensity of fluorescence from the green QD and bacteria concentration was used for quantification with a detection limit of 10 cfu mL^{-1}. Results from tests with egg shell wash samples showed that the FRET assay gave the same results as a conventional plate counting assay.[89]

FRET assays with QDs have also been utilized for the detection of genes associated with foodborne pathogens. In one example, classic molecular beacons for the detection of *Listeria monocytogenes*, *Bacillus thuringiensis* and *Salmonella typhimurium* were designed with a QD in place of a fluorescent dye.[90] The molecular beacons were complementary to the *hlyA* (*L. monocytogenes*), *Bt cry1Ac* (*B. thuringiensis*) and *invA* (*S. Typhimurium*) genes. Each beacon was modified with an amine group at its 5′-terminus and conjugated to carboxylated QDs using carbodiimide chemistry. A black hole quencher (BHQ) dye was located on the 3′-end to complete the FRET pair. In the absence of the gene target, the beacon was in the closed stem-loop form and the BHQ quenched the PL of the QD donor. Upon incubation with target, the stem-loop structure was opened by probe-target hybridization, restoring the PL from the QD. Bacterial genomic DNA targets were detected directly in water, milk and plant tissue.[90]

Fluorescence measurements are not the only possibility with QDs. Although less common, QDs are also useful in electrochemical assays. Upon dissolution, QDs yield a large quantity of metal ions that can be detected using sensitive techniques such as anodic stripping voltammetry (ASV). The size of the QD determines the number of metal ions and thus the amplification factor *versus* a molecular redox label, and different compositions of QD provide some capacity for multiplexing.[91] For example, a multiplexed square-wave anodic stripping voltammetry assay with three different types of QDs (PbS, CdS and ZnS) was developed for the detection of gene targets associated with the enteric pathogens *Vibrio cholera* (VrrA), *Salmonella* sp. (StyR-36) and *Shigella* sp. (CssrB) (Figure 2.10).[92] Each QD had a resolved

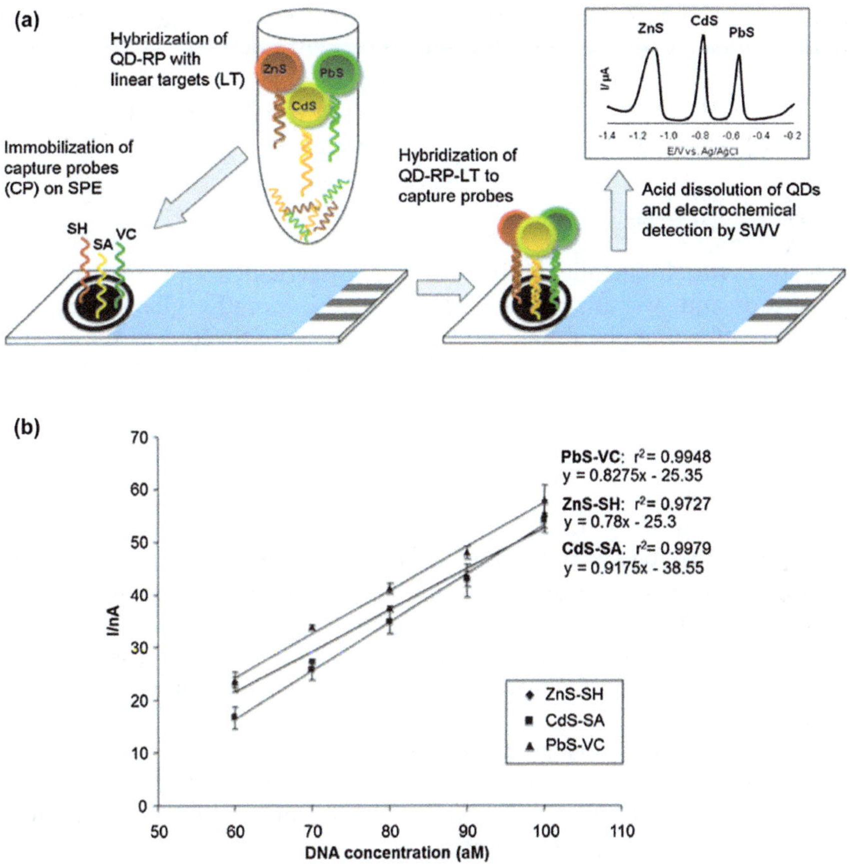

Figure 2.10 (a) Schematic of a sandwich hybridization assay for the simultaneous detection of *Vibrio cholerae* (VC), *Salmonella* sp. (SA) and *Shigella* sp. (SH) with electrochemical detection. The bound ZnS, CdS and PbS QDs are dissolved in acid and analyzed by square-wave ASV. (b) Calibration plot for a mixture of PCR products for the three pathogens.
Figure reproduced from *Biosens. Bioelectron.*, 77, D. Vijian, S. V. Chinni, L. S. Yin, B. Lertanantawong, W. Surareungchai, Non-protein coding RNA-based genosensor with quantum dots as electrochemical labels for attomolar detection of multiple pathogens, 805–811. Copyright 2016 with permission from Elsevier B.V.

anodic stripping peak (0.5 V for PbS, 0.75 V for CdS and 1.1 V for ZnS), which allowed for simultaneous electrochemical detection of three different targets. In this case, the target nucleic acids were PCR products corresponding to non-protein coding RNA sequences. DNA capture probes were electrochemically immobilized onto screen-printed carbon electrode surfaces and thiolated DNA reporter probes were immobilized onto the QD surfaces. A sandwich assay format was used to detect the three pathogens in less than 2 h, with attomolar detection limits for both oligonucleotide targets and PCR amplicons. Specific detection limits were 22 aM for *Vibrio cholera* (PbS QD), 34 aM for *Salmonella* (CdS QD) and 38 aM for *Shigella* (ZnS QD) with oligonucleotide targets. The dynamic range for a single target was between 10 and 60 aM, whereas multiplex detection had a dynamic range between 50 and 100 aM. The LODs for each pathogen using PCR products were 60 aM for *Vibrio cholera* (91 bp PCR amplicon), 59 aM for *Salmonella* (150 bp PCR amplicon) and 53 aM for *Shigella* (96 bp amplicon).[92]

2.3.2 Pesticides

Pesticides, which include herbicides and insecticides, are important for agriculture and are estimated to save one-third of the global crop production. Unfortunately, only 1% of pesticides are delivered to the crop of interest; the majority is accumulated in other plants, soil and water, and, because of their long half-life, pose a risk of exposure for animals and humans. As such, quantitation of pesticides in the environment is important for public safety. The most well-known class of pesticides are organophosphates (OP), which were developed in the 1940s and have been used heavily in agriculture since then. Nitro-OPs such as parathion, paraoxon and chloropyrifos are widely used (with declining use in high-income countries but still widespread use in low-income countries) and are known to have serious toxic effects.[93] Another class of pesticides are carbamates and, like OPs, they inhibit the enzyme activity of acetylcholinesterase (AChE) in organisms, including both insects and humans, causing malfunction of the nervous system.[94]

Conventional methods of detection of pesticides utilize liquid or gas chromatography coupled with mass spectroscopy (LC-MS, GC-MS). Although very useful, the limitations of chromatographic-mass spectrometric methods include laborious sample preparation time, large and expensive equipment and the need for highly trained personnel. There remains a need for inexpensive, portable and user friendly detection tools that will allow on-site environmental and biological monitoring. Electrochemical and optical assays have the potential to meet this demand. Nanomaterials such as metal nanoparticles, carbon nanotubes and QDs can enhance the sensitivity of both of these types of assays.[95–97]

A competitive immunoassay for 2,4-dichlorophenoxyacetic acid (2,4-D), a herbicide, was developed using QDs as a fluorescent reporter.[98] Capture antibodies were immobilized within a Sepharose column, and a sample with

2,4-D was spiked with QD-2,4-D conjugates. The intensity of the fluorescence of the column eluent was directly proportional to the amount of 2,4-D in the sample, with a detection limit of 250 pg mL^{-1}.[98] This type of methodology is expected to be applicable with a wide range of pesticides.

Another set of detection strategies for pesticides that utilize QDs rely on modulation of QD fluorescence intensity. Although materials such as CdSe/ZnS QDs tend to have robust fluorescence properties, they can also engage in FRET and charge transfer (CT), leading to quenching of their fluorescence. The fluorescence of other QD materials, such as core-only CdTe, is often sensitive to the presence of molecules adsorbed or coordinated to their surface, for example, as a result of hole trapping. The foregoing quenching processes are all strongly distance dependent and, for this reason, large receptor molecules such as antibodies are often non-ideal for these detection strategies, notwithstanding limitations in the availability of antibodies and aptamers for many pesticides.

In one example of a fluorescence modulation assay for a pesticide, QDs were prepared with dithizone ligands that quenched their fluorescence, putatively through FRET. Hydrolysis of chlorpyrifos, a common pesticide, resulted in two products, diethylphosphorothioate (DEP) and trichloro-2-pyridinol (TCP). DEP was able to bind to the QDs and displace dithizone, thus restoring the fluorescence of the QD, which was used for the quantification of the chlorpyrifos. When tested with spiked apple juice samples, this assay had a detection limit of 5.5 ppb, which was below the maximum residue limit (10 ppb).[5] In many cases, target pesticides will not have intrinsic binding to QDs and the QDs will need to be functionalized with receptor molecules that can bind the target pesticide. Macrocyclic host molecules like crown ethers, cyclodextrin and calixarenes are known to complex certain ions and small molecules. QDs in a sol–gel derived silica composite functionalized with calix(4)arene have been demonstrated for the fluorescent detection of methomyl, a carbamate pesticide.[99] Binding of methomyl enhanced QD fluorescence intensity, which was used for quantitation. Other pesticides elicited small or negligible enhancements in QD fluorescence intensity.[99] Presumably, binding of target disrupted charge transfer between the QD and calixarene. Another approach is to create artificial receptor sites around a QD through molecular imprinting. Molecularly imprinted polymers (MIPs; the topic of another chapter in this book) have binding cavities that are complementary to a target analyte. MIPs are much more robust and economical to produce than antibodies, but have lower specificity and affinity. To develop an assay for cyphenothrin, a synthetic pyrthroid insecticide, QDs were coated with a silica shell then further coated with MIP targeted to cyphenothrin. Binding of cyphenothrin to the MIP resulted in quenching of QD fluorescence that was linear with concentration in the range between 0.1 and 80 μM.[100] In a similar example, QDs were embedded within MIP nanospheres templated for diazinon, an OP pesticide (Figure 2.11).[101] Binding of diazinon quenched QD fluorescence, whereas other pesticides and structural analogues did not

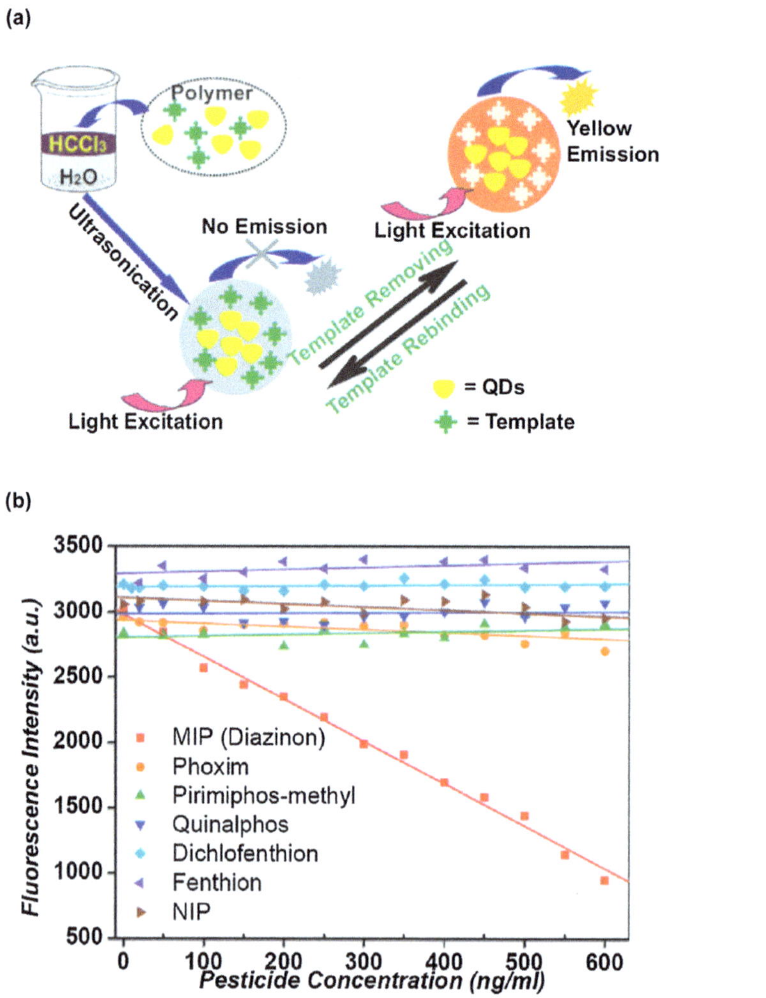

Figure 2.11 (a) Scheme for the preparation of molecularly imprinted polymer (MIP) nanospheres with embedded QDs. After template removal, selective binding of diazinon in a sample quenches the fluorescence of the QDs. (b) Calibration curves for diazinon and for several analogues, illustrating good selectivity. NIP is a non-imprinted polymer.
Reprinted with permission from Y. Zhao, Y. Ma, H. Li, L. Wang, Composite QDs@MIP Nanospheres for Specific Recognition and Direct Fluorescent Quantification of Pesticides in Aqueous Media, *Anal. Chem.*, 2012, **84**, 386–395. Copyright 2012 American Chemical Society.

(*e.g.*, dichlofenthion, fenthion, pirimphos-methyl, phoxim, quinalphos). The reported mechanism of quenching was FRET, albeit difficult to attribute to pure diazinon (which is colourless), suggesting the possibility of a different mechanism.

The quenching effect of hydrogen peroxide on the fluorescence intensity of QDs can also be used for the measurement of OPs. One such assay was a two enzyme approach with AChE and choline oxidase, where there was sequential enzymatic conversion of acetylcholine to choline to hydrogen peroxide, which is thought to quench QDs through oxidation (and loss) of ligands and possibly oxidation of the QD surface itself. Inhibition of AChE by OPs limits conversion of acetylcholine to choline and thus hydrogen peroxide, such that the OP concentration is directly proportional to QD fluorescence intensity.[102] This assay format has also been implemented in bulk solution,[102,103] and as a layer-by-layer film on a glass substrate,[104,105] for detection of pesticides such as paraoxon, dichlorvos and parathion. Detection limits were in the picomolar range with a dynamic range spanning 5–6 orders of magnitude, and the assays were tested with apple-derived samples.[104,105]

As noted earlier in this chapter, fluorescence assays are most common with QDs, but electrochemical assays are also possible. Pesticide detection is perhaps the application in food analysis that has most frequently used QDs in electrochemical assays. A somewhat unusual example adopted QDs as a passive, high-surface-area scaffold for the immobilization of a bacterial phosphotriesterase, methyl parathion degrading enzyme (MPDE), which can be used for the detection of methyl parathion. MPDE hydrolyzes the P–S bond of the OP agent and forms *p*-nitrophenol, which can be oxidized on a working electrode at fixed potential to generate a current proportional to its concentration. A glassy carbon electrode was sequentially modified with multi-walled carbon nanotubes (MWCNTs), gold nanoparticles and QD-MDPE conjugates. The gold NPs and MWCT increased the electrochemical response of the electrode, and the QDs permitted immobilization of a greater number of enzymes to increase the assay sensitivity. The detection limit was 1.0 $ng\,mL^{-1}$ for methyl parathion, and the enzyme provided selectivity to pesticides with a P–S bond. The platform also had the potential to be reused.[106] More often, QDs have active roles in generating measurable signals; for example, photoelectrochemical assays take advantage of the optical and electronic properties of QDs. In a photochemical assay for OPs, an indium tin oxide (ITO) electrode was modified with a composite layer-by-layer film of graphene, poly(ethylene imine), QDs and AChE.[107] The combination of materials was designed to improve the photocurrent response *versus* QDs and AChE alone. Hydrolysis of acetylthiocholine by AchE generated thiocholine, which acted as an electron donor for the photogenerated holes in the QDs, resulting in an enhancement of the observed photocurrent. The presence of pesticide inhibited AChE activity with a reduction in thiocholine generation, such that the photocurrent was inversely proportional to the OP concentration. Limits of detection were 10^{-14} M for paraoxon and 10^{-12} M for dichlorvos, and were comparable to those for established GC-MS methods.[107]

In addition to the detection of pesticides in food, water and environmental samples, assay methods are needed for the assessment of exposure to OPs.

These latter assays typically target the metabolites of pesticides or their inhibitory effects on choline esterase enzymes. For example, trichloropyridinol (TCP) is a metabolite of chlorpyrifos and has been detected with a competitive lateral flow immunochromatographic assay.[108] A sample with TCP was spiked with TCP-QD conjugates and applied to a lateral flow strip with a test line of immobilized anti-TCP antibodies. The QD fluorescence intensity from the test line was inversely proportional to the amount of TCP in the sample, with a detection limit of 1.0 $ng\,mL^{-1}$. The measurement time was only 15 minutes and the method was inexpensive, user-friendly and adaptable to different environments.[108]

QDs have also been used as components of electrochemical assays for assessment of pesticide exposure. Many methods measure cholinesterase (ChE) activity in blood, where exposure is determined from the degree of inhibition of esterase activity.[109] A drawback of these assays is the need to assume an average normal activity for ChE activity, even though there is variation between individuals in a population. This challenge has been addressed with a method that measures both ChE activity and total enzyme concentration.[110] Magnetic beads were modified with anti-butyrylcholinesterase (BChE) antibodies to capture BChE, a bloodborne enzyme that is phosphorylated by OPs. To complete the immunomagnetic assay, the bound BChE was labelled with QD-antibody conjugates and analyzed by anodic stripping voltammetry (ASV). To measure activity, butylthiocholine was added and its BChE-catalyzed conversion to redox-active choline was measured by amperometry. The difference between the signals for activity and the total amount of BChE was used to estimate the amount of phosphorylated BChE.[110]

An alternative approach is to directly target phosphorylated cholinesterase enzymes. Such measurements can be facilitated by zirconia (ZrO_2) nanoparticles, which have high affinity for OPs and are easy to produce. In one example of an assay utilizing these materials, zirconia nanoparticles were pre-coated on a screen printed electrode (SPE) and used as a selective sorbent to capture phosphorylated AChE, which is one of the products of exposure to OP pesticides such as paraoxon.[111] Antibodies conjugated with Cd-based QDs bound to the captured phosphorylated AChE and were measured by dissolution with acid followed by ASV. The voltammetric response for paraoxon, which was used as a model OP, showed a linear trend between 10 pM and 4 nM with a detection limit of 8 pM. This sensitive method was robust enough to work with complex biological samples like human blood plasma.[111] Two variations of this method have been reported. In one variation, the detection of phosphorylated BChE was demonstrated (detection limit of 0.03 nM).[112] In the other variation, the zirconia nanoparticle-coated screen printed electrode was replaced with magnetic particles modified with antibodies for capturing phosphorylated AChE, which were then labelled with QD-antibody conjugates. Once isolated, the QDs were dissolved in acid and the ASV signal measured.[113]

2.3.3 Antibiotics

Antibiotics are used in modern animal agriculture to prevent infections and promote growth in animals. Several thousand tonnes of antibiotics are used annually in poultry, hog and cattle farming, as well as in aquaculture. Ambitions of high-profit farming have also led to the use of illegal antibiotics, making it imperative to monitor vestigial antibiotics in farmed animal products,[114] especially meat and dairy.[115] The consumption of antibiotic contaminated food can cause allergic reactions and reduce the efficacy of antibiotics for treatment of infections in humans as microbes develop resistance.

As highlighted earlier in this chapter, the use of QDs as fluorescent reporters for immunoassays is well established for the analysis of food quality and safety, and this methodology extends to the detection of antibiotics. A microplate-based, competitive immunoassay has been reported for the high-throughput detection of streptomycin (SM), tetracycline (TC) and penicillin G (PC-G) in milk. Conjugates of these antibiotics with bovine serum albumin (BSA) were immobilized in wells and samples were added along with QD-antibody conjugates. The QDs had peak fluorescence at 520 nm, 565 nm and 610 nm, and were conjugated with antibodies targeting SM, TC and PC-G, respectively. The measured QD fluorescence decreased with increasing antibiotic concentration with a detection limit of 5 $pg\,mL^{-1}$ and was validated against the standard ELISA assay format.[115] A competitive immunoassay for screening of sulfonamides and quinolones in milk has been implemented in a flow-through column format using a Sepharose support for capture antibodies.[116] Multiple QDs were encapsulated in liposomes as a method of increasing fluorescence signals, and the liposomes were conjugated with antibiotics carried by BSA. The column was divided into detection zones with different capture antibodies, permitting the use of one colour of QDs, with an increasing concentration of antibiotics signalled by less fluorescence in the column. Compared to an analogous assay with horseradish peroxidase and a chromogen, QDs provided a 10-fold improvement of the limit of detection.[116] QDs have also been shown to provide five-fold better sensitivity, decreased rates of false positive results and better specificity than gold nanoparticles in a lateral flow immunochromatographic assay for the detection of gentamicin in milk samples.[117]

A competitive binding lateral flow immunochromatographic assay for parallel detection of ofloxacin (OFL), chloramphenicol (CAP) and streptomycin (STM) has also been reported, and was enabled by the use of three different colours of QDs as labels (525 nm, 585 nm, 625 nm), with each colour of QD binding a particular antibiotic through conjugation with the appropriate antibody (Figure 2.12).[118] With increasing concentration of an antibiotic the fluorescence intensity in the corresponding test zone decreased. The calculated limits of detection for ofloxacin, chloramphenicol and streptomycin were 0.3, 0.1 and 0.2 $ng\,mL^{-1}$, respectively, which were

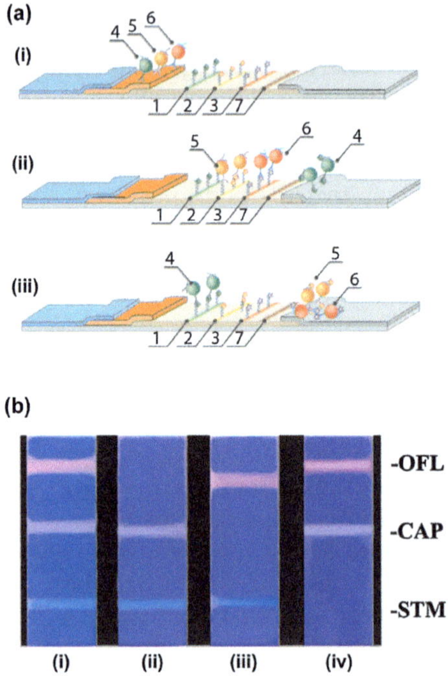

Figure 2.12 (a) Principle of a competitive binding lateral flow assay. (i) Prior to addition of the sample to the sample pad (blue). The membrane has three lines (1–3) of immobilized target antibiotics: STM, CAP and OFL. Three different colours of QD-antibody conjugate (4–6) are adsorbed on the conjugate pad (orange), where the antibody for each conjugate targets a different antibiotic. A control line is also shown (7). (ii) Expected assay result for a sample that is positive for STM. (iii) Expected assay result for a sample that is positive for OFL and CAP. (b) Actual assay results for various samples: (i) no antibiotics, (ii) OFL (220 ng mL^{-1}), (iii) CAP (10 ng mL^{-1}) and (iv) STM (500 ng mL^{-1}). Reprinted from *Biosens. Bioelectron.*, **63**, N. A. Taranova, A. N. Berlina, A. V. Zherdev, B. B. Dzantiev, 'Traffic light' immunochromatographic test based on multicolour quatum dots for the simultaneous detection of several antibiotics in milk, 255–261. Copyright 2015 with permission from Elsevier B.V.

between 80 and 200 times lower than those achievable with ELISA using the same antibodies.[118]

Similar to what was summarized earlier for pesticide detection, the adsorption of some antibiotics on the surface of certain QDs can lead to quenching or enhancement of QD fluorescence. This modulation of fluorescence intensity can be used for quantitative detection if proportional to the concentration of antibiotics. Enhancement of the fluorescence intensity of CdTe QDs has been reported for the detection of lincomycin[119] and sparfloxacin (SPF).[120] In the latter example, the enhancement effect was attributed to strong hydrogen bonding interactions between SPF and the silica-encapsulated QDs, but the actual mechanism of enhancement was not determined. The fluorescence method produced results that were in agreement with those from an HPLC method when tested with SPF-spiked human serum samples, suggesting that a low-cost, fluorescence-based assay may be able to compete with more expensive and time-consuming HPLC methods.[120] In other assays, interactions between antibiotics and QDs quench fluorescence. In one example, iron oxide magnetic nanoparticle-ZnS:Mn QD composites were synthesized and used for the detection of tetracycline hydrochloride (TC), which is a broad spectrum antibiotic against Gram positive and negative bacteria.[121] In presence of TC, QD fluorescence was quenched with a dynamic range between 10 and 700 nM and a detection limit of 1.2 nM. Magnetic isolation of the nanoparticles from the sample was utilized prior to fluorescence measurements.[121] In another example, vancomycin, a glycopeptide antibiotic for Gram positive bacteria, was found to dynamically quench QD fluorescence through putative electron transfer (Figure 2.13).[122] The method had a detection limit of 0.46 ng mL^{-1} and was validated by testing environmental water samples and human serum spiked with vancomycin.[122] There was no engineered interaction between the QDs and vancomycin, similar to other examples where putative dynamic quenching of QD fluorescence was used to detect TC,[123] doxycycline[124] and Rifampicin and Rifaximin.[125] The method for Rifampicin and Rifaximin used an automated flow system that enabled assays of 70 samples within one hour.[125]

FRET-type assays with QDs have also been developed for the detection of antibiotics and are more general than approaches that rely on the specific properties of an antibiotic to quench QD fluorescence. One such example is an aptamer-based assay for the detection of chloramphenicol (CAP) (Figure 2.14).[126] QDs were embedded in liposomes and aptamer-functionalized Au NPs were attached to the liposome using a single-stranded DNA binding protein. The Au NPs quenched the fluorescence of the QDs. Upon addition of CAP, the aptamer preferentially bound CAP and the Au NP-aptamer conjugate was released from the liposome surface, leading to restoration of QD fluorescence. The assay was able to detect CAP in diluted milk samples, with a linear response between 1 pM and 10 nM and a detection limit of 0.3 pM.[126]

Chemiluminescent and electrochemiluminescent systems with QDs have also been proposed for the detection of antibiotics. In the former, a system utilized permanganate, carminic acid and CdS QDs, where the QDs were proposed to act as a catalyst for the chemiluminescent reaction between permanganate and carminic acid. Cloxacillin was found to quench the chemiluminescence in a flow injection assay, with a detection limit of 5.8 μg L^{-1}.[127] In the latter, a common tris(2,2′-bipyridyl)ruthenium(II) complex ($Ru(bpy)_3$) electrochemiluminescence (ECL) system was combined with QDs for the detection of nitrofuran antibiotics.[128] These antibiotics have been banned in many countries because of concerns about their carcinogenicity and mutagenicity, but are still used in some developing countries. A screen-printed electrode was coated with $Ru(bpy)_3$, which, when oxidized,

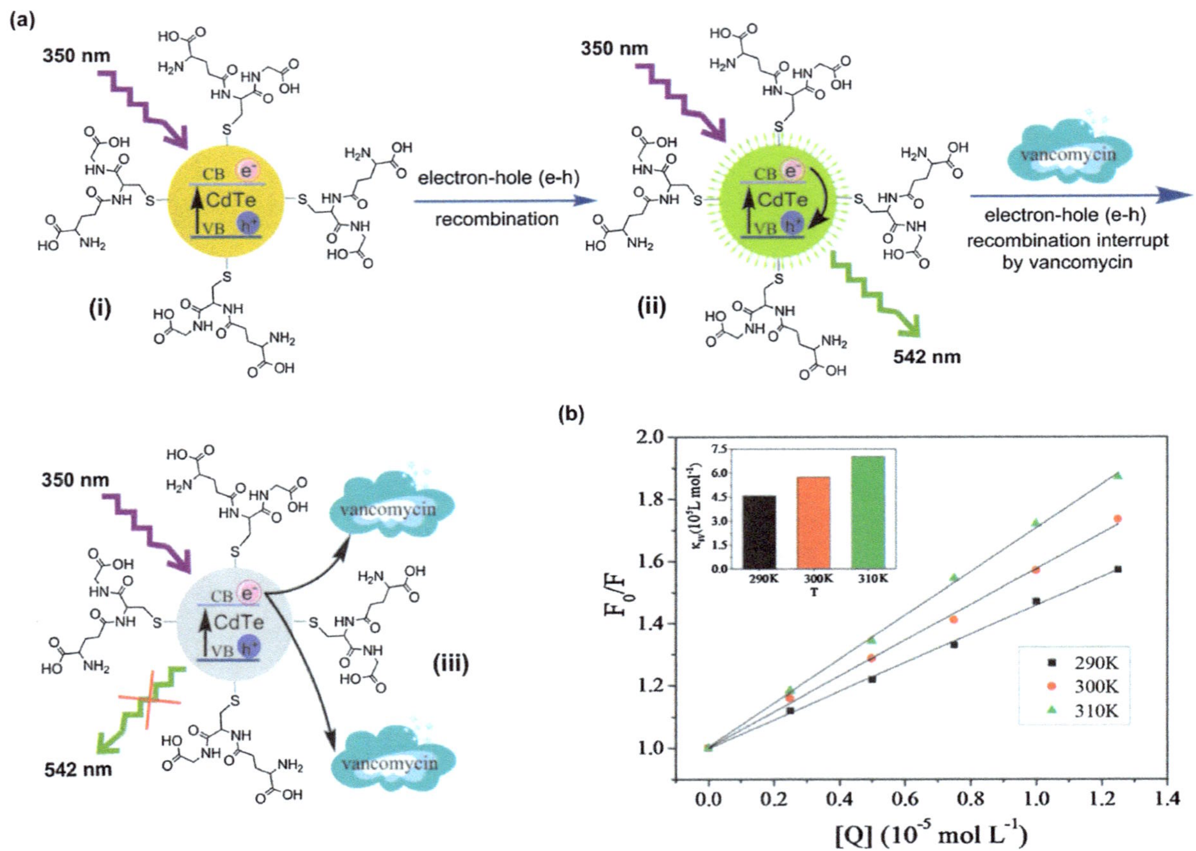

Figure 2.13 (a) Scheme showing detection of vancomycin through quenching of QD fluorescence. (b) Stern–Volmer plots for GSH-CdTe QDs with different concentrations of vancomycin at three different temperatures. The inset shows the relationship between the Stern–Volmer quenching constant (k_{SV}) and the temperature (T).
Reproduced from ref. 122 with permission from the Royal Society of Chemistry and the Centre National de la Recherche Scientifique.

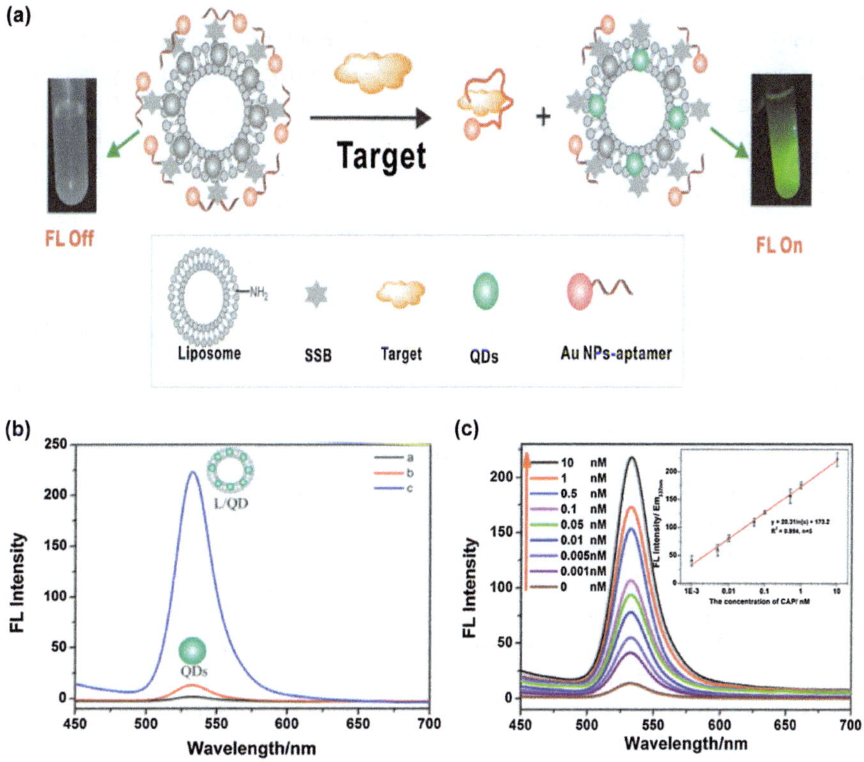

Figure 2.14 Aptamer-based fluorescence assay for the detection of chloramphenicol (CAP). (a) Schematic showing QDs embedded in liposomes and aptamer-functionalized Au NPs attached to the liposome using a single-stranded DNA binding protein. The Au NPs quench the fluorescence of the liposome-embedded QDs. Upon addition of CAP, the aptamer preferentially binds CAP and releases the Au NP from the liposome surface. The increase in QD fluorescence is the analytical signal. (b) Fluorescence spectra of liposome-embedded QDs before addition of CAP (black trace), and after addition of 10 mM CAP to only QDs (red trace) and liposome-embedded QDs (blue trace). (c) Calibration of the assay showing the linear response of the fluorescence between 0 and 10 mM CAP.
Reprinted from *Anal. Chim. Acta*, **929**, Y. B. Miao, H. X. Ren, N. Gan, Y. Zhou, Y. Cao, T. Li, Y. Chen, A homogeneous and "off-on" fluorescence aptamer-based assay for chloramphenicol using vesicle quantum dot-gold colloid composite, 49–55. Copyright 2016 with permission from Elsevier B. V.

was able to sensitize emission from the QDs. Increasing concentrations of nitrofurans resulted in progressive quenching of the ECL signal, which was used for quantitative measurement. Detection limits were between 0.4 and 0.7 μM for furaltadone (FTD), furazolidone (FZD) and nitrofurantoin (NFT). Assays with animal feed were successfully validated against HPLC methods.[128]

A purely electrochemical assay has been developed for detection of CAP and polychlorinated biphenyl 72 (PCB72), a persistent organic pollutant, in food samples.[129] As in earlier examples, stripping voltammetry was the method of detection. Thiol-functionalized aptamers for CAP and PCB72 were immobilized on magnetic iron oxide/Au NP composites through the well-established thiol/gold interaction. Oligonucleotide probes complementary to the CAP and PCB72 aptamer sequences were modified with thiol linkers and attached the PbS QDs and CdS QDs, respectively, through coordination. To enhance the sensitivity of the assay, clusters of QDs were created on a dendritic complex that comprised a polymer skeleton conjugated at multiple points with horseradish peroxidase (HRP) and secondary antibodies. The complex is typically used for immunohistochemical staining applications; however, in this assay, the HRP was modified with the QD-oligonucleotide conjugates. Upon introduction of the CAP and PCB72 targets, the aptamers present on the iron oxide/Au NP composites bound their targets and the QD clusters were displaced. After magnetic separation, the released clusters of QD were dissolved and measured using square wave stripping voltammetry. CAP and PCB72 were extracted from actual fish samples and analyzed. The dynamic range was between 0.001 and 100 ng mL^{-1}, and the detection limits for CAP and PCB72 were 0.33 and 0.35 pg mL^{-1}, respectively. The assay outperformed a commercially available ELISA assay for the same targets by between two and three orders of magnitude.[129] In another example, a multiplexed assay was developed for the simultaneous detection of streptomycin, CAP and TC in milk samples using aptamers and QD-oligonucleotide conjugates.[130] Aptamers specific to each antibiotic were first hybridized with a complementary oligonucleotide to create double-stranded DNA. Upon incubation with the antibiotics, the aptamers bound their targets and displaced the small DNA strands. An indirect sandwich assay was then used to detect the displaced oligonucleotides. The solution was exposed to a gold electrode that had immobilized capture probes to bind the displaced oligonucleotides, and the sandwich assay was completed with QD-oligonucleotide conjugates of different materials. PbS QDs were used for streptomycin, CdS QDs were used for CAP and ZnS QDs were used for TC. The QDs captured at the gold electrode were then dissolved using nitric acid and analyzed using square wave ASV. The detection limits for streptomycin, CAP and TC were 10 nM, 5 nM and 20 nM, respectively.[130]

2.3.4 Genetically Modified Organisms (GMOs)

A fluorescence-based method for the multiplexed detection of genetically modified corn, soybean, colza (rapeseed) and cotton using QD-encoded silica nanoparticles and MNPs has been reported.[131] The SiO_2 nanoparticles were encoded by incorporating different ratios of different colours of QDs to create a set of optical barcodes for a sandwich assay with MNPs. (Optical barcodes were first proposed for biomedical applications.[132]) QDs with peak fluorescence at 525 nm and 585 nm were combined in 1 : 1 and 1 : 3 ratios for the corn and soybean barcodes, and QDs with peak fluorescence at 665 nm and 740 nm were combined in 1 : 1 and 1 : 3 ratios for the colza and cotton barcodes. Probe oligonucleotides corresponding to each GMO were conjugated to the encoded silica nanoparticles, and capture oligonucleotides for each GMO were conjugated to MNPs. Upon target binding, a sandwich structure between the uniquely encoded silica nanoparticles and the MNPs was created. The conjugates were then collected using magnetic separation, and their fluorescence spectra were recorded and analyzed. Using this method, four different targets could be detected simultaneously. Quantitative analysis of only the soybean barcode yielded a detection limit of 4.48 fM.[131]

2.4 Summary and Perspective

This chapter provided an overview of the state-of-the art in the utilization of QDs for food quality and safety analyses. The unique combination of physical, optical and electrochemical properties of QDs have been shown to offer several advantages. Assays that utilize QDs for the detection of pathogens, toxins, pesticides, antibiotics, GMOs and other analytes relevant to food safety have been reported in the scientific literature. The majority of the assays have been fluorescence-based and can be categorized as relying on one of two general concepts: heterogeneous binding assays that measure the intensity of QD fluorescence that remains after washing or separation steps to remove unbound materials; and homogeneous assays that do not require washing steps and instead utilize energy transfer or other quenching processes to modulate the intensity of QD fluorescence in response to binding events. The most exploited fluorescence properties of QDs in these assays are their outstanding brightness, which can increase the sensitivity and detection limits of assays, and their amenability to multiplexing. Quenching processes, such as FRET and charge transfer, can also be utilized with QDs to great advantage; however, it is imperative that the mechanism of detection is both well controlled and well understood to ensure reproducibility and selectivity. Electrochemical assays have also been developed and most commonly exploit the size and chemical composition of QDs, which can offer advantages in terms of sensitivity and multiplexing.

Cumulatively, the examples of assays presented in this chapter show that QDs can be useful for both the improvement of existing assay methods and the development of new assay methods. A critical reader may have already wondered whether the magnitude of the improvement of existing assay methods or the analytical merit of new methods will actually impact the practice of food safety and quality analysis in agriculture, industry and society. That is, whether or not QD-based assays will ever be translated beyond research laboratories. Of course, the answers to this question must be considered on a case-by-case basis, but it is nonetheless clear that QDs have not immediately revolutionized food safety and quality analyses. Rather, QDs have significant potential to addresses specific weaknesses in current methodologies. To overcome the status quo, justify the time and cost of rigorous validation and regulatory approval and ultimately replace or modify an established method, it will be necessary for QD-based methods to provide improvements of an order of magnitude or more in relevant analytical figures of merit (*e.g.*, detection limit, assay time) and/or reductions in cost per test of similar magnitude. Such improvements are not trivial, even with the unique properties of QDs.

Context is also important when justifying the use of new QD-based assays. ELISAs and PCR are two examples of established laboratory assays that are often sufficient in their analytical performance. To illustrate, if PCR can amplify a single copy of a gene, it may not matter that QDs are an order of magnitude brighter than conventional fluorescent dyes because amplification makes up for the shortcoming of the latter. The same logic will often apply to ELISAs as well. Revolutionary impact will come from detection of a single copy of a gene without PCR amplification, but QD materials cannot enable such detection on their own. New measurement technologies must also be developed within the contexts and restrictions of food safety and quality analyses. For example, instrumentation for the detection of a single DNA molecule labelled with a fluorescent dye is established, but its cost and complexity make it less attractive to a food analysis laboratory than well-established PCR methods.

The best value and greatest impact will likely come from utilizing QDs to greatly relax the technical requirements of assays; for example, so that the assays can be made more portable and lower-cost, require less (if any) sample preparation and/or require less (if any) specialized training, while still having relevant detection limits and practical assay times. Some of these advances may leverage combinations with current diagnostic technologies, research technologies and consumer technologies, examples of which include lateral flow test strips, microfluidic chips, and consumer electronics such as smartphones and smartwatches. Rapid and low-cost screening assays for monitoring food safety and quality at agricultural, production and packaging sites, during transportation and at points of sale to consumers have excellent potential for societal impact and commercialization, even if established technologies in specialized and accredited laboratories continue to be used for confirmatory analyses.

In summary, many of the examples of QD-based assays highlighted in this chapter represent interesting proof-of-concept studies, but only some of the examples have longer-term potential for *bona fide* applications in food safety and quality analysis. Future R&D in this area should evolve to focus more on innovative and potentially disruptive methods and technologies that are newly enabled by the special properties of QDs. There should be less focus on permutations of existing assays where QDs simply substitute other materials. Any new methods and technologies should be measured against current benchmarks for detection limits, assay times, cost and deployability. Given the ingenuity of scientists, a bright future can be envisioned for QDs in food safety and quality analysis, with the potential for global benefits.

References

1. P. L. Pingali, *Proc. Natl. Acad. Sci. U. S. A.*, 2012, **109**, 12302–12308.
2. R. E. Evenson, *Science*, 2003, **300**, 758–762.
3. G. S. Khush, *Nat. Rev. Genet.*, 2001, **2**, 815–822.
4. A. C. Vinayaka and M. S. Thakur, *Anal. Bioanal. Chem.*, 2010, **397**, 1445–1455.
5. K. Zhang, Q. Mei, G. Guan, B. Liu, S. Wang and Z. Zhang, *Anal. Chem.*, 2010, **82**, 9579–9586.
6. M. A. Hahn, J. S. Tabb and T. D. Krauss, *Anal. Chem.*, 2005, **77**, 4861–4869.
7. G. Kim, S. B. Park, J.-H. Moon and S. Lee, *Anal. Method*, 2013, **5**, 5717.
8. B. Stephen Inbaraj and B. H. Chen, *J. Food Drug Anal.*, 2016, **24**, 15–28.
9. Government of Canada, *Report of the Independent Investigator into the 2008 Listeriosis Outbreak*, July 2009.
10. D. R. O'Connor, *Report of the Walkerton Inquiry: The Events of May 2000 and Related Issues*, Published by Ontario Ministry of the Attorney General, Queen's Printer for Ontario, 2002.
11. Centers for Disease Control and Prevention. *Multistate Outbreak of Listeriosis Linked to Whole Cantaloupes from Jensen Farms, Colorado (FINAL UPDATE)*, August 27, 2012, http://www.cdc.gov/listeria/outbreaks/cantaloupes-jensen-farms/.
12. U. Buchholz, H. Bernard, D. Werber, M. M. Böhmer, C. Remschmidt, H. Wilking, Y. Deleré, M. an der Heiden, C. Adlhoch, J. Dreesman, J. Ehlers, S. Ethelberg, M. Faber, C. Frank, G. Fricke, M. Greiner, M. Höhle, S. Ivarsson, U. Jark, M. Kirchner, J. Koch, G. Krause, P. Luber, B. Rosner, K. Stark and M. Kühne, *N. Engl. J. Med.*, 2011, **365**, 1763–1770.
13. Public Health England, *E. coli O157 national outbreak update*, 11 August 2016, https://www.gov.uk/government/news/update-as-e-coli-o157-investigation-continues.
14. World Health Organization Media Centre, *Drinking-water, Fact sheet No. 391*, June 2015, http://www.who.int/mediacentre/factsheets/fs391/en/.
15. United States Department of Agriculture, *Summary of Recall Cases in Calendar Year 2015*, http://www.fsis.usda.gov/wps/portal/fsis/topics/recalls-and-public-health-alerts/recall-summaries.
16. N. Swaminathan, *Scientific American*, "Were Our Pets Deliberately Poisoned?" April 25, 2007, https://www.scientificamerican.com/article/pets-deliberately-poisoned/.
17. C. Xiu and K. K. Klein, *Food Policy*, 2010, **35**, 463–470.
18. S. Wunderlich and K. A. Gatto, *Adv. Nutr.*, 2015, **6**, 842–851.
19. P. Koedrith, T. Thasiphu, K. Tuitemwong, R. Boonprasert and P. Tuitemwong, *Sens. Mater.*, 2014, **26**, 711–736.
20. P. Tallury, A. Malhotra, L. M. Byrne and S. Santra, *Adv. Drug Delivery Rev.*, 2010, **62**, 424–437.
21. C. Kaittanis, S. Santra and J. M. Perez, *Adv. Drug Delivery Rev.*, 2010, **62**, 408–423.

22. S. Rebe Raz and W. Haasnoot, *TrAC, Trends Anal. Chem.*, 2011, **30**, 1526–1537.
23. R. W. Peeling and D. Mabey, *Clin. Microbiol. Infect.*, 2010, **16**, 1062–1069.
24. B. Pérez-López and A. Merkoçi, *Trends Food Sci. Technol.*, 2011, **22**, 625–639.
25. H. Bouwmeester, S. Dekkers, M. Y. Noordam, W. I. Hagens, A. S. Bulder, C. de Heer, S. E. C. G. ten Voorde, S. W. P. Wijnhoven, H. J. P. Marvin and A. J. A. M. Sips, *Regul. Toxicol. Pharmacol.*, 2009, **53**, 52–62.
26. M. G. Valdés, A. C. Valdés González, J. A. García Calzón and M. E. Díaz-García, *Microchim. Acta*, 2009, **166**, 1–19.
27. I. L. Medintz, H. T. Uyeda, E. R. Goldman and H. Mattoussi, *Nat. Mater.*, 2005, **4**, 435–446.
28. J. B. Blanco-Canosa, M. Wu, K. Susumu, E. Petryayeva, T. L. Jennings, P. E. Dawson, W. R. Algar and I. L. Medintz, *Coord. Chem. Rev.*, 2013, **263–264**, 101–137.
29. W. R. Algar, K. Susumu, J. B. Delehanty and I. L. Medintz, *Anal. Chem.*, 2011, **83**, 8826–8837.
30. N. Hildebrandt, C. M. Spillmann, W. R. Algar, T. Pons, M. H. Stewart, E. Oh, K. Susumu, S. A. Díaz, J. B. Delehanty and I. L. Medintz, *Chem. Rev.*, 2017, **117**, 536–711.
31. F. F. P. X. Michalet, L. A. Bentolila, J. M. Tsay, S. Doose, J. J. Li, G. Sundaresan, A. M. Wu, S. S. Gambhir and S. Weiss, *Science*, 2005, **307**, 538–544.
32. W. C. W. Chan, D. J. Maxwell, X. Gao, R. E. Bailey, M. Han and S. Nie, *Curr. Opin. Biotechnol.*, 2002, **13**, 40–46.
33. A. P. Alivisatos, *J. Phys. Chem.*, 1996, **100**, 13226–13239.
34. M. Dasog, J. Kehrle, B. Rieger and J. G. C. Veinot, *Angew. Chem., Int. Ed.*, 2016, **55**, 2322–2339.
35. C. M. Gonzalez and J. G. C. Veinot, *J. Mater. Chem. C*, 2016, **4**, 4836–4846.
36. S. Tamang, C. Lincheneau, Y. Hermans, S. Jeong and P. Reiss, *Chem. Mater.*, 2016, **28**, 2491–2506.
37. H. Mattoussi, G. Palui and H. B. Na, *Adv. Drug Delivery Rev.*, 2012, **64**, 138–166.
38. E. Petryayeva and W. R. Algar, *Appl. Spectrosc.*, 2012, **67**, 215–252.
39. K. D. Wegner and N. Hildebrandt, *Chem. Soc. Rev.*, 2015, **44**, 4792–4834.
40. E. R. Goldman, A. R. Clapp, G. P. Anderson, H. T. Uyeda, J. M. Mauro, I. L. Medintz and H. Mattoussi, *Anal. Chem.*, 2004, **76**, 684–688.
41. J. Zhou, Y. Yang and C. Y. Zhang, *Chem. Rev.*, 2015, **115**, 11669–11717.
42. R. Rossetti, *J. Chem. Phys.*, 1984, **80**, 4464.
43. C. B. Murray, D. J. Norris and M. G. Bawendi, *J. Am. Chem. Soc.*, 1993, **115**, 8706–8715.
44. M. A. Hines and P. Guyot-Sionnest, *J. Phys. Chem.*, 1996, **100**, 468–471.
45. Z. A. Peng and X. Peng, *J. Am. Chem. Soc.*, 2001, **123**, 183–184.

46. F. Wang, R. Tang and W. E. Buhro, *Nano Lett.*, 2008, **8**, 3521–3524.
47. Y. Yin and A. P. Alivisatos, *Nature*, 2005, **437**, 664–670.
48. M. Green, *J. Mater. Chem.*, 2010, **20**, 5797–5809.
49. S. J. Rosenthal, J. McBride, S. J. Pennycook and L. C. Feldman, *Surf. Sci. Rep.*, 2007, **62**, 111–157.
50. P. Samokhvalov, M. Artemyev and I. Nabiev, *Chem. - Eur. J.*, 2013, **19**, 1534–1546.
51. K. L. Sowers, B. Swartz and T. D. Krauss, *Chem. Mater.*, 2013, **25**, 1351–1362.
52. A. L. Rogach, T. Franzl, T. A. Klar, J. Feldmann, N. Gaponik, V. Lesnyak, A. Shavel, A. Eychmüller, Y. P. Rakovich and J. F. Donegan, *J. Phys. Chem. C*, 2007, **111**, 14628–14637.
53. B. Dubertret, P. Skourides, D. J. Norris, V. Noireaux, A. H. Brivanlou and A. Libchaber, *Science*, 2002, **298**, 1759–1762.
54. D. Janczewski, N. Tomczak, M. Y. Han and G. J. Vancso, *Nat. Protoc.*, 2011, **6**, 1546–1553.
55. K. Susumu, H. T. Uyeda, I. L. Medintz, T. Pons, J. B. Delehanty and H. Mattoussi, *J. Am. Chem. Soc.*, 2007, **129**, 13987–13996.
56. K. E. Sapsford, W. R. Algar, L. Berti, K. Boeneman Gemmill, B. J. Casey, E. Oh, M. H. Stewart and I. L. Medintz, *Chem. Rev.*, 2013, **113**, 1904–2074.
57. A. D. Keefe, S. Pai and A. Ellington, *Nat. Rev. Drug Discovery.*, 2010, **9**, 537–550.
58. Y. Xing, Q. Chaudry, C. Shen, K. Y. Kong, H. E Zhau, L. W. Chung, J. A Petros, R. M. O'Regan, M. V. Yezhelyev, J. W. Simons, M. D. Wang and S. Nie, *Nat. Protoc.*, 2007, **2**, 1152–1165.
59. W. R. Algar, D. E. Prasuhn, M. H. Stewart, T. L. Jennings, J. B. Blanco-Canosa, P. E. Dawson and I. L. Medintz, *Bioconjugate Chem.*, 2011, **22**, 825–858.
60. D. V. Talapin, J.-S. Lee, M. V. Kovalenko and E. V. Shevchenko, *Chem. Rev.*, 2010, **110**, 389–458.
61. P. Zrazhevskiy, M. Sena and X. Gao, *Chem. Soc. Rev.*, 2010, **39**, 4326.
62. F. Todescato, I. Fortunati, A. Minotto, R. Signorini, J. Jasieniak and R. Bozio, *Materials*, 2016, **9**, 672.
63. N. Lewinski, V. Colvin and R. Drezek, *Small*, 2008, **4**, 26–49.
64. F. M. Winnik and D. Maysinger, *Acc. Chem. Res.*, 2013, **46**, 672–680.
65. M. Bottrill and M. Green, *Chem. Commun.*, 2011, **47**, 7039.
66. J. Li, X. Chang, X. Chen, Z. Gu, F. Zhao, Z. Chai and Y. Zhao, *Biotechnol. Adv.*, 2014, **32**, 727–743.
67. K.-T. Yong, W.-C. Law, R. Hu, L. Ye, L. Liu, M. T. Swihart and P. N. Prasad, *Chem. Soc. Rev.*, 2013, **42**, 1236–1250.
68. K. M. Tsoi, Q. Dai, B. A. Alman and W. C. W. Chan, *Acc. Chem. Res.*, 2013, **46**, 662–671.
69. M. Z. Hu and T. Zhu, *Nanoscale Res. Lett.*, 2015, **10**, 469.
70. B. Wang, Q. Wang, Z. Cai and M. Ma, *LWT—Food Sci. Technol.*, 2015, **61**, 368–376.

71. X. L. Su and Y. Li, *Anal. Chem.*, 2004, **76**, 4806–4810.
72. L. Wang, C.-S. Wu, X. Fan and A. Mustapha, *Int. J. Food Microbiol.*, 2012, **156**, 83–87.
73. P.-L. Guo, M. Tang, S.-L. Hong, X. Yu, D.-W. Pang and Z.-L. Zhang, *Biosens. Bioelectron.*, 2015, **74**, 628–636.
74. H. Kuang, G. Cui, X. Chen, H. Yin, Q. Yong, L. Xu, C. Peng, L. Wang and C. Xu, *Int. J. Mol. Sci.*, 2013, **14**, 8603–8610.
75. L. Yang and Y. Li, *Analyst*, 2006, **131**, 394–401.
76. C.-Y. Wen, J. Hu, Z.-L. Zhang, Z.-Q. Tian, G.-P. Ou, Y.-L. Liao, Y. Li, M. Xie, Z.-Y. Sun and D.-W. Pang, *Anal. Chem.*, 2013, **85**, 1223–1230.
77. R. Jackeray, C. K. V. Zainul Abid, G. Singh, S. Jain, S. Chattopadhyaya, S. Sapra, T. G. Shrivastav and H. Singh, *Talanta*, 2011, **84**, 952–962.
78. H. Kurt, M. Yüce, B. Hussain and H. Budak, *Biosens. Bioelectron.*, 2016, **81**, 280–286.
79. F. Tang, D.-W. Pang, Z. Chen, J.-B. Shao, L.-H. Xiong, Y.-P. Xiang, Y. Xiong, K. Wu, H.-W. Ai, H. Zhang, X.-L. Zheng, J.-R. Lv, W.-Y. Liu, H.-B. Hu, H. Mei, Z. Zhang, H. Sun, Y. Xiang and Z.-Y. Sun, *Nanoscale*, 2016, **8**, 4688–4698.
80. A. Huang, Z. Qiu, M. Jin, Z. Shen, Z. Chen, X. Wang and J. W. Li, *Int. J. Food Microbiol.*, 2014, **185**, 27–32.
81. Y. Zhao, M. Ye, Q. Chao, N. Jia, Y. Ge and H. Shen, *J. Agric. Food Chem.*, 2009, **57**, 517–524.
82. I. L. Medintz, D. Farrell, K. Susumu, S. A. Trammell, J. R. Deschamps, F. M. Brunel, P. E. Dawson and H. Mattoussi, *Anal. Chem.*, 2009, **81**, 4831–4839.
83. H. Zhu, U. Sikora and A. Ozcan, *Analyst*, 2012, **137**, 2541.
84. J. G. Bruno, *Pathogens*, 2014, **3**, 341–355.
85. H. Yang, D. Li, R. He, Q. Guo, K. Wang, X. Zhang, P. Huang and D. Cui, *Nanoscale Res. Lett.*, 2010, **5**, 875.
86. G. Kim, J.-H. Moon, C.-Y. Moh and J.-G. Lim, *Biosens. Bioelectron.*, 2015, **67**, 243–247.
87. R. Wang, Y. Ni, Y. Xu, Y. Jiang, C. Dong and N. Chuan, *Anal. Chim. Acta*, 2015, **853**, 710–717.
88. S. Agrawal, A. Morarka, D. Bodas and K. M. Paknikar, *Appl. Biochem. Biotechnol.*, 2012, **167**, 1668–1677.
89. B.-B. Wang, Q. Wang, Y.-G. Jin, M.-H. Ma and Z.-X. Cai, *J. Photochem. Photobiol., A*, 2015, **299**, 131–137.
90. K. P. Burris, T. C. Wu, M. Vasudev, M. A. Stroscio, R. J. Millwood and C. N. Stewart, *IEEE Trans. Nanobiosci.*, 2013, **12**, 233–238.
91. J. Wang, G. Liu and A. Merkoçi, *J. Am. Chem. Soc.*, 2003, **125**, 3214–3215.
92. D. Vijian, S. V. Chinni, L. S. Yin, B. Lertanantawong and W. Surareungchai, *Biosens. Bioelectron.*, 2016, **77**, 805–811.
93. P. Kumar, K.-H. Kim and A. Deep, *Biosens. Bioelectron.*, 2015, **70**, 469–481.
94. N. Xia, Q. Wang and L. Liu, *Sensors*, 2014, **15**, 499–514.
95. W. Zhang, A. M. Asiri, D. Liu, D. Du and Y. Lin, *TrAC, Trends Anal. Chem.*, 2014, **54**, 1–10.

96. S. Su, W. Wu, J. Gao, J. Lu and C. Fan, *J. Mater. Chem.*, 2012, **22**, 18101.
97. G. Aragay, F. Pino and A. Merkoçi, *Chem. Rev.*, 2012, **112**, 5317–5338.
98. A. C. Vinayaka, S. Basheer and M. S. Thakur, *Biosens. Bioelectron.*, 2009, **24**, 1615–1620.
99. H. Li and F. Qu, *Chem. Mater.*, 2007, **19**, 4148–4154.
100. X. Ren and L. Chen, *Biosens. Bioelectron.*, 2015, **64**, 182–188.
101. Y. Zhao, Y. Ma, H. Li and L. Wang, *Anal. Chem.*, 2012, **84**, 386–395.
102. X. Meng, J. Wei, X. Ren, J. Ren and F. Tang, *Biosens. Bioelectron.*, 2013, **47**, 402–407.
103. X. Gao, G. Tang and X. Su, *Biosens. Bioelectron.*, 2012, **36**, 75–80.
104. Z. Zheng, X. Li, Z. Dai, S. Liu and Z. Tang, *J. Mater. Chem.*, 2011, **21**, 16955.
105. Z. Zheng, Y. Zhou, X. Li, S. Liu and Z. Tang, *Biosens. Bioelectron.*, 2011, **26**, 3081–3085.
106. D. Du, W. Chen, W. Zhang, D. Liu, H. Li and Y. Lin, *Biosens. Bioelectron.*, 2010, **25**, 1370–1375.
107. X. Li, Z. Zheng, X. Liu, S. Zhao and S. Liu, *Biosens. Bioelectron.*, 2015, **64**, 1–5.
108. Z. Zou, D. Du, J. Wang, J. N. Smith, C. Timchalk, Y. Li and Y. Lin, *Anal. Chem.*, 2010, **82**, 5125–5133.
109. A. P. Periasamy, Y. Umasankar and S.-M. Chen, *Sensors*, 2009, **9**, 4034–4055.
110. D. Du, J. Wang, L. Wang, D. Lu, J. N. Smith, C. Timchalk and Y. Lin, *Anal. Chem.*, 2011, **83**, 3770–3777.
111. G. Liu, J. Wang, R. Barry, C. Petersen, C. Timchalk, P. L. Gassman and Y. Lin, *Chem. - Eur. J.*, 2008, **14**, 9951–9959.
112. D. Lu, J. Wang, L. Wang, D. Du, C. Timchalk, R. Barry and Y. Lin, *Adv. Funct. Mater.*, 2011, **21**, 4371–4378.
113. H. Wang, J. Wang, C. Timchalk and Y. Lin, *Anal. Chem.*, 2008, **80**, 8477–8484.
114. T. Yang, H. Huang, F. Zhu, Q. Lin, L. Zhang and J. Liu, *Sensors*, 2016, **16**, 1118.
115. E. Song, M. Yu, Y. Wang, W. Hu, D. Cheng, M. T. Swihart and Y. Song, *Biosens. Bioelectron.*, 2015, **72**, 320–325.
116. W. Jiang, N. V. Beloglazova, Z. Wang, H. Jiang, K. Wen, S. de Saeger, P. Luo, Y. Wu and J. Shen, *Biosens. Bioelectron.*, 2015, **66**, 124–128.
117. N. V. Beloglazova, P. S. Shmelin and S. A. Eremin, *Talanta*, 2016, **149**, 217–224.
118. N. A. Taranova, A. N. Berlina, A. V. Zherdev and B. B. Dzantiev, *Biosens. Bioelectron.*, 2015, **63**, 255–261.
119. B. Ge, Z. Li, Y. Xie, L. Yang and R. Wang, *Optics Spectrosc.*, 2015, **118**, 748–755.
120. D. Li, S. Jia, E. K. Fodjo, H. Xu, C. Kong and Y. Wang, *RSC Adv.*, 2015, **5**, 99454–99460.
121. X. Li, C. Li and L. Chen, *New J. Chem.*, 2015, **39**, 9976–9982.

122. W. Liang, S. Liu, Z. Liu, D. Li, L. Wang, C. Hao and Y. He, *New J. Chem.*, 2015, **39**, 4774–4782.
123. Z. Wang, J. Lin, J. Gao and Q. Wang, *Mater. Chem. Phys.*, 2016, **178**, 82–87.
124. J. Tashkhourian, G. Absalan, M. Jafari and S. Zare, *Spectrochim. Acta, Part A*, 2016, **152**, 119–125.
125. J. Jimenez-López, L. Molina-García, S. S. M. Rodrigues, J. L. M. Santos, P. Ortega-Barrales and A. Ruiz-Medina, *J. Lumin.*, 2016, **175**, 158–164.
126. Y. B. Miao, H. X. Ren, N. Gan, Y. Zhou, Y. Cao, T. Li and Y. Chen, *Anal. Chim. Acta*, 2016, **929**, 49–55.
127. A. Khataee, A. Hasanzadeh, R. Lotfi and S. W. Joo, *Talanta*, 2016, **152**, 171–178.
128. N. Taokaenchan, T. Tangkuaram, P. Pookmanee, S. Phaisansuthichol, S. Kuimalee and S. Satienperakul, *Biosens. Bioelectron.*, 2015, **66**, 231–237.
129. M. Chen, N. Gan, H. Zhang, Z. Yan, T. Li, Y. Chen, Q. Xu and Q. Jiang, *Microchim. Acta*, 2016, **183**, 1099–1106.
130. J. Xue, J. Liu, C. Wang, Y. Tian and N. Zhou, *Anal. Methods*, 2016, **8**, 1981–1988.
131. Y. Zhao, C. Hao, W. Ma, Q. Yong, W. Yan, H. Kuang, L. Wang and C. Xu, *J. Phys. Chem. C*, 2011, **115**, 20134–20140.
132. M. Han, X. Gao, J. Z. Su and S. Nie, *Nat Biotechnol.*, 2001, **19**, 631–635.

CHAPTER 3

Microfluidic "Lab-on-a-Chip" Sensing in Food Safety and Quality Analysis

XIAN HUANG,*[a] CHONGYUE TANG,[a] QINGMEI XU,[b] YICONG ZHAO[a] AND DACHAO LI[b]

[a] Tianjin University, Biomedical Engineering, School of Precision Instrument and Opto-electronics Engineering, Tianjin 300072, China; [b] Tianjin University, State Key Laboratory of Precision Measuring Technology and Instruments, Tianjin 300072, China
*Email: huangxian@tju.edu.cn

3.1 Introduction

Food essentially comprises various natural or synthetic chemicals that supply energy, flavours or extended preservation time. Among these food compositions, excessive food additives, toxins, bacteria, pathogens, heavy metals, pesticide residues, antibiotics and biogenic amines may lead to negative effects on the human body, leading to serious issues in almost all physiological systems,[1–3] such as digestive systems, neurosystems and reproductive systems. In addition, overconsumption of fats, proteins, peptides and carbohydrates in foods may lead to health issues, such as obesity, diabetes and cardiovascular diseases. In terms of national security, contaminated food products can easily cause large spread of disease, causing tremendous threat to the consumers. In China, there have been numerous outbreaks related to foods due to contaminated food sources or misconduct of

Food Chemistry, Function and Analysis No. 2
Sensing Techniques for Food Safety and Quality Control
Edited by Xiaonan Lu

Published by the Royal Society of Chemistry, www.rsc.org

the food producers.[4–6] In the United States, many incidents caused by bacterial infections originating from contaminated poultry, meat, produce and water have also been reported.[7–9] Thus, analysis of the food contents to ensure their safety for consumption is critical. Food contamination can be monitored using analytical equipment. However, using bench-top instruments makes it either inconvenient to conduct field tests or costly to equip food production facilities. An approach to overcome the problems associated with conventional equipment is to use microfluidic "lab-on-a-chip" systems, which offer several advantages over conventional methods.

Lab-on-a-chip (LOC) is an emerging technology that aims at miniaturizing chemical and biological processes and conducting various operation procedures, including sample sorting, bioanalysis and flow manipulation.[10–12] LOC technology integrates multiple microfluidic components, such as sensors, microvalves, micromixers, microchannels and micropumps, into a single chip with miniaturized footprint, allowing high detection sensitivity and speed with a reduced cost. As a result, microfluidic LOC technology has been used in many applications, such as clinical analysis,[13–16] drug discovery,[17–19] environmental monitoring[1,20,21] and forensics,[22–24] showing great promise in replacing the bulky equipment in the laboratory.

LOC devices possess many advantages over conventional analytical instruments and are very suitable to be used to analyze different food compositions. First, their high surface area-to-volume ratio allows surface modification to increase detection sensitivity and rapid antibody–antigen reaction. Second, sample consumption can be largely reduced due to the miniaturized dimension of microfluidic devices. Furthermore, detection throughput can be increased by integrating additional parallel sensing channels and introducing automated fluid handling systems. Lastly, the cost of food analysis can be minimized by reducing reagent consumption and chemical waste. In this chapter, food analysis approaches based on LOC technology are introduced. The major materials, structures and main operation principles of LOC devices will be introduced, followed by some representative applications of LOC devices in determining various food compositions to ensure food safety and quality (Figure 3.1).

3.2 Materials, Structures and Fabrication Methods of LOC Devices

3.2.1 Major Materials Used in Microfluidic LOC Devices

Microfluidic LOC devices can be made of a wide range of materials, such as glass, silicon, polymer and paper, all of which can be used to construct structural components (*e.g.*, device substrates, microchannels, microvalves and micromixers). In addition, metals such as gold (Au), platinum (Pt), copper (Cu), aluminium (Al) and silver (Ag), inorganic compounds (*e.g.*, silver chloride (AgCl) and zinc oxide (ZnO)), composite materials (*e.g.*, carbon paste) and other nanoparticle pastes can be used to build electrodes and

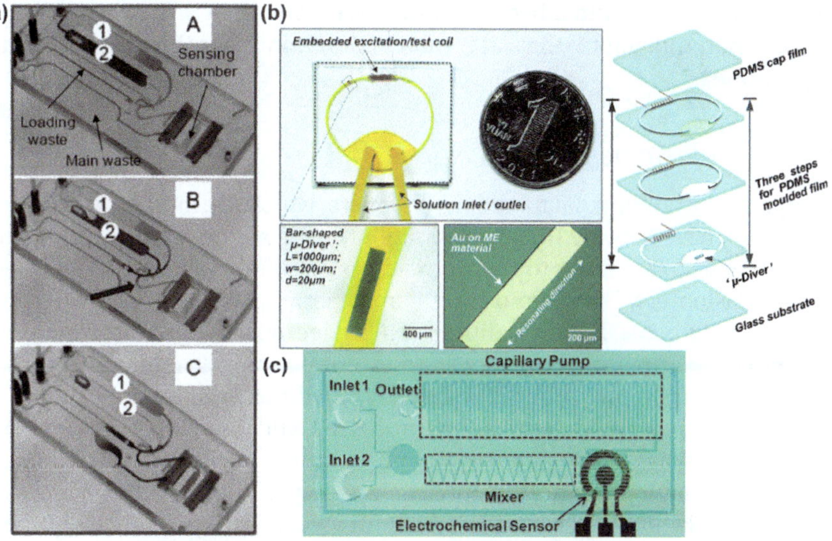

Figure 3.1 Examples of integrated microfluidic LOC devices used for food safety and quality analysis. (a) A multiplexed biosensor for the detection of antibiotics in milk. (Reproduced from ref. 25 with permission from the Royal Society of Chemistry). (b) A fully integrated system to determine foodborne pathogens. (Reproduced from ref. 26 with permission from the Royal Society of Chemistry). (c) An integrated device for rapid detection of toxins. (Reproduced from ref. 27 with permission from the Royal Society of Chemistry).

sensors. Other active components based upon organic and inorganic semiconductor materials can be used to construct active components, such as photo emitters, detectors and transistors in LOC devices.

3.2.1.1 *Glass and Silicon*

Glass and silicon can be patterned to form microstructures or used directly as the substrates to construct microfluidic chips.[11,28] The advantages of glass and silicon include resistance to organic solvents, ease in metal deposition, high thermal conductivity and stable electro-osmotic mobility. Their applications include on-chip reaction, droplet formation, solvent extraction and *in situ* fabrication. However, high fabrication cost and complex fabrication processes (etching and bonding) limit the use of glass and silicon as structural materials in LOC devices. In addition, as silicon is not transparent, it is not suitable to act as the substrate material when optical measurement is involved. Furthermore, as glass and silicon are not gas permeable, chips constructed on the basis of glass and/or silicon channels and chambers are not suitable for cell cultivation in the long term. The aforementioned limitations of glass and silicon urge the development of other chip materials based upon polymers that are more compatible for broader biological applications.

3.2.1.2 Elastomers and Plastics

Compared with inorganic materials, polymers, which are inexpensive and easy to process, have become the most commonly used material for the production of microchips. According to their physical properties, polymers can be classified as elastomers,[29,30] thermosets[31,32] and thermoplastics.[33,34]

Elastomers consist of cross-linked, entangled polymer chains, which offer reversible deformation in response to external forces. The most commonly used elastomer in microfluidic systems is polydimethylsiloxane (PDMS),[35–38] which features simple fabrication processes and low cost. Liquid PDMS precursor can be thermally cured at temperatures below 120 °C, and can be moulded into micro-/nanostructures using templates obtained by patterning photosensitive polymers or other simplified fabrication methods, such as laser milling, micromachining and 3D printing. PDMS can be bonded with glass or other PDMS layers through plasma or UV-assisted surface treatment, followed by direct contact bonding. This eventually can form permanent chemical bonds to allow device sealing. However, absorption of small molecules onto the walls of PDMS channels and water evaporation due to gas permeability of PDMS yield negative influence on the applications of PDMS-based devices. Various modification strategies, including chemical vapour deposition, silanization and surfactants may be used to minimize molecule absorption and water evaporation.

The thermosetting materials represented by SU-8 photoresist[39,40] and polyimide[41,42] can crosslink to form an irreversible rigid polymer network when they are heated or radiated. These thermosetting materials are stable at high temperature and possess high resistance to solvents and high optical transparency. The high strength and capability to conduct photopolymerization enable the materials to fabricate free-standing or three-dimensional structures with high aspect ratios. Thermosetting materials can be used to fabricate the entire microfluidic chips or used as the templates for other elastomers and thermoplastics.

Thermoplastics can reversibly change their geometry by heating the materials to glass transition temperature, and can be moulded using metal or silicon templates, allowing mass production for commercial applications. Typical thermoplastics used in LOC include polymethyl methacrylate (PMMA),[43,44] polycarbonate (PC),[45,46] polystyrene (PS)[47,48] and polyethylene terephthalate (PET).[49,50] These materials show minimum gas permeability, making them unsuitable for cell cultivation and characterization in the long term. However, their improved solvent resistance as compared with PDMS allows them to be used in the situation where the analytes are dissolved in solvents.

3.2.1.3 Hydrogels

Hydrogels contain 3D networks of hydrophilic polymer chains that span the aqueous media. They are highly porous with controllable pore sizes, allowing small molecules and particles to diffuse through. The combination of aqueous nature and high permeability makes hydrogels perfect for encapsulating cells for 3D culture. Hydrogels can be categorized as synthetic hydrogels (*e.g.*, polyethylene glycol (PEG)[51] and polyacrylamide (PAM)[52]) and natural hydrogels. Natural hydrogels include gelatine[53] and collagen[54] derived from animals as well as agarose[55] and cellulose[56] derived from plants. Animal-derived hydrogels contain factors that promote cell adhesion and proliferation. In contrast, plant-derived hydrogels and synthesized hydrogels lack cell adhesion sites and have to introduce extra grafting structures in order to improve cell adhesion. Fabrication of hydrogels can be achieved by a direct writing approach followed by moulding in the concave templates. Hydrogels commonly act as cell culture media,[57] delivery agents[58] and reaction chambers[59] in LOC devices.

3.2.1.4 Paper

Paper contains highly porous structures made of cellulose, and achieves liquid wicking through capillary forces without using any external force. Selective areas of the paper can be modified hydrophobically, allowing liquid to flow through the hydrophilic regions in a controlled manner. Fabrication of paper-based microfluidic devices can be achieved either by lithographic approaches to yield high resolution patterns or by inkjet printing methods in which functional regions can be defined by ink printing. Typical detection methods for paper-based LOC include colorimetry,[60,61] luminescence[62,63] and electrochemical detection.[64,65] Using paper to construct LOC devices offers several advantages. First, paper can act as a passive pump to transport liquid solution and as a filter to remove large particles without the need for external power and components. Second, the large surface-to-volume ratio of paper allows sufficient capability of reagent storage by simply soaking the paper, followed by drying. Moreover, the low cost and ease of fabrication makes paper very attractive as a material for massive application of LOC devices.

3.2.2 Major Structures and Components

A comprehensive LOC platform can perform standard laboratory functions, which include sample transportation, mixing, reaction, separation and recycling. These functions can be realized by a combination of various components, such as microvalves (Figure 3.2a), microseparators (Figure 3.2b), micropumps (Figure 3.2c) and micromixers (Figure 3.2d), all of which are connected by microchannels to construct functional devices. The following

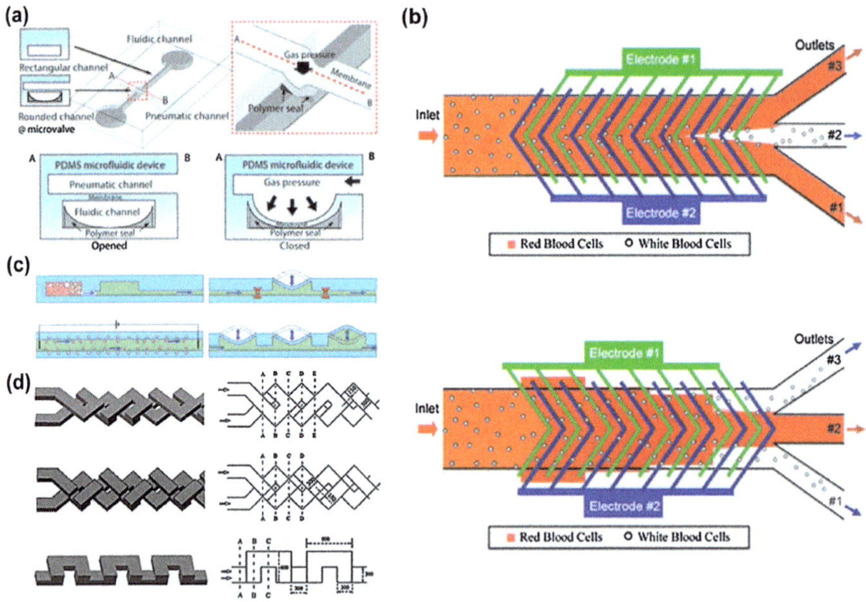

Figure 3.2 Representative images of the major components in a microfluidic chip, including (a) microvalves (reproduced from ref. 70 with permission from the Royal Society of Chemistry), (b) microseparators (reproduced from ref. 71 with permission from the Royal Society of Chemistry), (c) micropumps (reproduced from ref. 72 with permission from the Royal Society of Chemistry), and (d) micromixers (reproduced from ref. 73 with permission from the Royal Society of Chemistry).

sections provide a brief overview of each of the aforementioned microfluidic components. As there are numerous formats and operation principles of individual components, detailed information about these components can be referred to more comprehensive and specific reviews by Zhang *et al.*[66] and Wu *et al.*,[67] as well as books by Nguyen *et al.*[68] and Geschke *et al.*[69]

3.2.2.1 Microvalves

Microvalves can control and limit flow in LOC devices. Ideal microvalves feature low leakage, low power consumption, fast response, linear operating capacity and wide adaptation. According to the functions of microvalves, they can be divided into one-way valves[74–76] and switching valves.[77,78] Check valves are typically only represented by one-way valves, which offer two working states, namely channel closed and open. However, there are many types of switching valves, such as three-way valves and multi-position selector valves. According to the actuation mode of valves, they can be categorized as active valves and passive valves. Active valves use external driving forces to achieve open and closed operation, and are typically realized by actuation methods, such as pneumatic actuation,[79] thermal expansion,[80] piezoelectric effect,[81] shape memory alloy,[82] magnetic and electrostatic[83] and electromagnetic[84] actuation. Passive valves do not require external power or control,[85] but they rely upon the change of flow and pressure of the fluid itself. Detailed introduction about the use of microvalves in LOC devices can be found in reviews by Kwang *et al.*[86] and Au *et al.*[87]

3.2.2.2 Microseparators

Microseparators are components that separate different biomolecules according to their physical properties such as mass, size, diffusion coefficient, magnetic susceptibility and polarity. Different molecules in mixtures are commonly separated by electrophoresis,[88,89] in which molecules with different mobility under an electric field can be separated according to their spatial location or dynamic speed. The use of electrophoresis in continuous flow devices can select charged molecules and particles in the samples and move them into a specific stream of flow that can be separated downstream.[90] Besides the electrophoresis approach, some separation methods apply force fields generated through electric, magnetic and acoustic approaches or with a rotational CD to create centrifugal force.[91,92] Other methods rely upon passive hydrodynamic phenomena in microchannels, such as sedimentation, cross-flow filtration, hydrodynamic filtration or centrifugation in curved channels.[93] In addition, microfilters can be considered as the passive type of microseparators that discriminate particles based upon their geometrical sizes. Reviews about different separation technology used in LOC devices have been written by Gossett *et al.*[94] and Bhagat *et al.*[95]

3.2.2.3 Micropumps

Micropumps are used in LOC to create a pressure difference in order to drive the motion of fluid within the microchannels. Common operation principles of micropumps include piezoelectric, thermopneumatic, electrostatic and electromagnetic actuation. Detailed reviews about the principles and operation of micropumps can be found in multiple literatures.[87,96,97] The micropumps can be categorized into two types, namely mechanical displacement micropumps and electro- or magnetokinetic micropumps. The former applies oscillatory or rotational forces on the fluid through moving solid–fluid or fluid–fluid boundaries using diaphragm displacement, fluid displacement and rotation, while the latter generates steady flows due to the continuous application of energy from electro-osmotic, electrohydrodynamic, magnetohydrodynamic and electrowetting effects.

Within the category of mechanical displacement pumps, diaphragm displacement pumps contain deformable diaphragms actuated by piezoelectric, electrostatic, electromagnetic, pneumatic and thermopneumatic approaches. The deflection of the diaphragm during the expansion and compression strokes causes volume and pressure changes within the microchambers, and achieves dynamic adjustment of the liquid amount within the chambers. Another type of mechanical displacement pump is the fluid displacement pump, which uses secondary fluid in direct contact with the working fluid and achieves working fluid manipulation without using a diaphragm. The secondary fluid is usually made of ferro-magnetic liquid or gas. Rotary micropumps consist of a toothed gear rotating in a fluid chamber with an inlet and an outlet port.

Electro- and magnetokinetic micropumps directly convert electrical and magnetic energy into the directional fluid motion. Electrokinetic pumps pull ions within the microchannels and, in turn, drag along the bulk fluid by momentum transfer due to viscosity. Magnetokinetic pumps typically apply the Lorentz force on the bulk fluid to drive the microchannel flow.

3.2.2.4 *Micromixers*

Fluid mixing in microchannels is important for many biological and chemical applications. Mixing purely by spontaneous particle diffusion within microchannels is slow, but the introduction of passive and active mixers can greatly reduce the distance over which particles travel, and increase the efficiency of mixing.[98,99] Active mixers demand the involvement of external energy, while passive mixers rely purely upon fluid dynamics in the specific geometry of microstructures in microfluidic channels. A type of passive mixer contains grooves on the floor of the microfluidic channels. These grooves align asymmetrically, and generate a transverse component to the flow within the microfluidic channels. This transverse component when combined with the axial pressure gradient along the direction of flow motion yields vigorous relative movement of fluid elements, causing increased contact area between the flowing streams and enhanced diffusion. Another passive mixing design uses serpentine channels, leading to both inertial forces and centrifugal forces when the fluid flows through curved channels. These two combined forces can also generate transverse flow to enable effective mixing. Active mixers use approaches such as electrowetting, nonlinear electrokinetic effects, acoustic streaming and bubble movement to facilitate mixing. However, active mixers often involve complex structures that demand complicated fabrication processes and external power sources. Thus, applications of active mixers in LOC devices are limited due to the technical challenges and cost efficiency. In contrast, passive mixers are easier to fabricate, and are more suitable for species that are sensitive to electrical, mechanical or thermal agitation. In addition, passive mixers also feature robustness and stability, and can be easily integrated with microfluidic systems. Thus, they are more favourable in LOC devices. Detailed reviews about the principles and operation of micromixers can be found in several recent literatures.[100–103]

3.2.3 Fabrication Approaches

Fabrication of LOC adopts techniques, including CMOS, MEMS and other micromachining processes (*e.g.*, moulding, milling and cutting). Recent development in additive manufacturing also yields direct pattern generation through 3D printing that includes stereolithography, fused deposition and inkjet printing. This section offers brief reviews of some typical fabrication approaches in microfluidic LOC devices. More state-of-the-art fabrication techniques can be found in reviews by Anderson *et al.*[104] and Abgrall *et al.*[105]

3.2.3.1 *Surface Micromachining*

Micromachining includes techniques such as film deposition, photolithography, etching and bonding of microchips.[106–108] These techniques can be used to process various materials, such as Si, glass, metals, polymers and semiconductors used in LOC devices to form structural components (*e.g.*, microchannels, valves, diaphragms, cantilevers and pumps) as well as electrical functional components (*e.g.*, electrodes, heaters and detectors). The deposition of materials is achieved through either chemical or physical processes. In chemical processes, reactions between gas phase and liquid phase chemicals under a certain temperature and pressure yield thin layers of polymers, inorganic materials or semiconductors in a controlled manner. While in physical processes, raw materials can be evaporated or sputtered onto the target surface without the involvement of chemical reactions. Photolithography defines selective regions of materials with a photoresist that can be spin-coated and exposed to form photoresist patterns. The underneath materials of the photoresist are further processed through either dry or wet etching using the photoresist as a mask layer, forming patterns on the materials in correspondence to the photoresist. The bonding of the processed materials onto different substrates (*e.g.*, glass and silicon) can be achieved by methods such as surface plasma treatment, UV ozone activation and anodic bonding.

3.2.3.2 *Soft Lithography*

Soft lithography (Figure 3.3) uses elastomeric stamps to replace hard stamps in photolithography for the fabrication of micro-patterns and structures.[109,110] It can be used to create complex 3D structures and can be applied to many materials including polymers, colloid materials, glass and ceramics. In general laboratory environment, its feature size can reach 30 nm to 1 μm with relatively simple setup as compared with the lithography method as aforementioned. The essential components of soft lithography are elastomeric stamps, which are usually made of PDMS through moulding. Typical techniques in soft lithography include microcontact printing, capillary moulding, replica moulding, micro transfer moulding and solvent-assisted moulding. Two representative techniques are contact printing and capillary moulding, which generate non-structured, chemically modified surfaces and a topographically modified physical micro-/nanostructure, respectively. Contact printing uses an elastomeric PDMS stamp, which is soaked in a molecular ink, to contact with a substrate, resulting in transfer printing of the ink onto the substrate surface. In contrast, capillary

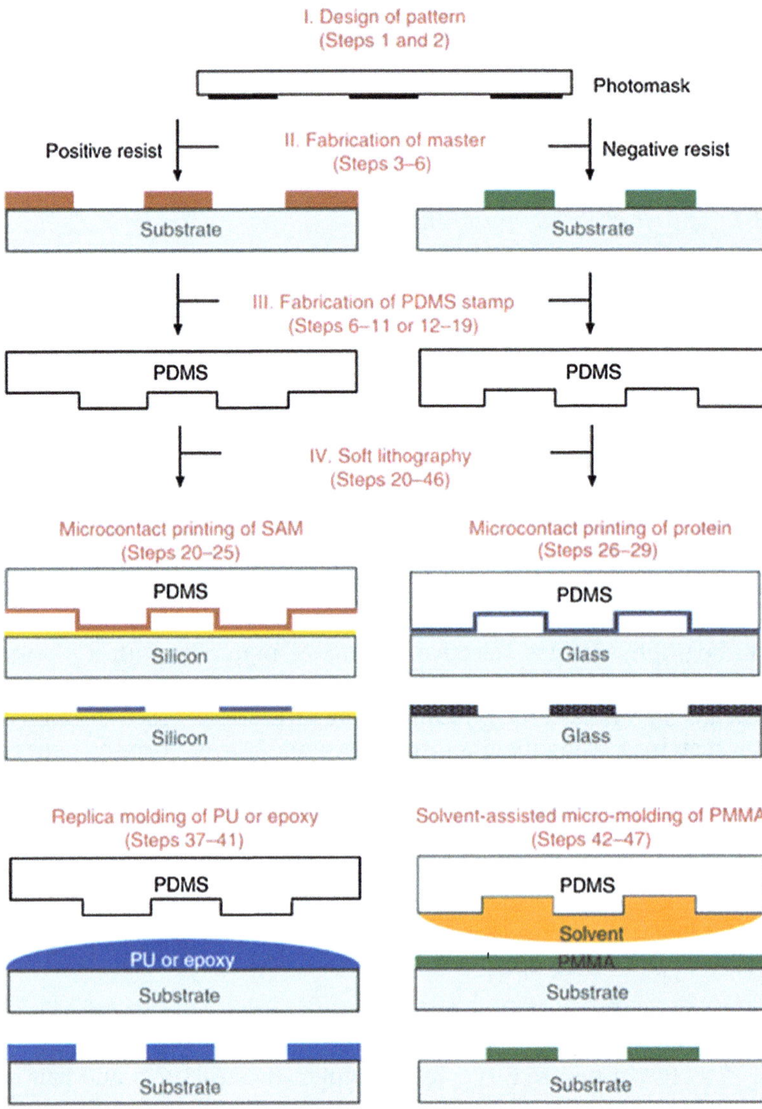

Figure 3.3 Schematic illustration of the four major steps involved in soft lithography and two major soft lithographic techniques – microcontact printing and replica moulding.
Reprinted by permission from Macmillan Publishers Ltd: *Nature Protocols*, (112). Copyright 2010.

moulding uses a patterned PDMS mould to contact with the polymer surface. Heating the polymer above its glass transition temperature allows the polymer to melt and fill the void space of the PDMS mould, generating a negative replica of the mould. Besides thermal approaches, pattern formation in capillary moulding can also be achieved by solvent-laden polymers of UV-curable polymers through solvent evaporation or UV exposure to achieve sub-100-nm soft lithography.[111]

3.2.3.3 3D Printing

3D printing is represented by a series of printing techniques that are capable of fabricating polymeric structures using additive manufacturing approaches (Figure 3.4). Popular 3D printing techniques include stereolithography,[113] two-photon polymerization,[114] fused deposition modelling[115] and inkjet printing,[116] among which the former two techniques are based upon photocurable resins. Photocurable resins solidify under light exposure supplied through a UV laser or near-infrared femtosecond laser in a certain spectral range. Fused deposition modelling uses a stacked layer of

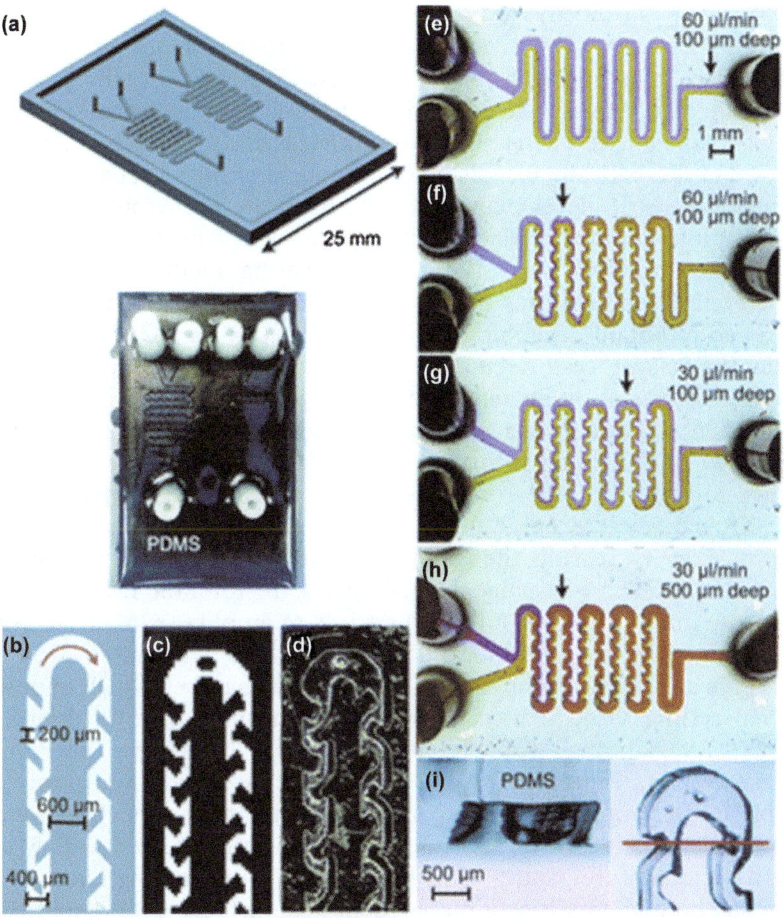

Figure 3.4 A 3D printed template that is used for cast moulding a PDMS microfluidic micromixer (reproduced from ref. 118 with permission from the Royal Society of Chemistry).

thermoplastic materials, which are melted by heaters and extruded through nozzles onto the stages. The extruded materials cool down and solidify on the stages to form models. Materials that can be inkjet printed require appropriate viscosity and surface tension, allowing the formation of droplets and the ability to flow through nozzles with diameters ranging from 5 to 50 μm. The inks of inkjet printing typically contain fully dissolved chemicals that can solidify under the exposure of UV light or heat, or contain nanoparticles that form a connective matrix when sintered by heat, microwave, laser or light.[117]

3.3 Methods Used in LOC Detection of Food Safety and Quality Analysis

3.3.1 PCR and Isothermal Amplification

The supreme specificity of nucleic acids allows sensitive detection of pathogens and recognition of some biomolecules in foods using DNA or RNA. However, biologically relevant samples typically contain very small amounts of nucleic acids, making direct detection almost impossible under many circumstances. Therefore, various nucleic acid amplification methods that aim at dramatically increasing the numbers of nucleic acids in the test samples have been developed. Currently, there are numerous nucleic acid amplification methods available, but the most widely used methods are the polymerase chain reaction[119] (PCR) and isothermal methods.[120]

The cycling temperatures in PCR-based microfluidics must be precisely controlled to obtain desirable temperature kinetics for DNA amplification. During the PCR process, a template of double-stranded DNA is first broken apart at the denaturation temperature (90 to 95 °C), resulting in single-stranded DNA for subsequent amplification. The temperature is further reduced to an annealing temperature (~55 °C), allowing annealing of DNA primers to serve as the starting molecules for DNA synthesis and flanking the DNA sequence to be amplified from both sides. Primer extension is performed by increasing the lower annealing temperature to a medium temperature, which is usually at 72 °C. Using microfluidic LOC technology can create miniaturized PCR (μPCR) devices to amplify DNA. μPCR devices can be categorized into static chamber (SC) and continuous flow (CF) devices. In an SC device, the sample is static in a chamber, which undergoes thermal cycling through precisely controlled heaters. In a CF device, the sample moves through fixed temperature zones to achieve the required thermal cycling, leading to faster DNA amplification and lower power consumption by eliminating the temperature ramping processes. Moreover, the comparable dimensions of μPCR devices and heating elements allow more effective temperature control, thus rapid heating and cooling rates.

The isothermal amplification approaches can achieve DNA amplification at a fixed temperature,[121–124] and involve multiple featured techniques, such as loop-mediated isothermal amplification (LAMP),[125] rolling circle amplification (RCA),[126] strand displacement amplification (SDA),[127] recombinase polymerase amplification (RPA)[128] and helicase dependent amplification (HDA).[129] A shift from PCR towards isothermal methods in microfluidic LOC devices is observed due to the simplicity and reduced thermal budget of the isothermal methods. The main advantage of isothermal methods over PCR is the elimination of the need for thermocycling, as isothermal methods are realized under a stable and constant temperature over time.[130]

3.3.2 Immunoassay

Immunoassay is a technique that exploits the sensitivity and specificity of antibody–antigen interactions for the detection of relevant analytes, and is the fundamental approach used in food analysis to detect specific particles and molecules. In immunoassay, the recognition site of a target antigen by a specific antibody-binding site is called the "epitope". Antigens contain numerous epitopes. However, antibodies can only bind to unique epitopes, resulting in highly specific interaction to achieve antigen recognition from other interfering biomolecules in test samples. Immunoassays can be classified into heterogeneous and homogeneous approaches. In heterogeneous immunoassays, antibodies immobilized on the solid surface interact with nearby antigens. Thus, influence from the unbound antibodies and other reagents can be eliminated. In homogenous immunoassays, both antibodies and antigens are freely suspended in solution. The physical or chemical changes due to bonding can be used to distinguish the bound and unbound antibodies.

Immunoassays have been utilized in a variety of sensor formats, resulting in the construction of immunosensors with different detecting techniques. An immunosensor generally detects immunoassay events through either optical or electrochemical signals. One of the widely used configurations for the capture of target antigens is known as the sandwich structure. The generation of an immunosensor signal from antigen capture is predominantly realized using some types of labels conjugated on a secondary antibody. The selection of labels depends upon the specific detection methodology, and generally includes fluorescent labels, enzymes, paramagnetic particles and metallic colloids. Immunoassay based upon label-free detection approaches have also been developed and offer many advantages, including reduced complexity of chemical reaction and capability to achieve more miniaturized systems. Some of the detection methods on the basis of the labelled or label-free immunoassays are introduced in the following section.

3.3.3 Detection Methods

The most common detection technique used in microfluidic immunoassays is optical detection, which includes approaches of laser-induced fluorescence (LIF) and chemiluminescence (CL) detection based upon labelled biomolecules. Another popular approach is electrochemical detection, which is capable of miniaturization with simplified external read-out.

In addition, some label-free detection methods, such as surface plasmon resonance (SPR), quartz crystal microbalance (QCM) and magnetoelectronic detection, have also been used in LOC devices.

3.3.3.1 Optical Approaches

The optical detection method is one of the most widely used methods in microfluidic detection. Some typical approaches include LIF, CL, spectroscopy and SPR. Microfluidic systems are mostly utilized for sample handling, such as loading, pre-treatment and separation. Such systems can be easily interfaced with external detection techniques or conventional optical detection instruments (*e.g.*, inverted fluorescence microscope, digital CCD cameras spectroscopes, light emitter and detectors and smartphones). However, the requirement of large detection equipment hinders the miniaturization and integration of entire detection systems, making on-line and in-field detection very challenging. Efforts in directly combining optical detection methods with microfluidics have been successfully demonstrated in many research works.

Fluorescence-based biosensors are by far the most prevalent type of biosensors encountered in microfluidic applications due to their ease of implementation. In the florescence process, a substance emits light as an effect of the absorption of light of a shorter wavelength. Specifically, a substance absorbs photons, which can lead to the shift of electrons to unstable higher energy levels. Subsequently, this can result in the return of the energy levels to the ground state and release energy in the form of photons. The benefit of fluorescence detection includes low detection limit, high selectivity and a wide array of fluorescence labels for tagging biomolecules. LOC devices based upon fluorescence detection have been developed with integrated microlenses, waveguides, light sources and detectors, resulting in highly compact systems that are suitable for portable detection. Additionally, miniaturized LOC devices with planar waveguides offer increased optical paths, leading to improved detection sensitivity. Fluorescence detection has been widely used in the determination of bacteria,[131] pathogens,[132–134] toxins[135] and biogenic amines[136] in different agri-food products.

Chemiluminescence is the emission of light as a result of a chemical reaction. In general, a chemiluminescent reaction can be generated by two basic mechanisms. In a direct reaction, two reagents, usually a substrate and an oxidant in the presence of some cofactors, react to form a product or intermediate with or without the presence of a catalyst. Then, some fractions of the product or intermediate will be formed in an electronically excited state, which can subsequently return to the ground state with emission of a photon. On the contrary, indirect or sensitized CL is based upon a process of energy transfer of the excited species to a fluorophore. This process makes it possible for those molecules that are unable to be directly involved in a CL reaction to transfer their energy to a fluorophore that is, in turn, excited followed by returning to its ground state with photon emission.

Surface plasmon resonance is a label-free optical detection method, which involves a resonant oscillation phenomenon that occurs at the interface between two media with distinct signs of dielectric constants. Such an interface can be formed by metals (Au or Ag) and liquids. SPR carries a mathematic form similar to that of the evanescent wave, and generates a decayed electrical field that is reduced exponentially as the field is moved away from the interface. SPR is a surface effect, which is sensitive to physical and chemical changes localized at the interface, while excluding other changes in the bulk regions. Beside the planar SPR sensors, metallic nanoparticles exhibit SPR effects and can serve as immunosensors. The surface plasmon present at a metal–ambient interface absorbs light at a specific wavelength, resulting in a resonance frequency that depends upon the shape of the nanoparticles and the ambient refractive index. In a general design, the surface of gold nanoparticles is immobilized with antibodies, which can bind with antigens and cause mass accumulation on the nanoparticles and shift of resonance frequency.

3.3.3.2 Electrochemical Approaches

Electrochemical detection (ED) is the most attractive alternative to optical detection because of its inherent sensitivity, capability to be miniaturized without loss of performance and high compatibility with the microfabrication technique. Electrochemical detection measures current, voltage, conductance or impedance changes in the process of affinity bonding between receptor/ligand or antigen/antibody systems or enzyme catalyzed chemical reactions. An electrochemical sensor typically contains a simple electrode configuration, including a working electrode, a counter electrode and a reference electrode. Functionalization of electrodes with enzymes can make use of their ability to selectively catalyze chemical reactions. Electrochemical detection offers less expensive read-out than optical systems and can be easily miniaturized and incorporated into microfluidic systems.

Among all electrochemical approaches in microfluidic systems, amperometric detection is one of the most common methods. It detects current changes due to generation and consumption of electroactive species during chemical reactions and affinity binding processes. Device miniaturization and high integration lead to fast response time and efficient collection of electroactive species due to shortened transportation distances. Conductometry is another widely used approach in ED systems. Such detection involves measurement of the conductance between two or four electrodes, through which an alternating current is passed. Conductometry allows convenient detection of ionic species, which are often not readily detected by other techniques, down to nM levels. Conductivity detection can be accomplished either by direct contact between test samples and sensing electrodes or by a contactless mode in which the electrodes do not come into contact with the solutions. Contactless detection is more favourable as it can avoid multiple issues (*e.g.*, current leakage, bubble generation and surface contamination) associated with the contact mode.[137]

In food microfluidics, ED has been successfully implemented in both amperometry[138,139] and conductometry,[140,141] as well as other ED approaches, such as impedance measurement[142,143] and voltammetry.[144,145] The large number of electroactive analytes in foods, the suitability of conductometry for detecting ionic food analytes and the capability to detect both transparent and opaque samples lead to the broad use of ED in food microfluidics.

3.3.3.3 Other Approaches

Two other non-optical and label-free detection approaches with high sensitivity and capability to conduct wireless detection are quartz crystal microbalance (QCM)[146,147] and magnetoelastic sensors[26,148] (ME). QCM measure mass deposition due to the absorption or release of small molecules onto the surface of the microbalance through changes in frequencies of a quartz crystal resonator. ME sensors are made of amorphous ferromagnetic alloys, which exhibit an ME resonance when excited by an external time-varying magnetic field. When target molecules make contact with the sensor surface, mass changes cause a shift of the resonance frequency, which can be measured remotely through a pick-up coil.

3.4 Applications in Food Safety and Quality Analysis

Microfluidic devices possess many unique features, such as low sample consumption, fast detection speed and miniaturized footprint, and these make microfluidic devices very attractive for food safety and quality analysis. There are many natural or synthesized compositions in foods that can generate positive or negative effects on the human body. In this section, LOC devices that measure or act on selective compositions that are important for the safety and quality of foods are presented. Additional information has also been provided in some systematic reviews.[149–151]

3.4.1 Food Additives

Food additives (*e.g.*, preservatives) can improve the preservation time of foods. In addition, food additives (*e.g.*, dyes, thickeners and whiteners) can improve food appeal. Although many food additives are applied under regulation and are typically safe in small quantity, extended use of food additives or intentional use of unregulated food additives may pose potential health risks, such as cancer, heart disease and bowel symptoms. Some typical food additives that have been measured by microfluidic LOC devices include nitrite[152,153] and food dyes.[154,155]

Nitrite exists in beverages and food products as a preservative because it provides highly effective protection against food poisoning microorganisms.

However, high concentrations of nitrite in the human body cause diabetes and nervous system disorders.[156] He and coauthors presented a microfluidic system that contained laser-etched microchannels and a microreactor of 1.8 μL in volume to determine nitrite in foods through the CL phenomenon.[157] Nitrite reacted with ferrocyanide to yield ferricyanide, which later reacted with luminol to generate CL light. CL light generated due to the chain reaction of ferrocyanide, nitrite and luminol in an acidic medium could be recorded by a luminescence analysis within 2 minutes, resulting in a linear detection range of the nitrite concentration from 8 to 100 $\mu g\,L^{-1}$ and a detection limit of 4 $\mu g\,L^{-1}$. Shiddiky and others developed two simple and fast methods for the extraction of nitrite in food samples.[153] The methods were characterized by UV-visible spectroscopy and electrochemical measurement. Microchip electrophoresis with electrochemical detection coupled with a copper/silane complex-modified carbon paste electrode has been employed to detect nitrite. In real sample detection, the sensor was demonstrated to successfully measure nitrite in ham and sausage samples with a wide concentration range from 1 to 150 ppm.

To detect food dyes in various products, Lee and colleagues developed a LOC device integrated with on-chip pre-concentration, separation and electrochemical detection.[154] The device contained three parallel channels: the first two were used for amplification and the last one was for electrochemical detection. Amperometric detection was performed using a glassy carbon electrode coupled horizontally with the microchip at the outlet of the separation channel. The sensitivity of this method was improved by approximately 10 800-fold compared to a conventional micellar electrokinetic chromatographic analysis. Reproducible response was observed during multiple injections of samples with detection limits between 1.0 and 5.0 nM for all food dyes.

3.4.2 Toxins

Toxins produced by microorganisms are chemical substances existing in foods contaminated by certain bacteria and fungi. Toxin contamination can occur throughout the entire food production chain, from processing to transportation and storage. Toxins are extremely hazardous to human health. Representative toxins include botulinum neurotoxins (BoNTs)[158] and mycotoxins.[159] BoNTs are considered as the most poisonous substances known to humans, and are the cause of the life-threatening neuroparalytic disease botulism. On the other hand, major fungal genera produce may contain mycotoxins, which are the secondary metabolites of fungi. There are various types of mycotoxins, such as aflatoxins (AFs), deoxynivalenol (DON), zearalenone (ZEA), fumonisin B1 (FB1), ochratoxin A (OTA) and citrinin (CIT).

There are many microfluidic LOC devices that have been developed to measure BoNTs.[27,160,161] A representative sensor measured botulinum neurotoxin type A (BoNT/A) using self-assembled monolayers (SAMs), which consisted of an immobilized synthetic peptide.[162] Microchannels in the sensor provided multiple functions, including facile fluid manipulation, sample incubation, analyte concentration and fluorescence detection, avoiding sample transfer and loss between different cells that achieve individual functions. Peptide SAMs were exposed to varying concentrations of BoNT/A or its catalytic light chain (ALC), resulting in enzymatic cleavage of the peptide substrate from the surface. Fluorescence detection was achieved down to 20 pg mL^{-1} for ALC and 3 pg mL^{-1} for BoNT/A in 3 h.

Several LOC devices have been developed to achieve rapid and accurate determination of low levels of mycotoxins in foods.[163–165] A research team developed a fully automated device with CL read-out system to detect OTA in green coffee extract.[166] Peptide-linked OTA and biotin conjugates were synthesized and immobilized in an array of 4×6 microspots, and used for an indirect competitive immunoassay. The device only required a CCD camera to detect CL signal. Thus, the complexity of the external detection system can be significantly reduced. The miniaturization and automation largely increased mass transport and reduced detection time (down to 12 mins), resulting in a detection limit of 0.3 g L^{-1} for OTA. Parker and others developed an electrochemical-based microarray for the detection of aflatoxin M1 (AFM1) in milk by using an antibody-modified screen-printed carbon working electrode with a carbon counter electrode and an Ag/AgCl reference electrode.[167] A simple pre-treatment technique of incorporating 18 mM calcium chloride in PBS to the milk samples eliminated the interfering signal in milk. The resulting immunosensor achieved a detection limit of 39 ng L^{-1} with a linear dynamic detection range up to 1 μg L^{-1}. Arévalo and colleagues developed a microfluidic electrochemical immunosensor coupled with flow injection system that can be used for rapid, sensitive and selective quantification of CIT at trace levels in rice samples using ED.[168] Due to the high selectivity of the antibody used in the detection, no sample pretreatment was required to purify the samples. The detection principle was based upon a competitive immunoassay. Specifically, CIT in rice and CIT deposited on a sensor electrode competed for the monoclonal mouse anti-CIT IgG antibody (mAb-CIT) present in solution, followed by a series of reactions that generate current inversely proportional to the amount of CIT present in the rice samples. The electrochemical detection was carried out within 2 min and the total analysis time did not exceed 45 min, leading to a detection limit of 0.1 ng mL^{-1}.

3.4.3 Bacterial and Foodborne Pathogens

Numerous cases of foodborne illness can be attributed to pathogens, such as Shiga-toxin producing *Escherichia coli*, *Salmonella enterica*, *Staphylococcus aureus*, *Campylobacter jejuni* and *Listeria monocytogenes*. LOC biosensors offer ideal tools to realize portable and real-time biosensing of pathogens in farms, packaging/processing facilities, delivery/distribution systems and at consumer levels.[169–171]

Several types of LOC biosensors, including immunoassay and PCR-based, have been developed and tested for the detection of foodborne pathogens. Strachan and coauthors developed a PMMA microfluidic chip integrated with on-board infrared-mediated PCR amplification. The device can be seamlessly integrated with a particle-based, visual DNA detection system for specific detection of *Salmonella enterica* in less than 35 minutes.[172] Microfluidic control was achieved using a capillary burst valve and a manual torque-actuated pressure system. The capillary burst valve that was fabricated using laser ablation can confine the PCR reagents to a chamber during thermal cycling. The manual pressure system mobilized the fluid from the PCR chamber to the detection reservoir containing oligonucleotide-adducted magnetic particles. Interaction of amplified products specific to the target organism with the beads in a rotating magnetic field allowed immediate detection ($<$30 s) based upon hybridization-induced aggregation of the particles and simple optical analysis.

A more integrated device for pathogen detection has been presented by Kim *et al.*[173] to determine *Salmonella* in milk. In this centrifugal microfluidic device, three main steps have been implemented in a single microfluidic disc, including DNA extraction, isothermal RPA and pathogen detection. The devices used a single laser diode to achieve wireless control of multiple functions (*e.g.*, valve actuation, cell lysis and non-contact heating) in the isothermal amplification step, resulting in a compact and miniaturized system. To achieve high detection sensitivity, *Salmonella* cells were first pre-enriched in phosphate-buffered saline (PBS) and milk samples before loading onto the disc through antibody-coated magnetic beads. The entire procedure, from DNA extraction through to detection, was completed within 30 min in a fully automated fashion with detection limits of 10 CFU mL^{-1} and 10^2 CFU mL^{-1} in PBS and in milk, respectively.

Impedance-based electrochemical detection has also been presented. Boehm and others have developed a simple and rapid method for the detection and identification of *E. coli.*[174] The developed device could detect impedance changes when *E. coli* in suspension passed through a microfluidic chamber immobilized with monoclonal antibodies. Continuous perfusion of bacteria suspension through the chamber not only identified specific bacteria but also enhanced the detection sensitivity of the chamber by accumulating bacteria on the chamber wall over time. The sensor could detect 9×10^5 CFU mL^{-1} *E. coli* in the solution by consecutive perfusions, and could increase its detection limit to 10^4 CFU mL^{-1} of *E. coli* when a shallower chamber (2 μm high) was used.

In addition to detecting pathogen, microfluidic devices can also be used for pathogen sorting by isolating pathogens from suspended particle mixture using dielectrophoresis.[175] By converging fluid flow through alternating current electro-osmotic flow in a microfluidic device,[176] the target pathogens can be directed towards the stagnation points, while the suspended particles can be swept towards the outlet along the fluidic flow. Using 3D electrodes to create a dielectrophoretic force field cage, bacterial cells inside a microfluidic channel can be captured more efficiently through tailoring the orientation of the 3D electrodes and by creating a dielectrophoretic force field cage.[177] In a device with 3D electrodes, whose orientation was tailored to achieve a spatial force field, the rate for sorting and collecting three different types of pathogens reached ~300 particles per second.

3.4.4 Antibiotics

Antibiotics are valuable tools for reducing animal disease and suffering from bacterial infections and pathogens as aforementioned. However, the extensive use of antibiotics has led to drug resistance of pathogens, which can be difficult to eliminate and affect human health in a prolonged time period. Applications of antibiotics must be used responsibly in both humans and animals to prevent the spread of drug-resistant bacteria.[178]

Serval microfluidic LOC devices have been developed with the capability to simultaneously detect multiple antibiotics.[179,180] For example, a multiplexed immunoassay-based antibiotic sensing device integrated in a LOC format has been developed to simultaneously determine three antibiotic families (*i.e.*, sulfonamides, fluoroquinolones and tetracyclines) in raw milk.[25] The device contained a polymer-based self-contained microfluidic cartridge and measured antibiotic levels based upon a multi-antibiotic competitive immunoassay. Immunoassay solution was pressure driven externally through a syringe pump and a multi-position valve. Pre-filled microfluidic cartridges were used for a positive/negative binary test for the simultaneous detection of three antibiotics. For result interpretation, any signal lower than the threshold value (100 $mg\,L^{-1}$) was considered to be negative for a given antibiotic. The reliability of the multiplexed detection system was assessed by a validation test conducted on a series of six blind milk samples. The test result had ~95% accuracy. The whole immunoassay procedure was fast (<10 minutes) and easy to handle (automated actuation). In another example, a microarray biosensor integrated with a commercialized Xantec HCX sensor and a flow cell has been demonstrated to quantitatively and simultaneously determine multiple antibiotic families.[181] The biosensor was based upon an imaging surface plasmon resonance (iSPR) platform, and can measure aminoglycosides, sulfonamides, fenicols and fluoroquinolones. By multiplexing seven immunoassays in a competitive format, all the target compounds can be detected in ppb levels in buffer and in 10-fold diluted milk, offering a promising alternative for multi-analyte food profiling.

3.4.5 Heavy Metals

Due to the pollution in water and soil, heavy metals can enter into the food chain through plants and animals, and accumulate within the food, leading to various health issues (*e.g.*, kidney damage, nervous system damage and cancers).[182] Heavy metals, such as cadmium (Cd), chromium (Cr), lead (Pb), arsenic (As), copper (Cu), nickel (Ni) and mercury (Hg), are among the elements that are toxic and can be found within contaminated foods. Therefore, the detection of heavy metal contents in food is necessary and has been achieved by multiple LOC devices.[183–186]

One of the simplest approaches to detect heavy metals in food is by conducting colorimetry detection. Wang and coauthors developed a 3D paper-based microfluidic device to measure Cu, Ni, Cd and Cr in water.[187] The device contained stacked layers of wax patterned paper and double-side adhesive tapes, and allowed fluid transportation through capillary force without the need of external pumps and power. The ions of the heavy metals can react with metal-selective chromogenic reagents and induce colour changes that can be captured by any portable cameras. The resolution of such a colorimetry LOC device was 0.29, 0.33, 0.19 and 0.35 ppm for Cu(II), Ni(II), Cd(II) and Cr(VI), respectively. Due to the miniaturized size and no demand for large external supporting equipment, this paper-based LOC device can be very useful for determination of heavy metals on a daily basis.

More quantitative results can be achieved using electrochemical detection. For example, an electrochemical sensor with a three-electrode configuration was developed to determine Mn, Zn, Cd and Pb by anodic stripping voltammetry. The sensor consisted of an Ag/AgCl reference electrode, an Au auxiliary electrode and a special bismuth working electrode, which could be used to minimize the hydrolysis effect of the device and extend the potential window to the −0.3 to −1.9 V range with respect to the Ag/AgCl reference electrode.[185] The sensor can detect highly electronegative metals (*e.g.*, Mn) and offers superior measurement capabilities for even mildly electronegative metals (*e.g.*, Pb and Cd). In addition, the sensor can detect multiple analytes simultaneously in the low μM concentration range that is relevant to the biophysiological levels of these heavy metals.

Optical detection approaches including fluorescence and CL have also been used in quantitative determination of heavy metals in water and foods. A miniaturized lead sensor has been developed by combining a lead-specific DNAzyme with a microfabricated device containing a network of microfluidic channels coupled *via* a nanocapillary array interconnect.[188] The interconnect can manipulate fluid flows and deliver small volumes of samples to DNAzyme in a spatially confined detection window. The DNAzyme was interrogated using LIF detection in this window, showing linear response over a Pb^{2+} concentration range of 0.1–100 μM, and a detection limit of 11 nM. Nogami and others developed a capillary electrophoresis (CE) micro-device with CL detection using the reaction of 1,10-phenanthroline and hydrogen peroxide for separation and determination of two metal ion groups, the platinum metal group (Ru(III), Rh(III), Pd(II), Os(VIII), Ir(III) and Pt(IV)) and the fourth periodic transition metal group (Cu(II), Fe(II), Co(II) and Ni(II)) in water.[189] The microchip consisted of two microchannels that crossed at the intersection and four reservoirs that were located at the ends of the channels. The metal ions in the sample solution migrated in the channel along with 1,10-phenanthroline in a solution. The solution mixed with hydrogen peroxide in a different reservoir to emit CL, which can be detected by a photomultiplier tube located just above the reservoir.

3.4.6 Pesticide Residues

Pesticide residue refers to pesticides, such as insecticide, fungicide, herbicide and nematocide, all of which remain on or in food after they are applied to food crops. Exposure of the general population to these residues may commonly occur through consumption of treated food sources (*e.g.*, meat, poultry, fish, nuts and vegetables). These pesticide residues exhibit bioaccumulation, leading to build-up of harmful materials to dangerous levels, causing either acute symptoms (*e.g.*, nausea, abdominal cramps, diarrhoea, dizziness, anxiety and confusion) or chronic issues (*e.g.*, respiratory problems, neurological disorders, skin conditions, depression, miscarriage, birth defects and cancer).

The herbicide glyphosate is widely used in a great variety of crops. In direct contact with humans, this species can generate dermal irritation and damage to the respiratory, ocular, endocrine and cardiovascular systems. Additionally, recent studies demonstrated that there is a direct relationship of long exposure to glyphosate with the formation of cancerous cells in human. Silva and colleagues reported a method for rapid, simple, direct and reproducible determination of glyphosate and its major metabolite aminomethylphosphonic acid (AMPA).[190] The microchip was made of polyester with the capability to conduct contactless capacitive detection and measure conductivity and electrophoresis separation of the analytes. The polyester microchip contained microchannels (150 μm in width and 12 μm in depth) that were used for sample injection and separation. Under an excitation voltage of 4.5 Vpp at 320 kHz and pH 8.8, the detection limits for glyphosate and AMPA were 45.1 and 70.5 $\mu mol\,L^{-1}$, respectively. A paper-based sensor for detecting the presence of organophosphate pesticides (*e.g.*, Malathion and Paraoxon) in water has been developed by Sicard and others.[191] The sensor applied the inhibition effect of the organophosphates to the hydrolysis reaction of indoxyl acetate with acetylcholinesterase (AChE). The blue-coloured reaction product indigo was used to determine the concentration of organophosphates in the tested samples. Using a cell phone camera and imaging analysis software, the blue pixels of the sensor images can be counted to quantify the concentration of the pesticides. The colour change in response to a concentration of 10 nM could be identified using this colorimetry approach. Electrochemical detection of pesticides using a LOC device has been demonstrated by Islam and others.[192] The device conducted electrophoresis separation, followed by in-channel pulsed amperometric detection of three common triazine herbicides (*i.e.*, simazine, atrazine and ametryn). Redox reaction of these herbicides generated current peaks in the amperometric curve at a distinct time. Both the amplitude of the peaks and the time were recorded to indicate different species and concentrations. The response time for the detection of these herbicides was less than 90 s, and the limit of detection for the sensor was determined to be 0.35, 0.45 and 0.55 nM for simazine, atrazine and ametryn, respectively.

3.4.7 Migrants from Packaging Materials

Synthetic chemicals used in food packaging may pose long-term health issues to the human body. Many of these packaging substances can leak into the food, causing chronic and continuous exposure to substances, whose long-term impact is either known to be harmful or still under investigation. Some known toxicants, such as formaldehyde, a cancer-inducing substance, are legally used in food packaging materials (*e.g.*, plastic bottles and tableware). Other chemicals such as bisphenol A (BPA) and tributyltin are known to disrupt hormone production, but they are still used in food packaging materials. Thus, to identify the trace of these packaging materials in foods is very critical.

Formaldehyde can be found in preservatives and food packaging materials. It can cause minor to serious problems, such as pain, vomiting, coma and possible death, when a large amount of formaldehyde is taken. Weng and colleagues developed a rapid and simple microfluidic analytical device to determine formaldehyde in eight different samples with only 2 μl sample consumption in one minute.[193] The device was based upon a heated PDMS microfluidic chip, which contained four reaction reservoirs and one substrate reservoir. Samples in the reservoirs were illuminated by 410 nm violet light, whose absorption through the samples can be measured to determine formaldehyde content. A novel integrated microfluidic chip with stacked layers of PMMA structures was developed for the detection of formaldehyde in Chinese herb samples.[194] The test sample and a fluorescence derivatization reagent were mixed in a circular mixing chamber and then flowed through a serpentine reaction channel heated to a temperature of 30 °C by an on-chip heater. The reaction could be completed in 4 min, and the device can be observed in a LIF detection system. The formaldehyde concentration (from 1 to 50 ppm) of the sample was then inversely derived from the measured value of the fluorescence intensity.

Using microfluidics LOC devices can also realize fast and simple detection of BPA with low sample consumption. Kubo and coauthors developed a disc-shaped LOC device using a soft lithography approach.[195] The device contained 32 microchannels and chambers. Anti-PBA antibody was immobilized on microbeads, and introduced into the microchambers through the microchannels in the device. The anti-BPA antibody reacted with horseradish peroxidase (HRP)-conjugated BPA based upon competitive immunoassay. The unbound HRP in each microchamber could be detected using CL. As a result, BPA can be determined at a concentration range between 3.9 and 250 ng mL^{-1}.

3.4.8 Biogenic Amines

Produced by the decarboxylation of tyrosine and histidine, tyramine and histamine are among the most harmful biogenic amines found in fermented beverages. These amines are produced as degradation products from microbial activities and are widely found in fermented foods and beverages, meat, fish and dairy products. Melamine is also a typical biogenic amine, and is used to produce a range of products such as plastics, laminates, coating agents, foams, pigments, glues and fire retardants.[196] However, multiple nationwide incidents associated with milk and beverage contamination in China have caused a national focus on melamine. In these incidents, melamine was illegally added to food products (*e.g.*, dairy products and animal feeds) to increase the apparent protein content, causing significant health effects on consumers (*e.g.*, blindness, kidney stones, reproductive damage and cancer).[197]

Detection of the aforementioned biogenic amines has been demonstrated using many devices. Jayarajah and others developed a portable microfabricated capillary electrophoresis (CE) instrument that was used for the determination of tyramine (3.3 to 30 $mg L^{-1}$) and histamine (1.0 to 3.0 $mg L^{-1}$) in fermented beverages.[198] The target molecules were labelled on their primary amino groups with fluorescamine in a 10 minute reaction, and the samples were analyzed directly, producing a detailed electropherogram in only 120 s on a glass CE device. A droplet-based microfluidic immunosensor has been reported to rapidly and accurately determine melamine in milk. The immunoassay was based upon the competitive reaction between native melamine and a melamine-fluorescein isothiocyanate conjugate against an anti-hapten antibody. The detection protocol provided a limit of detection of 300 ppb, which was below the maximum allowable melamine levels (2.5 ppm) defined by the US Food and Drug Administration and the European Commission.[199] Another detection method for melamine is based upon ultra-violet (UV) detection,[200] which also allows sensitive and fast screening of melamine in milk. This method uses the fracture sampling technique to directly detect sample milk through a narrow fracture and conduct separation through electrophoresis. The concentration of the melamine was later determined by UV absorption spectroscopy. This method can detect melamine in milk within 75 s. At the detection wavelength of 202 nm, the linear range for melamine was from 1.0 to 100 $g mL^{-1}$ with a detection limit of 0.23 $g mL^{-1}$.

3.4.9 Food Allergens

With the rising incidence of people with food hypersensitivity such as food allergy or food intolerance, there is a need for highly sensitive, low-cost analysis methods based upon LOC devices.[201–204] Heyries and others developed a microfluidic device for CL detection of allergenic proteins in peanut and milk.[205] Three different proteins (*i.e.*, β-lactoglobulin, peanut lectin and human IgG) were immobilized in microchambers for detecting specific antibodies. The device can concomitantly detect three specific antibodies in pM levels with only 300 μL of sample consumption and

6 minute sample incubation time. An electrochemical LOC device was also presented to determine food allergens. In this device, detection of food allergen-induced changes in cell morphology and cell metabolism measurement can be performed simultaneously. Two types of cells have been used to observe their allergic response under antigen stimulus. Two microfluidic channels integrated with gold electrodes can be used to introduce cells and measure cell-secreted inflammatory cytokines through ELISA and cell impedance changes through cell-based electrochemical assay.[206]

3.4.10 Antioxidants

Antioxidants are present naturally in foods or manually added in dietary supplements to provide nutrient and health benefits to humans. For example, phenolic acids can be readily absorbed through the walls of the intestinal tract, and can be beneficial to human health conditions due to their antioxidant effect that prevents cellular damage due to free-radical oxidation reactions. They may also promote anti-inflammatory conditions in the human body when they are consumed regularly.

Many plant food-based antioxidants can be efficiently and rapidly determined using a microfluidic system based upon a peroxyoxalate (PO)-CL assay.[207] This assay uses a 9,10-bis-(phenylethynyl)anthracene (BPEA) as the fluorescent probe and hydrogen peroxide as the oxidant. Antioxidant plugs injected into the hydrogen peroxide stream result in inhibition of the CL emission, which can be quantified and correlated with antioxidant capacity. In the tested plant-food-based antioxidants, β-carotene was found to be the most efficient hydrogen peroxide scavenger, followed by α-tocopherol and quercetin. Although the method is inherently simple and rapid, excellent analytical performance is achieved with high sensitivity, large dynamic range and high precision. A microfluidic device has been investigated to estimate the total phenolic contents or antioxidant levels in honey based upon CL detection.[208] Several phenolic antioxidants such as quercetin, catechin, gallic acid, caffeic acid and ferulic acid can emit analytically useful CL signals in an enhanced potassium permanganate CL system. The detection limits can be achieved in a range between 2.4 nmol L^{-1} for gallic acid and 34 nmol L^{-1} for *o*-coumaric acid.

3.4.11 Food Authentication

LOC devices can also be used to analyze certain food compositions associated with religious beliefs, dietary habits and authentication to prevent fraud and misconduct to pursue maximized profits.[209] For example, detection of canine species in foods is important from the perspectives of health, religion and the fair-trade food business. Rahman and others used PCR and restriction fragment length polymorphism (RFLP) assay with LOC detection platform for the authentication of canine DNA in processed foods to ensure Halal food regulation and animal rights protection.[210] In their work, a 100-bp fragment of canine mitochondrial Cytochrome b (*cytb*) gene was selected and amplified. The assay successfully detected 0.0001-ng canine DNA under pure state and 0.01% (w/w) canine meat spiked in chicken and beef burger formulations. The use of a LOC technique for the separation and quantification of milk proteins to prevent any impurity doping using low quality or diluted milk has also been presented.[211] A microfluidic chip with the capability to separate all major milk proteins (*e.g.*, a-lactalbumin, b-lactoglobulin, as-casein, b-casein and k-casein) was demonstrated. The separation and quantification of different proteins was realized by electrophoresis that was able to achieve a detection range from 0 to 0.3 mg mL^{-1} within 30 minutes.

3.5 Conclusions and Perspective

Current microfluidic LOC devices have demonstrated their use in determining various components in foods, allowing quantitative analysis of food safety and quality. The LOC devices feature miniaturized sizes and low sample consumption, and use optical, electrochemical and other electrical detection methods to specifically determine the concentrations of target molecules. The choice of materials and fabrication methods of LOC devices are flexible, including different combinations of various techniques according to the application needs.

However, it is also notable that the majority of the aforementioned LOC devices can only conduct single functions, while relying upon other external components to achieve comprehensive functions. More integrated microfluidic systems are still in high demand to achieve fully portable systems for daily use. Constructing miniaturized LOC systems with different components fabricated using various approaches is still challenging, and operation of these components with minimum external support still requires further research effort. It is worth mentioning that the use of cell phone cameras to assist optical detection and miniaturized batteries to achieve electrochemical detection represent two promising trends for LOC devices. In addition, use of porous structures, such as paper and sponge, can eliminate the conventional micropumps and microchannels in LOC devices for fluid handling, showing promising potential to achieve high integrated and miniaturized devices with simplified design. By resolving the aforementioned issues in terms of system miniaturization and simplification of device design and external setup, it can be expected that more portable LOC systems can be achieved to improve the efficiency and convenience of food safety and quality analysis. These portable LOC systems can facilitate improvement of personal health conditions by reducing intake of problematic foods and quantifying health compositions in foods, allowing more scientific dietary monitoring and fewer undesirable incidents with health risks.

References

1. A. Jang, Z. Zou, K. K. Lee, C. H. Ahn and P. L. Bishop, *Meas. Sci. Technol.*, 2011, **22**, 032001.
2. M. J. McLaughlin, D. R. Parker and J. M. Clarke, *Field Crops Res.*, 1999, **60**, 143–163.
3. F. P. Carvalho, *Environ. Sci. Policy*, 2006, **9**, 685–692.
4. X. Pei, A. Tandon, A. Alldrick, L. Giorgi, W. Huang and R. Yang, *Food Policy*, 2011, **36**, 412–420.
5. H.-M. Lam, J. Remais, M.-C. Fung, L. Xu and S. S.-M. Sun, *Lancet*, 2013, **381**, 2044–2053.
6. J. R. Ingelfinger, *N. Engl. J. Med.*, 2008, **359**, 2745–2748.
7. P. S. Mead, L. Slutsker, V. Dietz, L. F. McCaig, J. S. Bresee, C. Shapiro, P. M. Griffin and R. V. Tauxe, *Emerging Infect. Dis.*, 1999, **5**, 607.
8. S. Sivapalasingam, C. R. Friedman, L. Cohen and R. V. Tauxe, *J. Food Prot.*, 2004, **67**, 2342–2353.
9. N. A. Daniels, L. Mackinnon, S. M. Rowe, N. H. Bean, P. M. Griffin and P. S. Mead, *Pediatr. Infect. Dis. J.*, 2002, **21**, 623–628.
10. H. A. Stone, A. D. Stroock and A. Ajdari, *Annu. Rev. Fluid Mech.*, 2004, **36**, 381–411.
11. D. Figeys and D. Pinto, *Anal. Chem.*, 2000, **72**, 330 A–335 A.
12. G. M. Whitesides, *Nature*, 2006, **442**, 368–373.
13. Y. Wang, H. Xu, J. Zhang and G. Li, *Sensors*, 2008, **8**, 2043–2081.
14. M. S. Kim, T. Kim, S.-Y. Kong, S. Kwon, C. Y. Bae, J. Choi, C. H. Kim, E. S. Lee and J.-K. Park, *PLoS One*, 2010, **5**, e10441.
15. Y. Xu, J. A. Phillips, J. Yan, Q. Li, Z. H. Fan and W. Tan, *Anal. Chem.*, 2009, **81**, 7436–7442.
16. C. H. Ahn, J.-W. Choi, G. Beaucage, J. H. Nevin, J.-B. Lee, A. Puntambekar and J. Y. Lee, *Proc. IEEE*, 2004, **92**, 154–173.
17. P. S. Dittrich and A. Manz, *Nat. Rev. Drug Discovery*, 2006, **5**, 210–218.
18. J. Pihl, M. Karlsson and D. T. Chiu, *Drug Discovery Today*, 2005, **10**, 1377–1383.
19. P. Neuži, S. Giselbrecht, K. Länge, T. J. Huang and A. Manz, *Nat. Rev. Drug Discovery*, 2012, **11**, 620–632.
20. J. Gardeniers and A. Van den Berg, *Anal. Bioanal. Chem.*, 2004, **378**, 1700–1703.
21. L. Marle and G. M. Greenway, *TrAC, Trends Anal. Chem.*, 2005, **24**, 795–802.
22. K. M. Horsman, J. M. Bienvenue, K. R. Blasier and J. P. Landers, *J. Forensic Sci.*, 2007, **52**, 784–799.
23. E. Verpoorte, *Electrophoresis*, 2002, **23**, 677–712.
24. A. J. Hopwood, C. Hurth, J. Yang, Z. Cai, N. Moran, J. G. Lee-Edghill, A. Nordquist, R. Lenigk, M. D. Estes and J. P. Haley, *Anal. Chem.*, 2010, **82**, 6991–6999.
25. G. Suarez, Y.-H. Jin, J. Auerswald, S. Berchtold, H. F. Knapp, J.-M. Diserens, Y. Leterrier, J.-A. E. Manson and G. Voirin, *Lab Chip*, 2009, **9**, 1625–1630.

26. C. Xue, C. Yang, T. Xu, J. Zhan and X. Li, *Lab Chip*, 2015, **15**, 2318–2326.
27. P. B. Lillehoj, F. Wei and C.-M. Ho, *Lab Chip*, 2010, **10**, 2265–2270.
28. R. B. Fair, *Microfluid. Nanofluid.*, 2007, **3**, 245–281.
29. D. C. Duffy, J. C. McDonald, O. J. Schueller and G. M. Whitesides, *Anal. Chem.*, 1998, **70**, 4974–4984.
30. H. Becker and L. E. Locascio, *Talanta*, 2002, **56**, 267–287.
31. G. S. Fiorini, G. D. Jeffries, D. S. Lim, C. L. Kuyper and D. T. Chiu, *Lab Chip*, 2003, **3**, 158–163.
32. G. S. Fiorini, R. M. Lorenz, J. S. Kuo and D. T. Chiu, *Anal. Chem.*, 2004, **76**, 4697–4704.
33. C. Chen, J. Liu, L. Hromada, C. Tsao, C. Chang and D. DeVoe, *Lab Chip*, 2009, **9**, 50–55.
34. M. Heckele and W. Schomburg, *J. Micromech. Microeng.*, 2003, **14**, R1.
35. A. A. S. Bhagat, P. Jothimuthu and I. Papautsky, *Lab Chip*, 2007, **7**, 1192–1197.
36. K. Khare, J. Zhou and S. Yang, *Langmuir*, 2009, **25**, 12794–12799.
37. S. Seethapathy and T. Górecki, *Anal. Chim. Acta*, 2012, **750**, 48–62.
38. J. C. McDonald and G. M. Whitesides, *Acc. Chem. Res.*, 2002, **35**, 491–499.
39. H. Lorenz, M. Despont, N. Fahrni, N. LaBianca, P. Renaud and P. Vettiger, *J. Micromech. Microeng.*, 1997, **7**, 121.
40. C.-H. Lin, G.-B. Lee, B.-W. Chang and G.-L. Chang, *J. Micromech. Microeng.*, 2002, **12**, 590.
41. S. Metz, R. Holzer and P. Renaud, *Lab Chip*, 2001, **1**, 29–34.
42. K. I. Min, T. H. Lee, C. P. Park, Z. Y. Wu, H. H. Girault, I. Ryu, T. Fukuyama, Y. Mukai and D. P. Kim, *Angew. Chem., Int. Ed.*, 2010, **49**, 7063–7067.
43. H. Suzuki, K. V. Tabata, H. Noji and S. Takeuchi, *Langmuir*, 2006, **22**, 1937–1942.
44. T.-F. Hong, W.-J. Ju, M.-C. Wu, C.-H. Tai, C.-H. Tsai and L.-M. Fu, *Microfluid. Nanofluid.*, 2010, **9**, 1125–1133.
45. M. A. Witek, S. D. Llopis, A. Wheatley, R. L. McCarley and S. A. Soper, *Nucleic Acids Res.*, 2006, **34**, e74.
46. M. A. Witek, M. L. Hupert, D. S.-W. Park, K. Fears, M. C. Murphy and S. A. Soper, *Anal. Chem.*, 2008, **80**, 3483–3491.
47. A. S. Johnson, K. B. Anderson, S. T. Halpin, D. C. Kirkpatrick, D. M. Spence and R. S. Martin, *Analyst*, 2013, **138**, 129–136.
48. A. Bubendorfer, X. Liu and A. V. Ellis, *Smart Mater. Struct.*, 2007, **16**, 367.
49. J. Li, C. Liu, H. Qiao, L. Zhu, G. Chen and X. Dai, *J. Micromech. Microeng.*, 2007, **18**, 015008.
50. C. Geismann, A. Yaroshchuk and M. Ulbricht, *Langmuir*, 2007, **23**, 76–83.
51. J. A. Burdick, A. Khademhosseini and R. Langer, *Langmuir*, 2004, **20**, 5153–5156.
52. H. Shibata, Y. J. Heo, T. Okitsu, Y. Matsunaga, T. Kawanishi and S. Takeuchi, *Proc. Natl. Acad. Sci. U. S. A.*, 2010, **107**, 17894–17898.

53. M. Yamamoto, Y. Ikada and Y. Tabata, *J. Biomater. Sci., Polym. Ed.*, 2001, **12**, 77–88.
54. Y. Tabata, M. Miyao, M. Ozeki and Y. Ikada, *J. Biomater. Sci., Polym. Ed.*, 2000, **11**, 915–930.
55. M. Mayer, J. Yang, I. Gitlin, D. H. Gracias and G. M. Whitesides, *Proteomics*, 2004, **4**, 2366–2376.
56. C. Chang and L. Zhang, *Carbohydr. Polym.*, 2011, **84**, 40–53.
57. Y. Ling, J. Rubin, Y. Deng, C. Huang, U. Demirci, J. M. Karp and A. Khademhosseini, *Lab Chip*, 2007, **7**, 756–762.
58. J. Wan, *Polymers*, 2012, **4**, 1084–1108.
59. D. Choi, E. Jang, J. Park and W.-G. Koh, *Microfluid. Nanofluid.*, 2008, **5**, 703–710.
60. A. K. Ellerbee, S. T. Phillips, A. C. Siegel, K. A. Mirica, A. W. Martinez, P. Striehl, N. Jain, M. Prentiss and G. M. Whitesides, *Anal. Chem.*, 2009, **81**, 8447–8452.
61. W. Dungchai, O. Chailapakul and C. S. Henry, *Anal. Chim. Acta*, 2010, **674**, 227–233.
62. J. L. Delaney, C. F. Hogan, J. Tian and W. Shen, *Anal. Chem.*, 2011, **83**, 1300–1306.
63. J. Yu, L. Ge, J. Huang, S. Wang and S. Ge, *Lab Chip*, 2011, **11**, 1286–1291.
64. W. Dungchai, O. Chailapakul and C. S. Henry, *Anal. Chem.*, 2009, **81**, 5821–5826.
65. Z. Nie, F. Deiss, X. Liu, O. Akbulut and G. M. Whitesides, *Lab Chip*, 2010, **10**, 3163–3169.
66. C. Zhang, D. Xing and Y. Li, *Biotechnol. Adv.*, 2007, **25**, 483–514.
67. J. Wu and M. Gu, *J. Biomed. Opt.*, 2011, **16**, 080901–080901-080912.
68. N.-T. Nguyen and S. T. Wereley, *Fundamentals and Applications of Microfluidics*, Artech House, 2002.
69. O. Geschke, H. Klank and P. Telleman, *Microsystem Engineering of Lab-on-a-chip Devices*, Wiley Online Library, 2008.
70. W. Park, S. Han and S. Kwon, *Lab Chip*, 2010, **10**, 2814–2817.
71. K.-H. Han and A. B. Frazier, *Lab Chip*, 2008, **8**, 1079–1086.
72. H. So, A. P. Pisano and Y. H. Seo, *Lab Chip*, 2014, **14**, 2240–2248.
73. H. M. Xia, S. Y. M. Wan, C. Shu and Y. T. Chew, *Lab Chip*, 2005, **5**, 748–755.
74. T. Wang, H. Chen, K. Liu, Y. Li, P. Xue, Y. Yu, S. Wang, J. Zhang, E. Kumacheva and B. Yang, *Nanoscale*, 2014, **6**, 3846–3853.
75. D. Kim and D. J. Beebe, *Sens. Actuators, A*, 2007, **136**, 426–433.
76. T. Pan, S. J. McDonald, E. M. Kai and B. Ziaie, *J. Micromech. Microeng.*, 2005, **15**, 1021.
77. J. Kim, H. Kido, R. H. Rangel and M. J. Madou, *Sens. Actuators, B*, 2008, **128**, 613–621.
78. K. Ikuta, T. Hasegawa, T. Adachi and S. Maruo, Micro Electro Mechanical Systems, 2000. MEMS 2000. The Thirteenth Annual International Conference on, 2000.
79. K. Hosokawa and R. Maeda, *J. Micromech. Microeng.*, 2000, **10**, 415.

80. K. Pitchaimani, B. C. Sapp, A. Winter, A. Gispanski, T. Nishida and Z. H. Fan, *Lab Chip*, 2009, **9**, 3082–3087.
81. D. C. Roberts, L. Hanqing, J. L. Steyn, O. Yaglioglu, S. M. Spearing, M. A. Schmidt and N. W. Hagood, *J. Microelectromech. Syst.*, 2003, **12**, 81–92.
82. M. E. Piccini and B. C. Towe, *Sens. Actuators, A*, 2006, **128**, 344–349.
83. D. Anjewierden, G. A. Liddiard and B. K. Gale, *J. Micromech. Microeng.*, 2012, **22**, 025019.
84. A. Meckes, J. Behrens, O. Kayser, W. Benecke, T. Becker and G. Müller, *Sens. Actuators, A*, 1999, **76**, 478–483.
85. Y. Feng, Z. Zhou, X. Ye and J. Xiong, *Sens. Actuators, A*, 2003, **108**, 138–143.
86. W. O. Kwang and H. A. Chong, *J. Micromech. Microeng.*, 2006, **16**, R13.
87. A. K. Au, H. Lai, B. R. Utela and A. Folch, *Micromachines*, 2011, **2**, 179–220.
88. D. Wu, J. Qin and B. Lin, *J. Chromatogr. A*, 2008, **1184**, 542–559.
89. D. J. Beebe, G. A. Mensing and G. M. Walker, *Annu. Rev. Biomed. Eng.*, 2002, **4**, 261–286.
90. A. Adami, A. Mortari, E. Morganti and L. Lorenzelli, *J. Sens.*, 2016, 9.
91. R. Gorkin, J. Park, J. Siegrist, M. Amasia, B. S. Lee, J.-M. Park, J. Kim, H. Kim, M. Madou and Y.-K. Cho, *Lab Chip*, 2010, **10**, 1758–1773.
92. M. Madou, J. Zoval, G. Jia, H. Kido, J. Kim and N. Kim, *Annu. Rev. Biomed. Eng.*, 2006, **8**, 601–628.
93. J. Marchalot, Y. Fouillet and J. L. Achard, *Microfluid. Nanofluid.*, 2014, **17**, 167–180.
94. D. R. Gossett, W. M. Weaver, A. J. Mach, S. C. Hur, H. T. K. Tse, W. Lee, H. Amini and D. Di Carlo, *Anal. Bioanal. Chem.*, 2010, **397**, 3249–3267.
95. A. A. S. Bhagat, H. Bow, H. W. Hou, S. J. Tan, J. Han and C. T. Lim, *Med. Biol. Eng. Comput.*, 2010, **48**, 999–1014.
96. M. De Volder and D. Reynaerts, *J. Micromech. Microeng.*, 2010, **20**, 043001.
97. F. Abhari, H. Jaafar and N. A. M. Yunus, *Int. J. Electrochem. Sci.*, 2012, 7, 9765–9780.
98. V. Hessel, H. Löwe and F. Schönfeld, *Chem. Eng. Sci.*, 2005, **60**, 2479–2501.
99. L. Capretto, W. Cheng, M. Hill and X. Zhang, in *Microfluidics*, Springer, 2011, pp. 27–68.
100. N. Nam-Trung and W. Zhigang, *J. Micromech. Microeng.*, 2005, **15**, R1.
101. L. Falk and J.-M. Commenge, *Chem. Eng. Sci.*, 2010, **65**, 405–411.
102. C.-Y. Lee, C.-L. Chang, Y.-N. Wang and L.-M. Fu, *Int. J. Mol. Sci.*, 2011, **12**, 3263–3287.
103. Y. K. Suh and S. Kang, *Micromachines*, 2010, **1**, 82–111.
104. J. R. Anderson, D. T. Chiu, H. Wu, O. Schueller and G. M. Whitesides, *Electrophoresis*, 2000, **21**, 27–40.
105. P. Abgrall and A. Gue, *J. Micromech. Microeng.*, 2007, **17**, R15.
106. J. M. Bustillo, R. T. Howe and R. S. Muller, *Proc. IEEE*, 1998, **86**, 1552–1574.

107. N. Tas, T. Sonnenberg, H. Jansen, R. Legtenberg and M. Elwenspoek, *J. Micromech. Microeng.*, 1996, **6**, 385.
108. B. Ziaie, A. Baldi, M. Lei, Y. Gu and R. A. Siegel, *Adv. Drug Delivery Rev.*, 2004, **56**, 145–172.
109. G. M. Whitesides, E. Ostuni, S. Takayama, X. Jiang and D. E. Ingber, *Annu. Rev. Biomed. Eng.*, 2001, **3**, 335–373.
110. J. A. Rogers and R. G. Nuzzo, *Mater. Today*, 2005, **8**, 50–56.
111. P. Kim, K. W. Kwon, M. C. Park, S. H. Lee, S. M. Kim and K. Y. Suh, *BioChip J.*, 2008, **2**, 1–11.
112. D. Qin, Y. Xia and G. M. Whitesides, *Nat. Protoc.*, 2010, **5**, 491–502.
113. A. K. Au, W. Lee and A. Folch, *Lab Chip*, 2014, **14**, 1294–1301.
114. W. Xiong, Y. S. Zhou, X. N. He, Y. Gao, M. Mahjouri-Samani, L. Jiang, T. Baldacchini and Y. F. Lu, *Light: Sci. Appl.*, 2012, **1**, e6.
115. E. J. McCullough and V. K. Yadavalli, *J. Mater. Process. Technol.*, 2013, **213**, 947–954.
116. K. Abe, K. Suzuki and D. Citterio, *Anal. Chem.*, 2008, **80**, 6928–6934.
117. C. M. B. Ho, S. H. Ng, K. H. H. Li and Y.-J. Yoon, *Lab Chip*, 2015, **15**, 3627–3637.
118. G. Comina, A. Suska and D. Filippini, *Lab Chip*, 2014, **14**, 424–430.
119. N. Arnheim and H. Erlich, *Annu. Rev. Biochem.*, 1992, **61**, 131–156.
120. T. Notomi, H. Okayama, H. Masubuchi, T. Yonekawa, K. Watanabe, N. Amino and T. Hase, *Nucleic Acids Res.*, 2000, **28**, e63.
121. C.-H. Wang, K.-Y. Lien, T.-Y. Wang, T.-Y. Chen and G.-B. Lee, *Biosens. Bioelectron.*, 2011, **26**, 2045–2052.
122. Y. Mori and T. Notomi, *J. Infect. Chemother.*, 2009, **15**, 62–69.
123. P. J. Asiello and A. J. Baeumner, *Lab Chip*, 2011, **11**, 1420–1430.
124. P. Craw and W. Balachandran, *Lab Chip*, 2012, **12**, 2469–2486.
125. X. Fang, Y. Liu, J. Kong and X. Jiang, *Anal. Chem.*, 2010, **82**, 3002–3006.
126. M. M. Ali, F. Li, Z. Zhang, K. Zhang, D.-K. Kang, J. A. Ankrum, X. C. Le and W. Zhao, *Chem. Soc. Rev.*, 2014, **43**, 3324–3341.
127. M. A. Burns, B. N. Johnson, S. N. Brahmasandra, K. Handique, J. R. Webster, M. Krishnan, T. S. Sammarco, P. M. Man, D. Jones and D. Heldsinger, *Science*, 1998, **282**, 484–487.
128. S. Lutz, P. Weber, M. Focke, B. Faltin, J. Hoffmann, C. Müller, D. Mark, G. Roth, P. Munday and N. Armes, *Lab Chip*, 2010, **10**, 887–893.
129. M. Mahalanabis, J. Do, H. ALMuayad, J. Y. Zhang and C. M. Klapperich, *Biomed. Microdevices*, 2010, **12**, 353–359.
130. G. D. Kaprou, G. Papadakis, D. P. Papageorgiou, G. Kokkoris, V. Papadopoulos, I. Kefala, E. Gizeli and A. Tserepi, *Microsyst. Technol.*, 2016, **22**, 1529–1534.
131. M. Ikeda, N. Yamaguchi, K. Tani and M. Nasu, *J. Microbiol. Methods*, 2006, **67**, 241–247.
132. Y. Li, C. Zhang and D. Xing, *Anal. Biochem.*, 2011, **415**, 87–96.
133. O. Strohmeier, N. Marquart, D. Mark, G. Roth, R. Zengerle and F. von Stetten, *Anal. Methods*, 2014, **6**, 2038–2046.
134. A. Rasooly and K. E. Herold, *J. AOAC Int.*, 2006, **89**, 873–883.

135. Y. Dong, K. S. Phillips and Q. Cheng, *Lab Chip*, 2006, **6**, 675–681.
136. L. Wang, J. Wu, Q. Wang, C. He, L. Zhou, J. Wang and Q. Pu, *J. Agric. Food Chem.*, 2012, **60**, 1613–1618.
137. M. Pumera, *Talanta*, 2007, **74**, 358–364.
138. A. G. Crevillén, M. Ávila, M. Pumera, M. C. González and A. Escarpa, *Anal. Chem.*, 2007, **79**, 7408–7415.
139. M. Scampicchio, D. Ballabio, A. Arecchi, S. M. Cosio and S. Mannino, *Microchim. Acta*, 2008, **163**, 11–21.
140. Z. Muhammad-Tahir and E. C. Alocilja, *Biosens. Bioelectron.*, 2003, **18**, 813–819.
141. M. Hnaiein, W. Hassen, A. Abdelghani, C. Fournier-Wirth, J. Coste, F. Bessueille, D. Leonard and N. Jaffrezic-Renault, *Electrochem. Commun.*, 2008, **10**, 1152–1154.
142. L. Yang and R. Bashir, *Biotechnol. Adv.*, 2008, **26**, 135–150.
143. F. Tan, P. H. Leung, Z.-b. Liu, Y. Zhang, L. Xiao, W. Ye, X. Zhang, L. Yi and M. Yang, *Sens. Actuators, B*, 2011, **159**, 328–335.
144. Y. Yardım, M. Gülcan and Z. Şentürk, *Food Chem.*, 2013, **141**, 1821–1827.
145. M. Regiart, S. V. Pereira, V. G. Spotorno, F. A. Bertolino and J. Raba, *Sens. Actuators, B*, 2013, **188**, 1241–1249.
146. M. A. Cooper and V. T. Singleton, *J. Mol. Recognit.*, 2007, **20**, 154–184.
147. H. Sun, Z. Mo, J. T. Choy, D. Zhu and Y. Fung, *Sens. Actuators, B*, 2008, **131**, 148–158.
148. J. Wan, M. L. Johnson, R. Guntupalli, V. A. Petrenko and B. A. Chin, *Sens. Actuators, B*, 2007, **127**, 559–566.
149. S. Neethirajan, I. Kobayashi, M. Nakajima, D. Wu, S. Nandagopal and F. Lin, *Lab Chip*, 2011, **11**, 1574–1586.
150. B. Van Dorst, J. Mehta, K. Bekaert, E. Rouah-Martin, W. De Coen, P. Dubruel, R. Blust and J. Robbens, *Biosens. Bioelectron.*, 2010, **26**, 1178–1194.
151. H. Bridle, B. Miller and M. P. Desmulliez, *Water Res.*, 2014, **55**, 256–271.
152. V. J. Sieben, C. F. Floquet, I. R. Ogilvie, M. C. Mowlem and H. Morgan, *Anal. Methods*, 2010, **2**, 484–491.
153. M. J. A. Shiddiky, L. Kyungsun, S. Jungik, P. Deogsu and S. Yoonbo, *J. Agric. Food Chem.*, 2009, **57**, 4051–4057.
154. K. S. Lee, M. J. A. Shiddiky, S. H. Park, D. S. Park and Y. B. Shim, *Electrophoresis*, 2008, **29**, 1910–1917.
155. S. Teerasong and R. L. McClain, *J. Chem. Educ.*, 2011, **88**, 465–467.
156. G. Ellis, I. Adatia, M. Yazdanpanah and S. K. Makela, *Clin. Biochem.*, 1998, **31**, 195–220.
157. D. He, Z. Zhang, Y. Huang and Y. Hu, *Food Chem.*, 2007, **101**, 667–672.
158. D. Dressler and F. Adib Saberi, *Eur. Neurol.*, 2005, **53**, 3–9.
159. H. S. Hussein and J. M. Brasel, *Toxicology*, 2001, **167**, 101–134.
160. S. Sun, M. Ossandon, Y. Kostov and A. Rasooly, *Lab Chip*, 2009, **9**, 3275–3281.

161. C.-Y. Koh, U. Y. Schaff, M. E. Piccini, L. H. Stanker, L. W. Cheng, E. Ravichandran, B.-R. Singh, G. J. Sommer and A. K. Singh, *Anal. Chem.*, 2015, **87**, 922–928.
162. M. L. Frisk, W. H. Tepp, E. A. Johnson and D. J. Beebe, *Anal. Chem.*, 2009, **81**, 2760–2767.
163. L. Guo, J. Feng, Z. Fang, J. Xu and X. Lu, *Trends Food Sci. Technol.*, 2015, **46**, 252–263.
164. Z. Olcer, E. Esen, T. Muhammad, A. Ersoy, S. Budak and Y. Uludag, *Biosens. Bioelectron.*, 2014, **62**, 163–169.
165. J. C. Vidal, L. Bonel, A. Ezquerra, S. Hernández, J. R. Bertolín, C. Cubel and J. R. Castillo, *Biosens. Bioelectron.*, 2013, **49**, 146–158.
166. J. C. Sauceda-Friebe, X. Y. Karsunke, S. Vazac, S. Biselli, R. Niessner and D. Knopp, *Anal. Chim. Acta*, 2011, **689**, 234–242.
167. C. O. Parker and I. E. Tothill, *Biosens. Bioelectron.*, 2009, **24**, 2452–2457.
168. F. J. Arévalo, A. M. Granero, H. Fernández, J. Raba and M. A. Zón, *Talanta*, 2011, **83**, 966–973.
169. J. Mairhofer, K. Roppert and P. Ertl, *Sensors*, 2009, **9**, 4804–4823.
170. J. C. Jokerst, J. A. Adkins, B. Bisha, M. M. Mentele, L. D. Goodridge and C. S. Henry, *Anal. Chem.*, 2012, **84**, 2900–2907.
171. D. L. Stokes, G. D. Griffin and T. Vo-Dinh, *Fresenius' J. Anal. Chem.*, 2001, **369**, 295–301.
172. B. C. Strachan, H. S. Sloane, E. Houpt, J. C. Lee, D. C. Miranian, J. Li, D. A. Nelson and J. P. Landers, *Analyst*, 2015, **141**, 947–955.
173. T.-H. Kim, J. Park, C.-J. Kim and Y.-K. Cho, *Anal. Chem.*, 2014, **86**, 3841–3848.
174. D. A. Boehm, P. A. Gottlieb and S. Z. Hua, *Sens. Actuators, B*, 2007, **126**, 508–514.
175. K. Khoshmanesh, S. Nahavandi, S. Baratchi, A. Mitchell and K. Kalantar-zadeh, *Biosens. Bioelectron.*, 2011, **26**, 1800–1814.
176. Z. Gagnon and H. C. Chang, *Electrophoresis*, 2005, **26**, 3725–3737.
177. I. F. Cheng, H. C. Chang, D. Hou and H. C. Chang, *Biomicrofluidics*, 2007, **1**, 21503.
178. M. Lee, H. Lee and P. Ryu, *Asian-Australas. J. Anim. Sci.*, 2001, **14**, 402–413.
179. P. Sun, Y. Liu, J. Sha, Z. Zhang, Q. Tu, P. Chen and J. Wang, *Biosens. Bioelectron.*, 2011, **26**, 1993–1999.
180. J. Choi, Y.-G. Jung, J. Kim, S. Kim, Y. Jung, H. Na and S. Kwon, *Lab Chip*, 2013, **13**, 280–287.
181. S. Rebe Raz, M. G. Bremer, W. Haasnoot and W. Norde, *Anal. Chem.*, 2009, **81**, 7743–7749.
182. L. Järup, *Br. Med. Bull.*, 2003, **68**, 167–182.
183. K. Hylton and S. Mitra, *Anal. Chim. Acta*, 2008, **607**, 45–49.
184. S. Z. Hossain and J. D. Brennan, *Anal. Chem.*, 2011, **83**, 8772–8778.
185. P. Jothimuthu, R. A. Wilson, J. Herren, E. N. Haynes, W. R. Heineman and I. Papautsky, *Biomed. Microdevices*, 2011, **13**, 695–703.
186. W. Yantasee, Y. Lin, K. Hongsirikarn, G. E. Fryxell, R. Addleman and C. Timchalk, *Environ. Health Perspect.*, 2007, 1683–1690.

187. H. Wang, Y.-j. Li, J.-f. Wei, J.-r. Xu, Y.-h. Wang and G.-x. Zheng, *Anal. Bioanal. Chem.*, 2014, **406**, 2799–2807.
188. I.-H. Chang, J. J. Tulock, J. Liu, W.-S. Kim, D. M. Cannon, Y. Lu, P. W. Bohn, J. V. Sweedler and D. M. Cropek, *Environ. Sci. Technol.*, 2005, **39**, 3756–3761.
189. T. Nogami, M. Hashimoto and K. T. Nogami, *J. Sep. Sci.*, 2009, **32**, 408–412.
190. E. R. Silva, T. P. Segato, W. K. T. Coltro, R. S. Lima, E. Carrilho and L. H. Mazo, *Electrophoresis*, 2013, **34**, 2107–2111.
191. C. Sicard, C. Glen, B. Aubie, D. Wallace, S. Jahanshahi-Anbuhi, K. Pennings, G. T. Daigger, R. Pelton, J. D. Brennan and C. D. M. Filipe, *Water Res.*, 2015, **70**, 360–369.
192. K. Islam, S. K. Jha, R. Chand, D. Han and Y.-S. Kim, *Microelectron. Eng.*, 2012, **97**, 391–395.
193. X. Weng, C. H. Chon, H. Jiang and L. Dongqing, *Food Chem.*, 2009, **114**, 1079–1082.
194. L.-M. Fu, Y.-N. Wang and C.-C. Liu, *Chem. Eng. J.*, 2014, **244**, 422–428.
195. I. Kubo, T. Kanamatsu and S. Furutani, *Sens. Mater.*, 2014, **26**, 615–621.
196. M. Rezai, B. Akbariadergani and M. Shekarchi, *Eur. J. Pharm. Sci.*, 2014, **57**, 99–151.
197. X. Duan, X. X. Dai, T. Wang, H. L. Liu and S. C. Sun, *Hum. Reprod.*, 2015, **30**, 1643–1652.
198. C. N. Jayarajah, A. M. Skelley, A. D. Fortner and R. A. Mathies, *Anal. Chem.*, 2007, **79**, 8162–8169.
199. J.-W. Choi, K.-M. Min, S. Hengoju, G.-J. Kim, S.-I. Chang, A. J. deMello, J. Choo and H. Y. Kim, *Biosens. Bioelectron.*, 2016, **80**, 182–186.
200. C. Zhai, W. Qiang, J. Sheng, J. Lei and H. Ju, *J. Chromatogr. A*, 2010, **1217**, 785–789.
201. M. Cretich, G. Di Carlo, C. Giudici, S. Pokoj, I. Lauer, S. Scheurer and M. Chiari, *Proteomics*, 2009, **9**, 2098–2107.
202. R. Schirhagl, A. Seifner, F. Husain, M. Cichna-Markl, P. Lieberzeit and F. Dickert, *Sens. Lett.*, 2010, **8**, 399–404.
203. X. Weng, G. Gaur and S. Neethirajan, *Biosensors*, 2016, **6**, 24.
204. W. G. Shreffler, *J. Allergy Clin. Immunol.*, 2011, **127**, 843–849.
205. K. A. Heyries, M. G. Loughran, D. Hoffmann, A. Homsy, L. J. Blum and C. A. Marquette, *Biosens. Bioelectron.*, 2008, **23**, 1812–1818.
206. H. Jiang, D. Jiang, P. Zhu, F. Pi, J. Ji, C. Sun, J. Sun and X. Sun, *Biosens. Bioelectron.*, 2016, **83**, 126–133.
207. M. Amatatongchai, O. Hofmann, D. Nacapricha, O. Chailapakul and A. J. Demello, *Anal. Bioanal. Chem.*, 2007, **387**, 277–285.
208. H. A. J. A. Lawati, B. A. Haddabi and F. E. O. Suliman, *Anal. Methods*, 2014, **6**, 7243–7249.
209. S. Primrose, M. Woolfe and S. Rollinson, *Trends Food Sci. Technol.*, 2010, **21**, 582–590.
210. M. M. Rahman, M. E. Ali, S. B. A. Hamid, S. Bhassu, S. Mustafa, M. Al Amin and M. A. Razzak, *Food Anal. Methods*, 2015, **8**, 1598–1606.
211. S. G. Anema, *Int. Dairy J.*, 2009, **19**, 198–204.

CHAPTER 4

Paper-fluidic Based Sensing in Food Safety and Quality Analysis

YANG LIN AND JIE XU*

Mechanical and Industrial Engineering, University of Illinois at Chicago, Chicago, IL 60607, USA
*Email: jiexu@uic.edu

4.1 Introduction

Food safety and quality have gained extensive attention from the public across the world in the past decade.[1,2] The safety and quality of food can be affected by various biological and chemical contaminants, such as food pathogens, herbicides, pesticides, heavy metals and toxins. These hazardous materials widely exist in different aspects of agricultural, industrial and food distributive processes.[3] Moreover, excessive food additives can also result in severe food safety concerns. Therefore, strict control of these foodborne hazardous substances has been intensively in demand by using reliable detection and analysis methods. However, traditional detection and analysis processes are costly and labour intensive.[4] Therefore, cheap, simple, portable and user-friendly alternatives are highly desirable.

Owing to the benefits of microelectromechanical system (MEMS) technology, microfluidics technology has undergone rapid growth in the past decade. Various sensing applications based upon microfluidics have emerged and microfluidics has been considered as a promising tool to replace the current detection methods that require special equipment and

Food Chemistry, Function and Analysis No. 2
Sensing Techniques for Food Safety and Quality Control
Edited by Xiaonan Lu

Published by the Royal Society of Chemistry, www.rsc.org

centralized laboratories.[5–7] The central principle of microfluidics lies in the miniaturization of analytical devices into the microscale; hence the consumption of samples and associated solutions, buffers and others can be reduced significantly compared to its macroscale counterparts. Another striking advantage of microfluidics is its capability of integrating all the essential components into one device. As a result, traditional procedures, such as mixing, purification and separation, can be achieved in a single chip resulting in tremendous savings in space and cost. However, these so-called "Lab on a chip" (LoC) devices still have discrepancies in terms of meeting the cost, operation and fabrication premises. Expensive cleanroom facilities are usually required for the fabrication of LoC chips.[8] For instance, photolithography is one of the most popular techniques involved in microfluidics fabrication and it consists of spin coating, prebake, UV exposure, postbake and associated cleaning and development steps.[9] All these procedures require strict operative conditions and expensive photoresists, thus inhibiting sufficiently commercial development of microfluidics especially in resource-limited settings. Given these concerns, the next-generation LoC devices, namely, microfluidic paper-based analytical devices (μPADs) have emerged in the past few years. Apart from the conventional silicon- or glass-based LoC chips, μPADs capitalize on commercial filter or chromatography papers to build various functional components (*e.g.*, mixing, filtration and dilution) or to conduct biological assays (*e.g.*, glucose detection).[10,11] The advantages of μPADs compared to conventional LoC chips can be categorized into five points: (i) μPADs apply capillary force to wick liquid instead of using additional devices, such as pumps or valves; (ii) μPADs can be simply fabricated by various methods, such as printing or cutting; (iii) μPADs are biocompatible, biodegradable and can be easily disposed of by burning; (iv) paper has high surface-to-volume-ratio that can enhance the performance of assays; and (v) paper is extremely cheap and ubiquitous all over the world. By virtue of these attractive merits, paper-based devices nowadays have been applied to diverse applications, such as point-of-care diagnosis,[12,13] environmental monitoring[14,15] and flexible electronics.[16] With respect to food safety and quality analysis, the use of μPADs has undergone burgeoning growth in recent years.[17,18] Although still at their infancy stage, the potential benefits of μPADs have already shaped them into promising substitutes for traditional equipment, especially for on-site or in-field sensing.

This chapter first focuses on μPADs fabrication techniques and the capabilities and detection mechanisms of μPADs. Then, representative applications of μPADs in food safety and quality analysis will be introduced and discussed. Finally, conclusions will be made and future perspectives will be provided.

4.2 Fabrication Techniques

Since the first modern μPAD was developed in 2007,[19] different types of paper substrates have been investigated. Typical ones consist of filters, chromatography papers and nitrocellulose that are made of or modified from pure cellulose.[20] Thereby the intrinsic hydrophilicity of cellulose enables them to be favourable materials for μPADs. Generally, in a μPAD, liquid samples or analyzing solutions are allowed to spread along the confined hydrophilic channels, which can be fabricated by various techniques as shown in Table 4.1.

The selection of proper substrates is of great importance when the influence of diverse parameters, such as flexibility, porosity, thickness and biocompatibility, is considered. Flexibility is one of the paramount advantages of μPADs, and consequently accidental bending or folding will cause little damage to the device.[21] Moreover, porosity is another pivotal parameter since it determines the size of particles that can be retained in the device.[22] In addition to these physical parameters, chemical and biological

Table 4.1 Fabrication techniques for μPADs.

Technique	Type	Characteristics	Substrate	Ref.
Wax printing	Printing	Only a wax printer and a hotplate or oven are required. Resolution is low.	Chromatography paper	24
Screen printing	Printing	A specific mask is required. Resolution is low.	Filter paper	25
Inkjet printing	Printing	Only a modified inkjet printer is required. Different types of ink could be replaced easily.	Filter paper	26
Flexographic printing	Printing	Low ink consumption.	Chromatography paper	27
Alkenyl ketene dimer printing	Printing	Extremely cheap.	Filter paper	28
PDMS printing	Printing	Simple and cheap.	Filter paper	29
Manual cutting	Cutting	Extremely simple. Resolution is very low.	Chromatography paper	30
Laser cutting	Cutting	Simple.	Chromatography paper	31
Drawing	Drawing	Simple and cheap.	Weighing paper	32
Plotting	Drawing	Suitable for different surfaces.	Chromatography paper	33
Embossing	Other	Suitable for 3D structures.	Filter paper	34
Stamping	Other	Fast and cheap.	Filter paper	35
Lamination	Other	Simple and cheap.	Chromatography paper	36
Wax dipping	Other	Simple and cheap.	Filter paper	37
Spraying	Other	A specific iron mask is required.	Filter paper	38
Plasma treatment	Other	Flexibility of paper can be kept.	Filter paper	21
Photolithography	Other	High resolution.	Chromatography paper	19
Fast lithographic activation	Other	Simple operation. High resolution.	Chromatography paper	39
Stereolithography	Other	Simple operation.	Filter paper	40
UV Patterning	Other	Simple and cheap.	Filter paper	41

parameters are also decisive. For instance, biocompatibility accounts for applications of nitrocellulose in lateral flow chromatographic immunoassay,[23] which has better ability in immobilizing proteins than others. However, these membranes also have some drawbacks due to their flammability and incompatibility with many organic solvents, such as SU-8 photoresist. Hence fabrication techniques for nitrocellulose are limited compared to filter and chromatography papers.

Currently, most techniques for fabricating μPADs are based upon printing due to its simplicity and user-friendliness. Among these printing techniques, wax printing and inkjet printing are two commonly used ones. To build a μPAD by wax printing, only a commercial wax printer and a hotplate are required. Similar to the conventional printing used in our daily lives, a wax printer prints patterns of wax directly on the surfaces of paper, then a hotplate or an oven is applied to heat up the solid wax and let it penetrate into the paper. One limitation is that the molten wax also spreads along the horizontal directions during heating, resulting in a decrease of the resolution of patterns (Figure 4.1A). Therefore, wax printing is widely exploited in those applications that do not require complicated designs and very high resolutions. The spreading of molten wax in a hydrophilic porous medium is governed by Washburn's equation: $L = (\gamma Dt/4\eta)^{1/2}$,[42] where L, γ, D, t, η denote spreading distance, surface tension, average pore diameter, spreading time and viscosity of molten wax, respectively. According to this equation, spreading distance is proportional to the square root of time; hence long channels are rarely applied in μPADs, especially for on-site sensing applications.

Inkjet printing is another ideal technique that has been exponentially explored for μPADs in recent years. At present, commercial inkjet printers are categorized into two different types, namely drop-on-demand (DOD) and continuous inkjet (CIJ).[26] As a non-contact method, CIJ technology has been widely used for marking and coding on almost any substrate (*e.g.*, packages) even with a curved surface. However, all common desktop inkjet printers, including laboratory printers, adopt DOD. In contrast to the continuous stream of ink droplets created by CIJ technology, the ejection of ink only occurs where it is needed. Additionally, the ink used in a DOD system is not cycled, thus preventing pollution of the ink and stabilizing the properties of the ink. It also avoids the usage of volatile organic solvents that are essential for a CIJ system, hence reducing the adverse impact on the environment. Owing to these obvious benefits, commercial inkjet printers with a DOD system are favourable in building μPADs. Besides the mechanism of inkjet printing, the properties of ink are also of great importance. The equation $Z = \mathrm{Re}/\mathrm{We}^{1/2} = (\gamma\rho d)^{1/2}/\eta$ is commonly used to evaluate the printing feasibility of a certain ink.[43] In the above equation, Re, We, γ, ρ, d, η denote Reynolds number, Weber number, surface tension, density, nozzle diameter and viscosity, respectively. Inks with a Z number in the range of 0.67 to 50 are suitable for conventional printers due to the fact that inks with a higher Z number will result in satellite droplets while smaller ones will lead to clogging. To date, inkjet printing has been applied not only to the

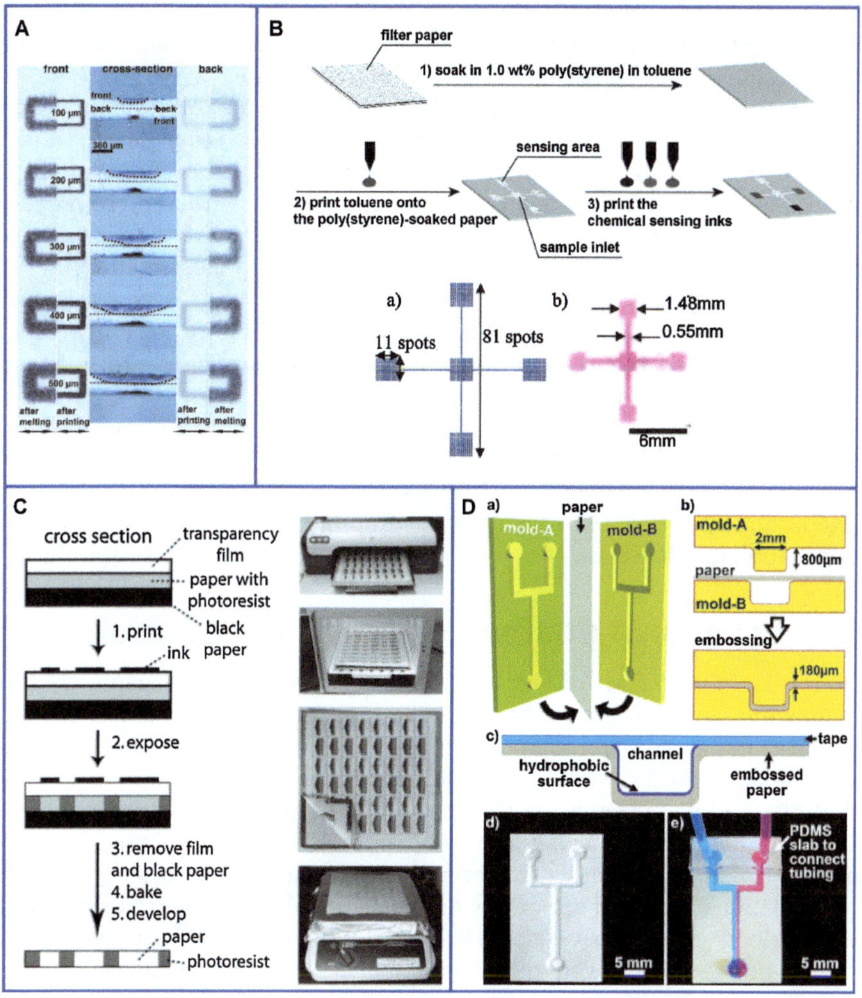

Figure 4.1 (A) Decrease of resolution of patterns after melting the wax. (Reproduced with permission from E. Carrilho, A. W. Martinez and G. M. Whitesides, *Anal. Chem.*, 2009, **81**, 7091–7095. Copyright 2009 American Chemical Society.) (B) Fabrication of μPAD and deposition of assay reagents by inkjet printing. (Reproduced with permission from K. Abe, K. Suzuki and D. Citterio, *Anal. Chem.*, 2008, **80**, 6928–6934. Copyright 2008 American Chemical Society.) (C) Fast lithographic method for fabrication of μPAD. (Reproduced from ref. 39 with permission from the Royal Society of Chemistry.) (D) Fabrication process of μPAD on the basis of embossing. (Reproduced with permission from M. M. Thuo, R. V. Martinez, W.-J. Lan, X. Liu, J. Barber, M. B. Atkinson, D. Bandarage, J.-F. Bloch and G. M. Whitesides, *Chemistry of Materials*, 2014, **26**, 4230–4237. Copyright 2014 American Chemical Society.)

fabrication of hydrophilic channels on paper, but also to the deposition of assay reagent directly (Figure 4.1B).[26,44] This convenience may realize the mass production potential of μPADs in the near future.

Other fabrication techniques, such as drawing, photolithography and embossing, also have their unique advantages compared to printing, thus it is worth introducing them briefly. Photolithography has been regarded as one of the techniques with relatively high resolution.[19] However, the process can be costly since cleanroom facilities and expensive solvent are required. To solve this problem, UV lamps and hotplates are applied to simplify the process and reduce the cost significantly (Figure 4.1C).[39] By using this technique, high resolution has been achieved (~200 μm). Recently, Thuo and coworkers developed a promising method to fabricate μPADs based upon embossing (Figure 4.1D).[34] They combined the technique of embossing and "cut and stack" to form three-dimensional structures from omniphobic papers. Moreover, He and collaborators capitalized on a commercial three-dimensional printer,[40] and achieved comparable high resolution to photolithography (*i.e.*, 367 ± 20 μm for hydrophilic channels and 400 ± 21 μm for hydrophobic barriers).

4.3 Functional Components and Flow Control

Aiming to integrate all of the essential steps onto a single paper device, various functional components have been intensively explored. With the aid of these components, requirements for user intervention and external equipment could be significantly reduced, thereby increasing the efficiency of μPADs and also decreasing the cost. Over the past decade, various functional elements have been realized on paper,[16] such as mixing,[11] separation,[45] timers,[46] switches,[21] batteries[47] and displays.[48]

For instance, Kar and coworkers have successfully separated plasma from whole blood on a paper-based H-channel (Figure 4.2A).[49] In this study, human blood and phosphate-buffered saline (PBS) were dispensed to the corresponding inlet R and B, respectively. Plasma was separated from outlet B1 and red blood cells (RBCs) were separated from R1. Aside from separation, uniform mixing has also been developed for μPADs by using surface acoustic wave (SAW).[11] Atomization was induced by a 30 MHz SAW in the end of the channel (Figure 4.2B), thus liquid can be drawn through the Y-shaped channel without backflow that occurs commonly in conventional μPADs. However, demand for the extra function generator casts a shadow on its further applications. Moreover, energy storage devices (*e.g.*, supercapacitors) and microbial fuel cells (MFCs) also pave the way for sophisticated designs that are self-powered. For instance, paper-based MFCs have been developed by Fraiwan and collaborators.[47] In this study, papers were adopted as reservoirs for anodes and cathodes, thus decreasing the cost for conventional counterparts. Electrochromic read-out associated with an internal battery has been adopted to sense glucose and hydrogen peroxide.[50] Besides read-out by electrochromic materials, another display method has

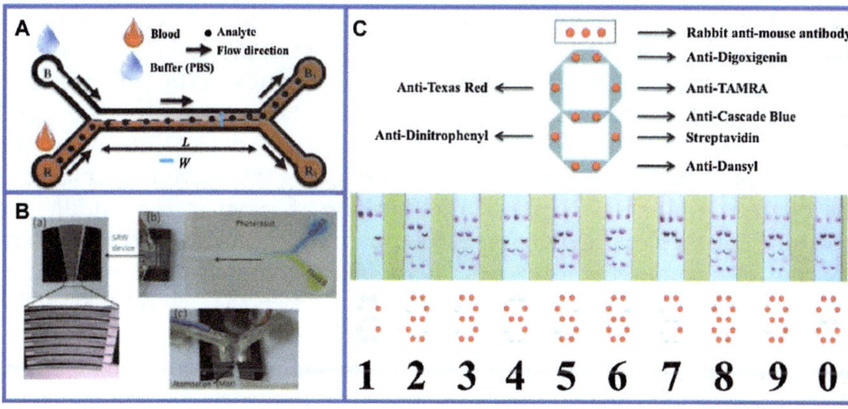

Figure 4.2 (A) Separation of plasma from blood on μPADs. (Reproduced from ref. 49 with permission from the Royal Society of Chemistry.) (B) Uniform mixing on paper based on SAW. (Reproduced from ref. 11 with permission from the Royal Society of Chemistry.) (C) 7-segment display for multiplex lateral flow detection. (Reproduced from ref. 48 with permission from the Royal Society of Chemistry.)

been recently developed by Li and Macdonald.[48] They capitalized on the conventional display method that is commonly used for a calculator (Figure 4.2C). This ingenious design has successfully expanded the detection of lateral flow chromatographic immunoassay without expanding the dimensions of devices and consumption of reagents.

Although the Lucas–Washburn equation accounts for the capillary flow in porous materials (*e.g.*, papers), its disadvantages still exist in a variety of aspects. This equation does not take ambient conditions, such as gravity, into account; therefore, it can only be used for rough estimation. Rogacs and coworkers have investigated the effect of gravity and found that the impact of gravity on the vertical capillary flow is unnoteworthy.[51] However, evaporation does cast a shadow on this equation, especially when applying μPADs in low humidity conditions.[52] Taking evaporation into account, a modified Lucas–Washburn equation has been developed:

$$L=\sqrt{\frac{\gamma D\phi h\cos\theta t}{8\eta q_0}\left(1-\mathrm{e}^{-\frac{2q_0}{\phi h}t}\right)} \tag{4.1}$$

where ϕ, h, θ, q_0 denote the porosity of paper, height of microchannel, contact angle and evaporation rate, respectively. It is worth noting that changing these parameters (*e.g.*, length, height and porosity) could result in control of the capillary flow velocity of μPADs; therefore, multi-step analysis can be realized. Recently, a myriad of developments in paper-based microfluidics have taken advantage of such benefits for flow control, including two-dimensional and three-dimensional devices. For instance, channels with different lengths have been demonstrated as a good example to show the feasibility of flow control in an easy way.[53] Additionally, baffle design was also applied to differentiate the length of microchannels, allowing for sequential reactions to take place.[54] In this study, different reactants were prepared in advance onto the corresponding zones of the paper (Figure 4.3A). After dipping the as-prepared paper device into a sample, antigen could be captured with second antibodies that have been placed in the bilateral channels. Owing to the delaying channel in the middle, the substrate mixture that generated a colorimetric signal finally reached both the control zone and the test zone, thus avoiding manual additions required for the conventional enzyme-linked immunosorbent assays (ELISA). Moreover, although fluid source pads with different sizes release fluids at a fixed rate, larger pads could provide more sustainable flow than the smaller ones. Therefore, Fu and collaborators capitalized on this property and generated sequential flow on the paper (Figure 4.3B).[55] Aside from the abovementioned techniques for flow control in μPADs, delay shunt, pressurized delay region and sugar deposited region are also preferred.[56–58] By integrating these particular regions in the microchannel, the flow rate could be decreased while other regions remain the same.

Over the last few years, not only have two-dimensional flow control systems been explored, but diverse three-dimensional valves have also shown an intensive potential for automatic sequential reactions on paper.[59] For example, a fluid-triggered valve has been developed recently.[60] A movable paper valve associated with fluid-triggered expanding sponge was used in one study (Figure 4.4). The status of the as-prepared valve can be controlled after the sponge expands by absorbing the liquid wicked through the timing channel. Consequently, changing the length of the timing channel could result in the control of the valve in a certain time, thus realizing sequential reactions.

4.4 Detection Mechanisms

Almost every μPAD that has emerged to date has capitalized on either single or multiple commonly used detection mechanisms, including colorimetric, fluorescent, electrochemical, chemiluminescent and electrochemiluminescent mechanisms.[61,62] Undoubtedly, the colorimetric mechanism has shown greater popularity compared with the others by virtue of its simplicity. The final results, *i.e.*, colour change, obtained in a colorimetric assay could be easily determined by the naked eye or image analysis software (*e.g.*, ImageJ) associated with image acquisitive apparatus such as cell phones. Moreover, with the aid of a calibration chart, semi-quantitative readouts can be realized. Generally, related paper devices are prepared by one of the fabrication techniques described above, and then reagents (*e.g.*, enzymes, antibodies) are spotted to the reaction zone in advance. After the samples are added to the devices, chemical reactions occur and the colour of the reaction zone changes, showing the existence of the corresponding targets. For instance, sensing of different concentrations of copper ions has

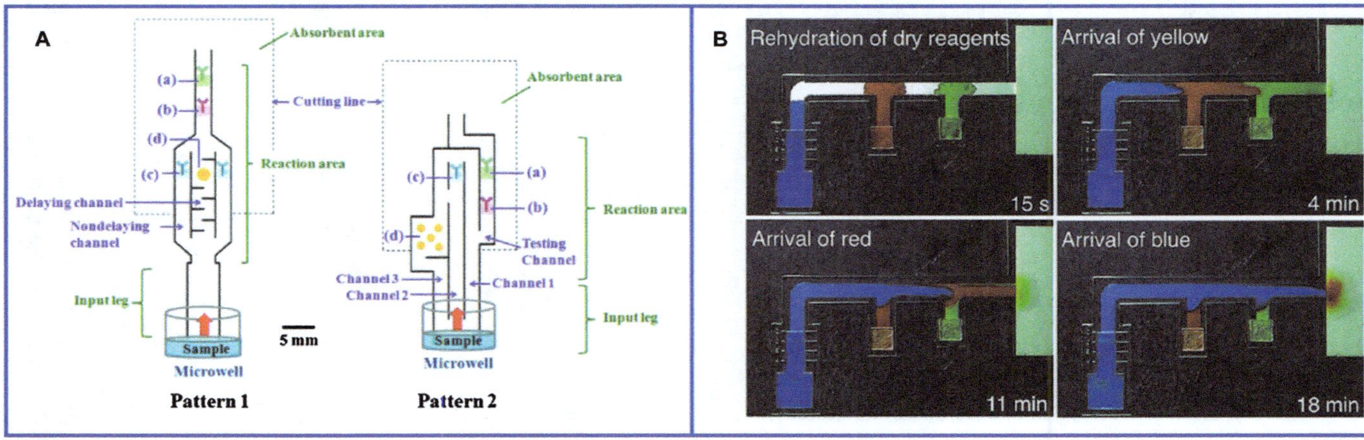

Figure 4.3 (A) Multi-step reactions on μPADs based upon the baffle design. (Reproduced from ref. 54 with permission from the Royal Society of Chemistry.) (B) Multi-step flow control by different size of sample source pads. (Reproduced with permission from E. Fu, T. Liang, P. Spicar-Mihalic, J. Houghtaling, S. Ramachandran and P. Yager, *Anal. Chem.*, 2012, **84**, 4574–4579. Copyright 2012 American Chemical Society.)

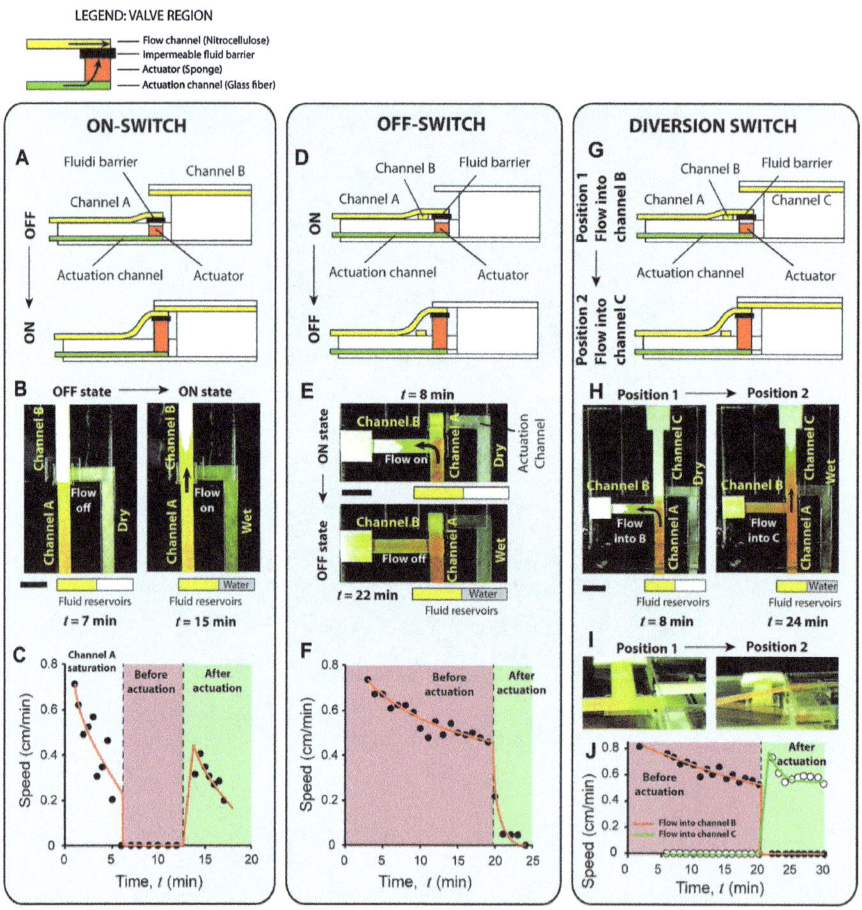

Figure 4.4 Different fluid-trigged valves for three-dimensional μPADs. (Reproduced from ref. 60 with permission from the Royal Society of Chemistry.)

been realized by Li and colleagues.[63] In this study, Cu^{2+} ions with concentrations from 33.6 nM to 39.8 μM were investigated and the limit of detection (LOD) was reported to be 33.6 nM (Figure 4.5A). The darker the paper showed, the higher the concentration of Cu^{2+} ions detected. Although colorimetric mechanism has still been considered as one of the most efficient methods, inevitable drawbacks still exist, thus postponing its commercialization worldwide. For instance, the cost for a specific reader and image processing software are indispensable for quantitative measurement.

Fluorescent mechanism is another method that has been widely adopted in μPADs. It enables higher sensitivity and selectivity compared with the colorimetric counterpart.[64] Similar to the colour change used in the colorimetric mechanism, the principle of fluorescent mechanism lies in the

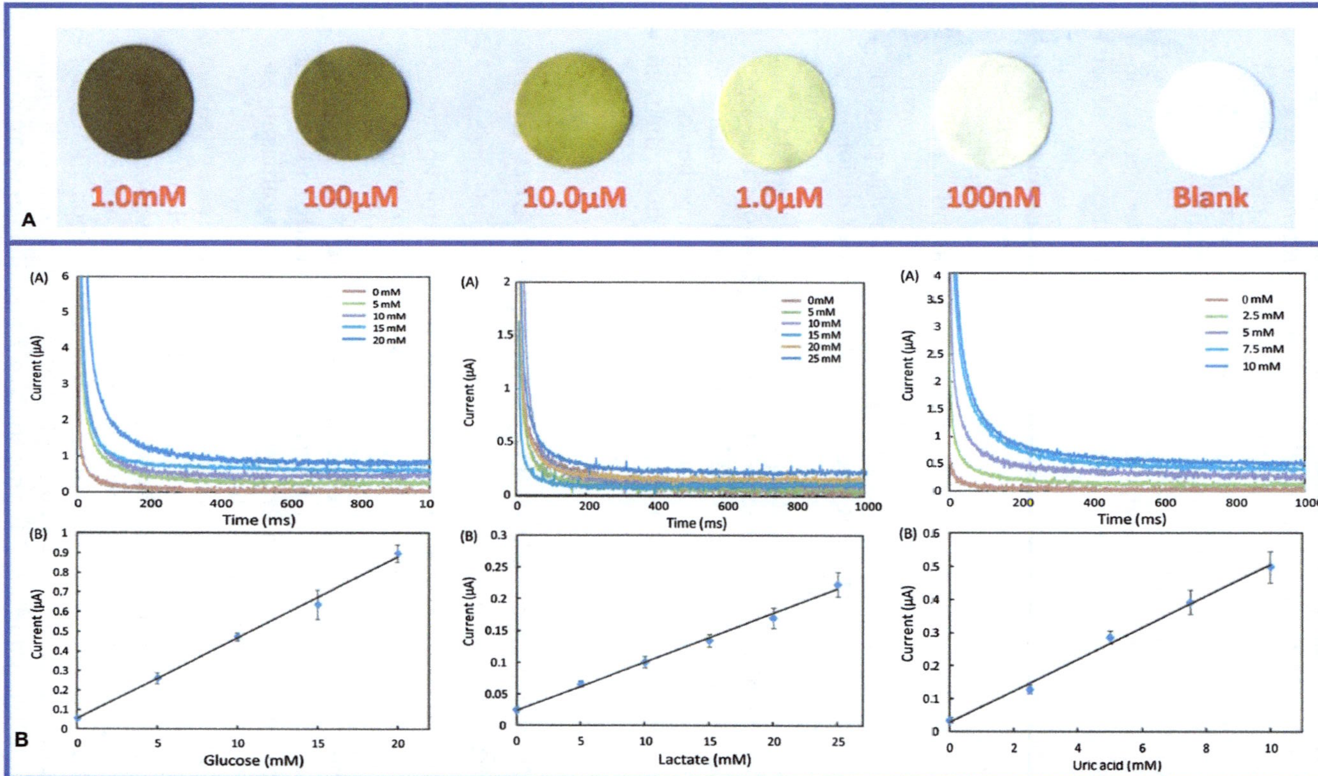

Figure 4.5 (A) Colorimetric detection of Cu^{2+} ions based upon μPADs. (Reproduced from *Sensors and Actuators B: Chemical*, **223**, J.-j. Li, C.-h. Ji, C.-j. Hou, D.-q. Huo, S.-y. Zhang, X.-g. Luo, M. Yang, H.-b. Fa and B. Deng, High efficient adsorption and colorimetric detection of trace copper ions with a functional filter paper, 853–860. Copyright 2016 with permission from Elsevier.) (B) Glucose, lactate and uric acid detection on the basis of electrochemical mechanism. (Reproduced with permission from A microfluidic paper-based electrochemical biosensor array for multiplexed detection of metabolic biomarkers, C. Zhao, M. M. Thuo and X. Liu, *Science and Technology of Advanced Materials*, 2013, **14**, 054402. Copyright © National Institute for Materials Science, reprinted by permission of Taylor & Francis Ltd, http://www.tandfonline.com on behalf of National Institute for Materials Science.)

changes of fluorescent intensity when chelating units associated with target analytes are connected to the fluorophore units.[65]

The electrochemical mechanism usually involves a three-electrode system, including working electrode, counter electrode and reference electrode. With the help of this system, many characteristics that are otherwise challenging to be obtained can be achieved by diverse electroanalytical methods, such as chronoamperometry, cyclic voltammetry, linear sweep voltammetry, square wave voltammetry, staircase voltammetry, anodic/cathodic stripping voltammetry and adsorptive stripping voltammetry.[67] For instance, chronoamperometry is one of the most popular methods applied for paper devices. Zhao and coworkers capitalized on this method to determine the concentration of glucose and demonstrate the linear relationship between the current and concentration of glucose, lactate and uric acid (Figure 4.5B).[66]

The chemiluminescent mechanism possesses a similar principle compared to the fluorescent method except that its light is generated from the chemical reactions.[68] Moreover, it can be divided into two categories by different mechanisms. The emitting species is created by oxidation of the reagents in one mechanism while being created by enhancing or inhibiting the performance of luminescent compounds in another.[69] Besides four conventional mechanisms applied in μPADs, other applications that adopt electrochemiluminescent, nanoparticle-based and hybrid mechanisms are also feasible. Interested readers can refer to the reference books and review articles listed here, but this chapter will not introduce them in detail.[20,61,70–73]

4.5 Representative Applications in Food Safety and Quality Analysis

As aforementioned, using paper as a platform is attractive for on-site detection due to its simplicity and cost-effectiveness. Therefore, a myriad of applications for food safety and quality analysis, including detection of foodborne pathogens, pesticides, herbicides, heavy metal ions and food additives, have been developed over the past decade. An early and efficient detection of these pathogens obtained by using μPADs can significantly diminish the possibilities of illnesses or even fatalities.

Bacterial contamination of foods is prevalent worldwide. It is reported that approximately 76 million cases of foodborne illnesses occur annually, most of which are caused by bacterial pathogens.[74] Given this situation, it is urgent to adopt early detection as a preventive strategy. Jokerst and coworkers have successfully developed a screening system based upon μPADs for simultaneous detection of *Escherichia coli* O157:H7, *Listeria* and *Salmonella enterica* serotype Typhimurium.[18] In this study, a 7-mm-diameter spot array was designed using the graphic software CorelDRAW and created on the paper by wax printing first. Then, hydrophobic barriers were obtained by

Table 4.2 Enzyme-substrate pairs for different bacterial pathogens.

Bacterial pathogen	Enzyme	Substrate	Colour Change
E. coli	β-galactosidase	Chlorophenol red β-galactopyranoside (CPRG)	Yellow to red-violet
L. monocytogenes	Phosphatidylinositol-specific phospholipase C (PI-PLC)	5-bromo-4-chloro-3-indolyl-myo-inositol phosphate (X-InP)	Colourless to blue
S. enterica	Esterase	Magenta caprylate	Colourless to purple

heating to melt the wax into the paper. Afterwards, packaging tape was placed onto the printed side of the paper to prevent leaking and enhance flow control. Three enzyme-substrate pairs were adopted to detect the corresponding bacteria as shown in Table 4.2.

After the colour of the substrates changed, a Xerox scanner was applied to acquire an image and then ImageJ was used for image analysis (Figure 4.6). The enrichment time required to decrease the minimal detectable amount in μPADs was significantly reduced compared with the conventional standard culture methods.

Lateral flow immunochromatographic assay, which is also known as lateral flow test (LFT), is another widely exploited type of μPAD. A typical LFT usually consists of four parts: sample pad, conjugate pad, reaction pad and absorbent pad. Additionally, a plastic cassette is commonly applied to enclose all the components inside (Figure 4.7A). Generally speaking, after the sample is added to the sample pad, a buffer solution is commonly used to accelerate the capillary action. When the sample reaches the conjugate pad, the prepared conjugates will recognize and bind to the targeted analytes in the sample if present. Afterwards, these target-conjugate couples arrive at the reaction pad, reacting with reactants (*e.g.*, antibodies), which are immobilized previously. According to different mechanisms, the results shown in the test line and control line may be different. Eventually, excessive samples and other solutions can be absorbed by the absorbent pad, affirming the success of assay performance.

Recently, Park and collaborators have developed a novel LFT on paper and successfully demonstrated its excellent performance in simultaneous detection of *E. coli* and *S. enterica*.[75] They have improved the performance of the conventional LFT by introducing pressed regions to delay fluid flow in specific channels, thus enabling sequential delivery of different reagents (Figure 4.7B). By virtue of this design, there is no need to design a particulate channel with different delivery times. In order to detect *E. coli* and *S. enterica* simultaneously, two types of capturing antibodies and the corresponding antibody-conjugated gold nanoparticles were utilized in this assay. Moreover, three types of gold enhancer solutions were also applied to increase the final performance. All these reagents were preloaded and dried on the

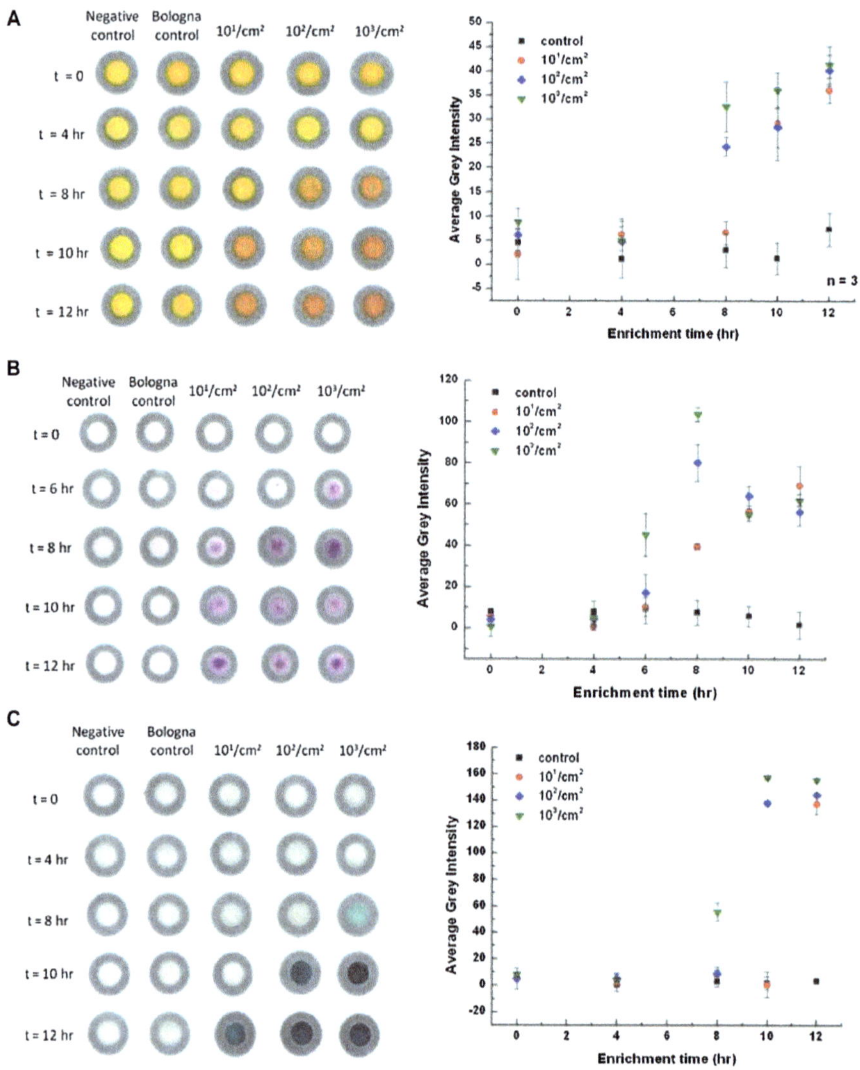

Figure 4.6 Determination of different concentrations of (A) *E. coli*, (B) *L. monocytogenes* and (C) *S. enterica* under different enrichment times. (Reproduced with permission from J. C. Jokerst, J. A. Adkins, B. Bisha, M. M. Mentele, L. D. Goodridge and C. S. Henry, *Anal. Chem.*, 2012, **84**, 2900–2907. Copyright 2012 American Chemical Society.)

nitrocellulose membrane in advance, and can be rehydrated after sample loading. As they were placed on the different regions of membrane, sequential reactions were carried out.

As a universal energy molecule, adenosine triphosphate (ATP) nowadays is regarded as a signalling molecule to determine the existence of living microbes.[76] Therefore, it has been adopted for the detection of pathogens

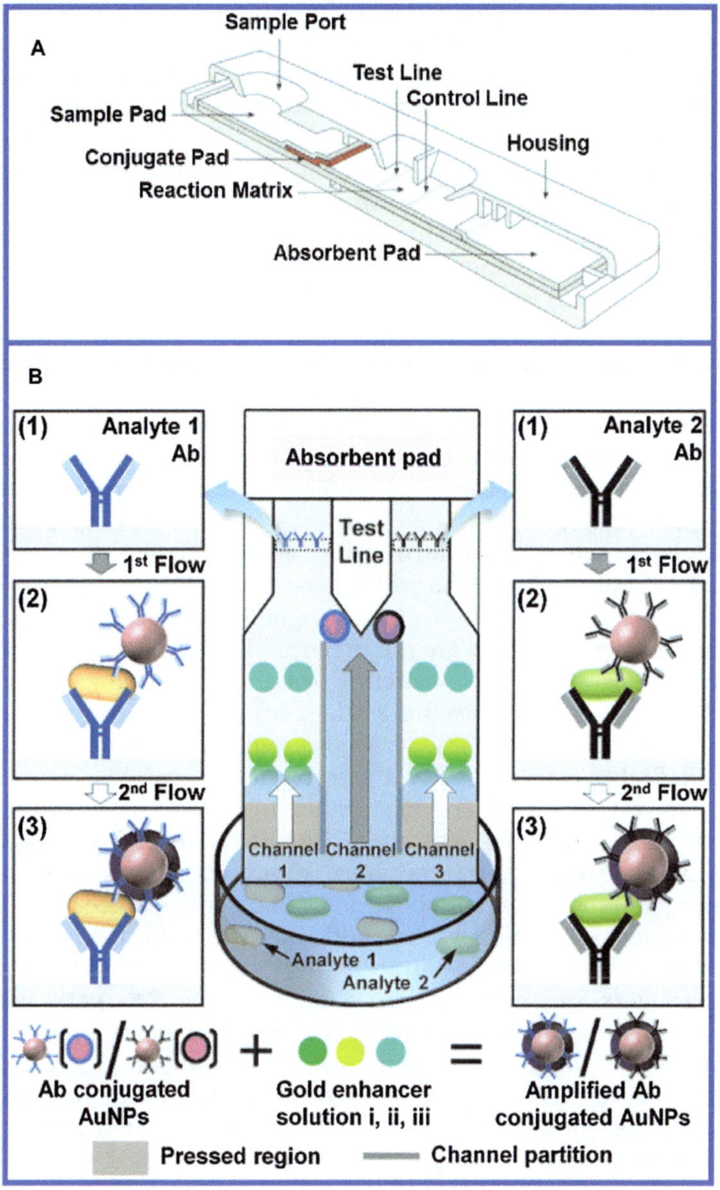

Figure 4.7 (A) Schematic illustration of the structure for lateral flow immunochromatographic assay. (Reproduced from ref. 20 with permission of the Royal Society of Chemistry.) (B) Sequential reactions on an LFT paper device by introducing pressed regions. (Reproduced with permission from J. Park, J. H. Shin and J.-K. Park, *Anal. Chem.*, 2016, **88**, 3781–3788. Copyright 2016 American Chemical Society.)

through a μPAD based upon chemiluminescent assay.[77] A simple Z-folding design was applied to control the flow and substrate transportation in this study. Specifically, conventional wax printing was applied to construct a hydrophobic pattern in a chromatographic paper, and then the surface of the paper was chemically modified, followed by attachment of an ATP aptamer in one reagent zone. After blocking non-immobilized regions with bovine serum albumin, horseradish peroxidase (HRP) labelled complementary DNA fragment was added to associate with the ATP aptamer. In another reagent zone, 3-amino-9-ethylcarbazole (AEC) associated with polyethylene glycol (PEG) was applied to treat the paper. When the sample was added to as-prepared μPAD, the paper device only needed folding twice to connect both prepared reagent zones. Afterwards, hydrogen peroxide was applied after transportation of the liquid between two reagent zones, and the final results were obtained by using an optical scanner. Finally, *Salmonella* bacteria were successfully detected through the colour change in the μPAD.

Besides using enzyme-substrate binding to determine the existence of the targeted pathogens in foods, a novel detection method based upon immunoagglutination associated with a smartphone was developed (Figure 4.8).[78] Since white light, which is now accessible to almost every cell phone, consists of light with a wide range of wavelengths, fluorescent and chemiluminescent methods are not preferred because they are wavelength-dependent. Therefore, Mie scattering was applied owing to its angle-dependent feature. When scanning Mie scattering intensity by varying the angles, sharp peaks of intensity can be observed. It is worth noting that these peaks are highly dependent upon the targeted materials, thus rendering itself a great tool for identifying certain materials. In this study, submicroparticles conjugated with anti-*S. enterica* were preloaded and dried on the paper. After dipping the paper device into the sample solution, immunoagglutination of antibody-conjugated particles occurred. Finally, such immunoagglutination was quantified by Mie scattering by using a cell phone with a well-designed smartphone application. The ingenious method introduced here also demonstrated a bright future for μPADs because a single cell phone can be considered as the terminal analysis device.

Apart from the μPADs described above, other works for the detection of foodborne pathogens have also emerged. To name a few, *E. coli* in a water distribution system was determined by a screening printed μPAD, and the grey value of silver stained spots on the paper was used to quantify the concentration of the bacteria.[79] Moreover, a bacteriophage that carries the *β*-galactosidase gene was also applied to detect *E. coli* by quantification of *β*-galactosidase when cell lysis was induced.[80]

Despite pesticides and herbicides having contributed to the maintenance of food safety for decades, their negative impacts still cast a shadow on human health.[81] Therefore, simple, cost-effective and efficient on-site μPADs are still imperative and are highly attractive for the determination of pesticide and herbicide residues in foods in the past decade. For instance, molecular imprinted polymers (MIPs) have been applied to detect pesticides

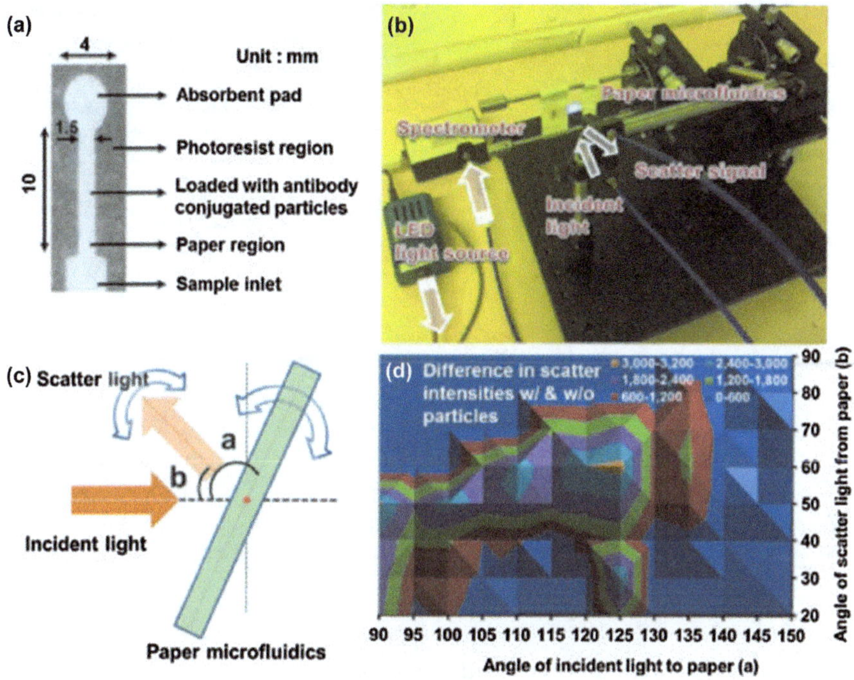

Figure 4.8 Quantification of pathogens through Mie scattering by using μPAD and a cell phone.
(Reproduced from ref. 78 with permission from the Royal Society of Chemistry.)

over a chemiluminescent mechanism.[82] As an alternative to antibodies, MIPs are formed by the arrangement and polymerization of monomers by using targeted molecules as the template. Thereby, MIPs have been demonstrated to have great memory towards target molecules in their shape, size and functional group, which is favourable for detection. Given their benefits, MIPs have been successfully adopted in μPADs to detect 2,4-dichlorophenoxyacetic acid (2,4-D), which is commonly used for broadleaf weeds. Specifically, an indirect competitive assay was carried out on the basis of competition between free 2,4-D in the target sample solution and MIPs conjugated 2,4-D-tobacco peroxidase. Finally, chemiluminescent intensity was utilized to quantify the analyte concentration. Additionally, MIPs have also been investigated for the detection of dichlorvos (DDV) *via* μPAD.[83] In this study, chromatography paper was cut into circles and then clipped between two glass slides (Figure 4.9). Afterwards, the glass slides were placed inside a flask, followed by the addition of DDV, methacrylic acid (MAA), ethylene glycol dimethacrylate (EGDMA), azobisisobutyronitrile (AIBN) and acetonitrile. After polymerization for 18 hours, the templates were removed by acetic acid. The final results showed that MIPs-based μPAD had a favourable selectivity of DDV and the limit of detection for DDV in vegetables could be as low as 0.8 ng mL^{-1}.

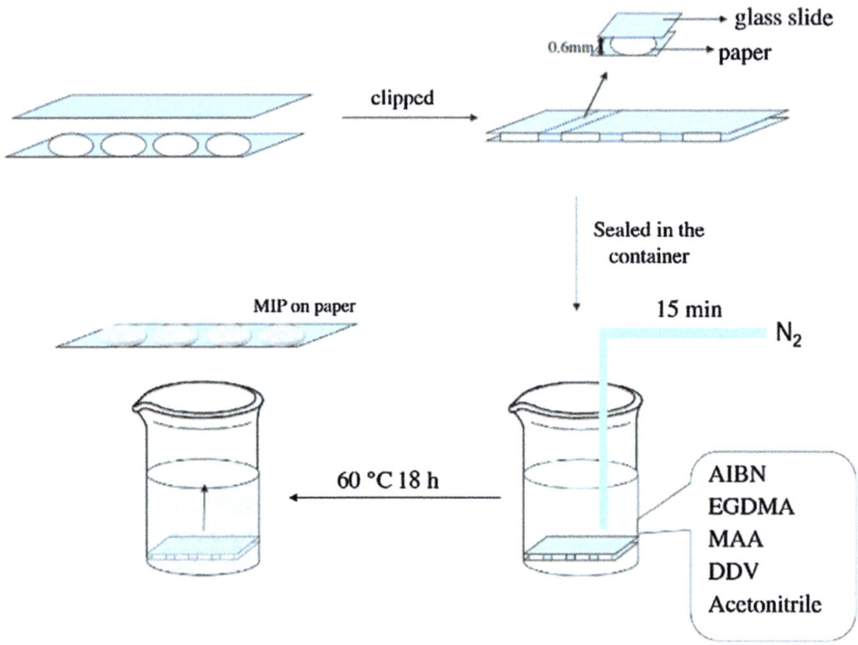

Figure 4.9 Preparation process of MIPs on paper.
(Reproduced from *Spectrochimica Acta Part A: Molecular and Biomolecular Spectroscopy*, **141**, W. Liu, Y. Guo, J. Luo, J. Kou, H. Zheng, B. Li and Z. Zhang, A molecularly imprinted polymer based a lab-on-paper chemiluminescence device for the detection of dichlorvos, 51–57. Copyright 2015 with permission from Elsevier)

Given that acetylcholinesterase (AChE) inhibitors (*e.g.*, organophosphate and carbamate pesticides) have been abused intensively in agriculture, monitoring their residues in foods is crucial. Badawy and collaborators recently developed a bioactive paper sensor to rapidly detect AChE by observing the colour change.[84] Similarly, Sicard and coworkers have determined the presence of pesticides based upon colour variations.[85] Specifically, colourless indoxyl acetate was applied as the colorimetric indicator because it can be converted to blue indigo product when the targeted pesticides exist. The reported experiment involved one μPAD for the sample test and another one for a clean water test as a control. A cell phone with an internal camera and well-designed image processing software was adopted to quantify colour intensity, thereby determining the concentration of the targeted pesticides. Furthermore, the internal GPS in cell phones also rendered them capable of recording the location of the test, which is beneficial to tracking, monitoring and mapping the water quality online. Another μPAD reported by Nouanthavong and collaborators, which took advantage of coating nanoceria onto paper, showed great promise in the detection of organophosphate pesticides.[86] Because nanoceria has the ability to catalyze the oxidation of H_2O_2 with peroxidase substrate, it has been demonstrated as a colorimetric indicator for μPAD, as the amount of H_2O_2 is correlated to the concentration of organophosphate pesticides associated with the colour change from colourless to yellow.[87] The authors declared that the limit of detection for methyl-paraoxon and chlorpyrifos-oxon in a spiked cabbage could be as low as 18 ng mL^{-1} and 5.3 ng mL^{-1}, respectively, which is comparable to conventional detection methods, such as LC-MS/MS.

Other pesticides such as pentachlorophenol (PCP) have also attracted growing attention over the past decade due to their potential carcinogenic effects towards public health. Sun and collaborators have devised a photoelectrochemical sensor to detect PCP on a paper device based upon the MIPs technique as aforementioned (Figure 4.10).[88] Specifically, a layer of gold nanoparticles was grown on the working electrode of a paper device, thus enhancing the conductivity and effective area of the working electrode. Afterwards, zinc oxide (ZnO) spheres were synthesized by a previous protocol,[89] and then deposited onto the surface of a gold-nanoparticles-modified electrode. After the template PCP was grafted onto the surface of ZnO layer by cyclic voltammetry, the electrode was treated by K_2HPO_4 solution to remove these template molecules. Another μPAD was fabricated as a control in the same way except using the template PCP. The limit of detection could be as low as 4 pg mL^{-1}.

Food additives are another type of foodborne contaminants that widely exist in food industries so as to improve the flavour or colour of the final products. However, excessive intake may result in health problems. For instance, vinegar, ascorbic acid and citric acid have been utilized as preservatives and antioxidants for decades. Food colouring additives have been adopted to compensate colour lost during production processes or make the

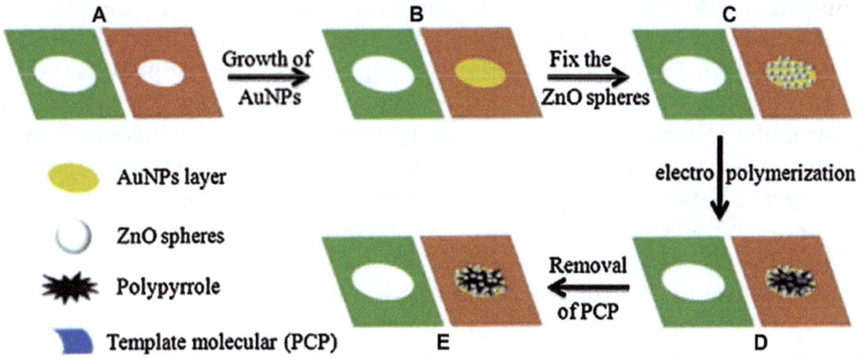

Figure 4.10 Schematic illustration of the fabrication process of μPAD for the detection of PCP.
(Reprinted from *Biosensors and Bioelectronics*, Volume 56, Guoqiang Sun, Panpan Wang, Shenguang Ge, Lei Ge, Jinghua Yu and Mei Yan, Photoelectrochemical sensor for pentachlorophenol on microfluidic paper-based analytical device based on the molecular imprinting technique, 97–103, Copyright 2014 with permission from Elsevier.)

final products more tempting. However, it is worth noting that these colouring additives could lead to behavioural issues among children, thereby their use is strictly restricted worldwide. Colletes and collaborators developed a novel μPAD with stamped paraffin barriers to detect sugars and related inhibitors in sugarcane bagasse liquors.[90] The authors declared that the quantification of glucose had high linearity (*i.e.*, $R^2 = 0.99$), and the limit of detection could be as low as 2.77 mmol L^{-1}. Moreover, multiplex determination of polyphenols, flavonoids and anthocyanins in wine was achieved by using a paper device.[91] The concentrations of these substances in various commercial wine samples were quantified by using the colorimetric assays on paper and the results were captured and analyzed by a cell phone and specific software. This reported μPAD had remarkable advantages compared with the conventional spectrophotometry technique. For instance, it only required 4 μL of reagents for each sample whereas the latter one required 1 mL. Furthermore, the total testing time was much less than for the spectrophotometry technique. In addition, no analytical equipment is required.

Nitrites have been widely used as a preservative in food industries for decades.[92] They can be easily ingested by humans *via* vegetables, meats and water, and they are considered to be toxic when a large amount is ingested, which may result in methemoglobinemia or even cancer.[93] Given these severe impacts on public health, various detection methods have emerged. For instance, Zhang and collaborators successfully determined the concentration of nitrite in water supplies, dairy products and meat *via* a colorimetric method.[94] The detection of nitrites was extended using μPADs. He and coworkers developed a novel fabrication method for μPAD by modifying conventional filter paper with octadecyltrichlorosilane.[41] This procedure converted the original hydrophilic paper into a hydrophobic substrate. Afterwards, UV light and a quartz mask were applied to turn the exposed region back to hydrophilic while other regions were still hydrophobic (Figure 4.11A). Finally, the as-prepared paper device was applied to determine nitrite *via* a colorimetric assay (Figure 4.11B), which also showed a great linear relationship between colour intensity and the concentration of nitrite. Other reported works for the determination of nitrite on paper devices were similar based upon the colorimetric mechanism; therefore, these are not elaborated here. Interested readers are referred to the articles cited here.[95–97]

Besides the μPADs for the determination of foodborne and waterborne contaminants described above, there are numerous other developments. For example, multiplex detection of chemical contaminants in foods in a simultaneous manner has been developed by using a graphene oxide sensor.[17] The graphene oxide sensor applied in this study was modified with different Cy5-labelled single-stranded DNA (ssDNA) as the corresponding sensitive probes for multiplex detection. Wax printing was also adopted to fabricate the paper devices developed here and different test zones were all connected to the central sample zone. As shown in

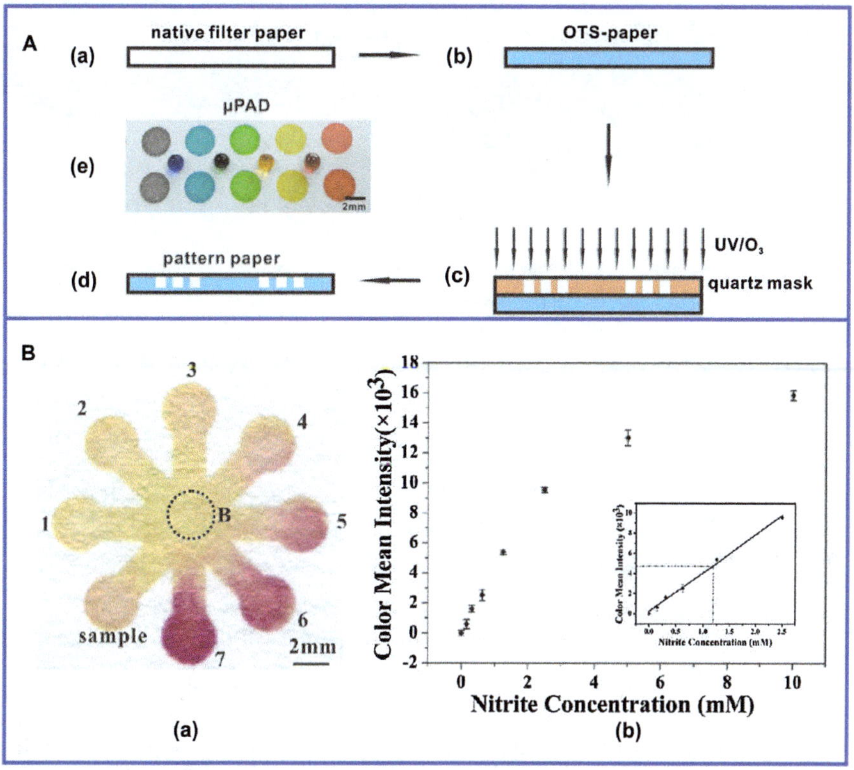

Figure 4.11 (A) Novel fabrication method for μPAD by UV exposure. (B) Determination of nitrite by the as-prepared μPAD.
(Reproduced with permission from Q. He, C. Ma, X. Hu and H. Chen, *Anal. Chem.*, 2013, **85**, 1327–1331. Copyright 2013 American Chemical Society.)

Figure 4.12, different ssDNA-modified graphene oxides were added to the corresponding test zone first, followed by drying in air. After the sample was added, it reached all the test zones by capillary action. All the well-designed ssDNA showed excellent affinity to the corresponding target contaminants (*e.g.*, Hg^{2+} ions) in foods. In the presence of the targeted contaminants, ssDNA was liberated from the surface of graphene oxide. Since graphene oxide is an outstanding distance-dependent fluorescent quencher, this liberation resulted in the recovery of fluorescence in the corresponding test zones. On the contrary, the presence of antibiotics allowed strong binding of graphene oxide and ssDNA, thus the fluorescence was quenched when the targeted contaminants existed. Similarly, another pathogen (*i.e.*, *Lactobacillus acidophilus*) was determined through a polydimethylsiloxane (PDMS)/paper hybrid device by using graphene oxide associated with an adapter.[98] The ready-to-use hybrid device has shown its sensitivity by a limit of detection of 11.0 CFU mL^{-1}.

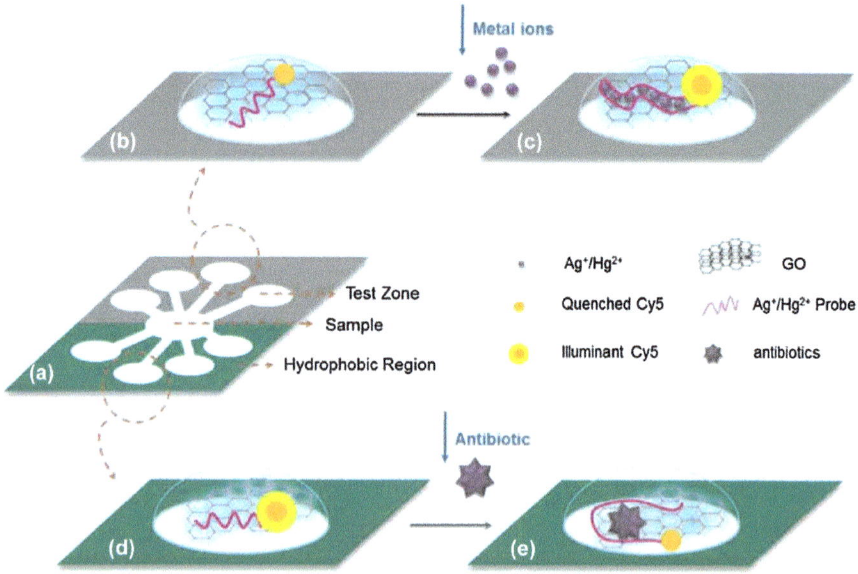

Figure 4.12 Simultaneous detection of various contaminants in foods through μPADs by using graphene oxide integrated with fluorescence labelled-ssDNA.
(Reproduced from *Biosens. Bioelectron.*, **68**, Y. Zhang, P. Zuo and B.-C. Ye, A low-cost and simple paper-based microfluidic device for simultaneous multiplex determination of different types of chemical contaminants in food, 14–19. Copyright 2015 with permission from Elsevier.)

4.6 Conclusions and Future Perspectives

Numerous applications of paper-based microfluidics to food safety and quality analysis were achieved in the past decade. These mainly include the determination of chemical and microbiological hazards in agri-food products. However, most of the current paper devices usually lack the capabilities to integrate all related essential functions together. For instance, food specimens usually require extensive pre-treatment before introduction to the paper devices because the collection of targets of interest is difficult to achieve on paper. To address this problem, integration of paper-based microfluidics and conventional lab-on-a-chip devices (*e.g.*, PDMS-based chip) may be one of the answers. For instance, extraction of target contaminants from vegetables, fruits, poultry and other foods can be realized by an individual chip, followed by delivering the intermediate products to the paper devices, which surpass the conventional chips in terms of simplicity, user-friendliness, cost-effectiveness and potential for mass production. Although most applications as aforementioned are basically reported as demonstrations, the development and commercialization of paper-based devices expanded massively during the past few years. It is expected that paper devices will make detection of foodborne and waterborne contaminants much easier and enable more effective prevention of human diseases, thus ensuring excellent food safety and quality.

References

1. K. G. Grunert, *Eur. Rev. Agric. Econ.*, 2005, **32**, 369–391.
2. N. L. Wilson and M. R. Worosz, *Food Policy*, 2014, **45**, 112–115.
3. L. S. A. Busa, S. Mohammadi, M. Maeki, A. Ishida, H. Tani and M. Tokeshi, *Micromachines*, 2016, **7**, 86.
4. S. Wang, *Food Chemical Hazard Detection: Development and Application of New Technologies*, John Wiley & Sons, 2014.
5. E. K. Sackmann, A. L. Fulton and D. J. Beebe, *Nature*, 2014, **507**, 181–189.
6. L. Guo, J. Feng, Z. Fang, J. Xu and X. Lu, *Trends Food Sci. Technol.*, 2015, **46**, 252–263.
7. J. Xu, *Food Chemical Hazard Detection: Development and Application of New Technologies*, 2014, pp. 263–289.
8. Y. Xia and G. M. Whitesides, *Annu. Rev. Mater. Sci.*, 1998, **28**, 153–184.
9. A. Folch, *Introduction to BioMEMS*, CRC Press, 2012.
10. X. Chen, J. Chen, F. Wang, X. Xiang, M. Luo, X. Ji and Z. He, *Biosens. Bioelectron.*, 2012, **35**, 363–368.
11. A. R. Rezk, A. Qi, J. R. Friend, W. H. Li and L. Y. Yeo, *Lab Chip*, 2012, **12**, 773–779.
12. S. Lathwal and H. D. Sikes, *Lab Chip*, 2016, **16**, 1374–1382.
13. D. Zang, L. Ge, M. Yan, X. Song and J. Yu, *Chem. Commun.*, 2012, **48**, 4683–4685.
14. Y. Lin, D. Gritsenko, S. Feng, Y. C. Teh, X. Lu and J. Xu, *Biosens. Bioelectron.*, 2016, **83**, 256–266.
15. S. Z. Hossain and J. D. Brennan, *Anal. Chem.*, 2011, **83**, 8772–8778.
16. Y. Lin, D. Gritsenko, Q. Liu, X. Lu and J. Xu, *ACS Appl. Mater. Interfaces*, 2016, **8**, 20501–20515.
17. Y. Zhang, P. Zuo and B.-C. Ye, *Biosens. Bioelectron.*, 2015, **68**, 14–19.
18. J. C. Jokerst, J. A. Adkins, B. Bisha, M. M. Mentele, L. D. Goodridge and C. S. Henry, *Anal. Chem.*, 2012, **84**, 2900–2907.
19. A. W. Martinez, S. T. Phillips, M. J. Butte and G. M. Whitesides, *Angew. Chem., Int. Ed.*, 2007, **46**, 1318–1320.
20. A. K. Yetisen, M. S. Akram and C. R. Lowe, *Lab Chip*, 2013, **13**, 2210.
21. X. Li, J. Tian, T. Nguyen and W. Shen, *Anal. Chem.*, 2008, **80**, 9131–9134.
22. M. A. Harvey, C. A. Audette and R. McDonogh, *IVD Technol.*, 1996, **2**, 34.
23. J. Credou, H. Volland, J. Dano and T. Berthelot, *J. Mater. Chem. B*, 2013, **1**, 3277–3286.
24. E. Carrilho, A. W. Martinez and G. M. Whitesides, *Anal. Chem.*, 2009, **81**, 7091–7095.
25. W. Dungchai, O. Chailapakul and C. S. Henry, *Analyst*, 2011, **136**, 77–82.
26. K. Yamada, T. G. Henares, K. Suzuki and D. Citterio, *Angew. Chem., Int. Ed.*, 2015, **54**, 5294–5310.

27. J. Olkkonen, K. Lehtinen and T. Erho, *Anal. Chem.*, 2010, **82**, 10246–10250.
28. X. Li, J. Tian, G. Garnier and W. Shen, *Colloids Surf., B*, 2010, **76**, 564–570.
29. D. A. Bruzewicz, M. Reches and G. M. Whitesides, *Anal. Chem.*, 2008, **80**, 3387–3392.
30. W. Wang, W.-Y. Wu and J.-J. Zhu, *J. Chromatogr. A*, 2010, **1217**, 3896–3899.
31. P. J. Bracher, M. Gupta, E. T. Mack and G. M. Whitesides, *ACS Appl. Mater. Interfaces*, 2009, **1**, 1807-1812.
32. Y.-L. Tai and Z.-G. Yang, *J. Mater. Chem.*, 2011, **21**, 5938–5943.
33. J. Nie, Y. Zhang, L. Lin, C. Zhou, S. Li, L. Zhang and J. Li, *Anal. Chem.*, 2012, **84**, 6331–6335.
34. M. M. Thuo, R. V. Martinez, W.-J. Lan, X. Liu, J. Barber, M. B. Atkinson, D. Bandarage, J.-F. Bloch and G. M. Whitesides, *Chem. Mater.*, 2014, **26**, 4230–4237.
35. V. F. Curto, N. Lopez-Ruiz, L. F. Capitan-Vallvey, A. J. Palma, F. Benito-Lopez and D. Diamond, *RSC Adv.*, 2013, **3**, 18811–18816.
36. C. L. Cassano and Z. H. Fan, *Microfluid. Nanofluid.*, 2013, **15**, 173–181.
37. T. Songjaroen, W. Dungchai, O. Chailapakul and W. Laiwattanapaisal, *Talanta*, 2011, **85**, 2587–2593.
38. T. Nurak, N. Praphairaksit and O. Chailapakul, *Talanta*, 2013, **114**, 291–296.
39. A. W. Martinez, S. T. Phillips, B. J. Wiley, M. Gupta and G. M. Whitesides, *Lab Chip*, 2008, **8**, 2146–2150.
40. Y. He, W.-b. Wu and J.-z. Fu, *RSC Adv.*, 2015, **5**, 2694–2701.
41. Q. He, C. Ma, X. Hu and H. Chen, *Anal. Chem.*, 2013, **85**, 1327–1331.
42. E. W. Washburn, *Phys. Rev.*, 1921, **17**, 273.
43. H. C. Nallan, J. A. Sadie, R. Kitsomboonloha, S. K. Volkman and V. Subramanian, *Langmuir*, 2014, **30**, 13470–13477.
44. K. Abe, K. Suzuki and D. Citterio, *Anal. Chem.*, 2008, **80**, 6928–6934.
45. T. Songjaroen, W. Dungchai, O. Chailapakul, C. S. Henry and W. Laiwattanapaisal, *Lab Chip*, 2012, **12**, 3392–3398.
46. X. Li, P. Zwanenburg and X. Liu, *Lab Chip*, 2013, **13**, 2609–2614.
47. A. Fraiwan, H. Lee and S. Choi, *Sens. J. IEEE*, 2014, **14**, 3385–3390.
48. J. Li and J. Macdonald, *Lab Chip*, 2016, **16**, 242–245.
49. S. Kar, T. K. Maiti and S. Chakraborty, *Analyst*, 2015, **140**, 6473–6476.
50. H. Liu and R. M. Crooks, *Anal. Chem.*, 2012, **84**, 2528–2532.
51. A. Rogacs, J. E. Steinbrenner, J. A. Rowlette, J. M. Weisse, X. L. Zheng and K. E. Goodson, *J. Colloid Interface Sci.*, 2010, **349**, 354–360.
52. C. K. Camplisson, K. M. Schilling, W. L. Pedrotti, H. A. Stone and A. W. Martinez, *Lab Chip*, 2015, **15**, 4461–4466.
53. E. Fu, P. Kauffman, B. Lutz and P. Yager, *Sens. Actuators, B*, 2010, **149**, 325–328.
54. A. Apilux, Y. Ukita, M. Chikae, O. Chailapakul and Y. Takamura, *Lab Chip*, 2013, **13**, 126–135.

55. E. Fu, T. Liang, P. Spicar-Mihalic, J. Houghtaling, S. Ramachandran and P. Yager, *Anal. Chem.*, 2012, **84**, 4574–4579.
56. B. J. Toley, B. McKenzie, T. Liang, J. R. Buser, P. Yager and E. Fu, *Anal. Chem.*, 2013, **85**, 11545–11552.
57. J. H. Shin, J. Park, S. H. Kim and J.-K. Park, *Biomicrofluidics*, 2014, **8**, 054121.
58. B. Lutz, T. Liang, E. Fu, S. Ramachandran, P. Kauffman and P. Yager, *Lab Chip*, 2013, **13**, 2840–2847.
59. S.-G. Jeong, S.-H. Lee, C.-H. Choi, J. Kim and C.-S. Lee, *Lab Chip*, 2015, **15**, 1188–1194.
60. B. J. Toley, J. A. Wang, M. Gupta, J. R. Buser, L. K. Lafleur, B. R. Lutz, E. Fu and P. Yager, *Lab Chip*, 2015, **15**, 1432–1444.
61. X. Li, D. R. Ballerini and W. Shen, *Biomicrofluidics*, 2012, **6**, 011301.
62. G. Aragay, H. Montón, J. Pons, M. Font-Bardía and A. Merkoçi, *J. Mater. Chem.*, 2012, **22**, 5978–5983.
63. J.-j. Li, C.-h. Ji, C.-j. Hou, D.-q. Huo, S.-y. Zhang, X.-g. Luo, M. Yang, H.-b. Fa and B. Deng, *Sens. Actuators, B*, 2016, **223**, 853–860.
64. M. Zheng, H. Tan, Z. Xie, L. Zhang, X. Jing and Z. Sun, ACS Appl. Mater. Interfaces, 2013, **5**, 1078–1083.
65. M. Formica, V. Fusi, L. Giorgi and M. Micheloni, *Coord. Chem. Rev.*, 2012, **256**, 170–192.
66. C. Zhao, M. M. Thuo and X. Liu, *Sci. Technol. Adv. Mater.*, 2013, **14**, 054402.
67. F. Scholz, *Electroanalytical Methods*, Springer, 2010.
68. M. Cormier, *Chemiluminescence and Bioluminescence*, Springer Science & Business Media, 2013.
69. H. Chen, L. Lin, H. Li and J.-M. Lin, *Coord. Chem. Rev.*, 2014, **263**, 86–100.
70. M. S. Akram, R. Daly, F. da Cruz Vasconcellos, A. K. Yetisen, I. Hutchings and E. A. Hall, *Lab-on-a-Chip Devices and Micro-Total Analysis Systems*, Springer, 2015, pp. 161–195.
71. D. D. Liana, B. Raguse, J. J. Gooding and E. Chow, *Sensors*, 2012, **12**, 11505–11526.
72. E. W. Nery and L. T. Kubota, *Anal. Bioanal Chem.*, 2013, **405**, 7573–7595.
73. Y. Xia, J. Si and Z. Li, *Biosens. Bioelectron.*, 2016, **77**, 774–789.
74. R. L. Scharff, Health-related costs from foodborne illness in the United States, 2010, http://citeseerx.ist.psu.edu/viewdoc/versions?doi=10.1.1.172.4518 (Accessed 5/15/2017).
75. J. Park, J. H. Shin and J.-K. Park, *Anal. Chem.*, 2016, **88**, 3781–3788.
76. F. Yu, L. Li and F. Chen, *Anal. Chim. Acta*, 2008, **610**, 257–262.
77. S.-Q. Jin, S.-M. Guo, P. Zuo and B.-C. Ye, *Biosens. Bioelectron.*, 2015, **63**, 379–383.
78. T. San Park, W. Li, K. E. McCracken and J.-Y. Yoon, *Lab Chip*, 2013, **13**, 4832–4840.
79. S. Ma, Y. Tang, J. Liu and J. Wu, *Talanta*, 2014, **120**, 135–140.

80. S. Burnham, J. Hu, H. Anany, L. Brovko, F. Deiss, R. Derda and M. W. Griffiths, *Anal. Bioanal. Chem.*, 2014, **406**, 5685–5693.
81. B. Gilbert-López, J. F. García-Reyes and A. Molina-Díaz, *Talanta*, 2009, **79**, 109–128.
82. S. Wang, L. Ge, L. Li, M. Yan, S. Ge and J. Yu, *Biosens. Bioelectron.*, 2013, **50**, 262–268.
83. W. Liu, Y. Guo, J. Luo, J. Kou, H. Zheng, B. Li and Z. Zhang, *Spectrochim. Acta, Part A*, 2015, **141**, 51–57.
84. M. E. Badawy and A. F. El-Aswad, *Int. J. Anal. Chem.*, 2014, **2014**, 536823.
85. C. Sicard, C. Glen, B. Aubie, D. Wallace, S. Jahanshahi-Anbuhi, K. Pennings, G. T. Daigger, R. Pelton, J. D. Brennan and C. D. Filipe, *Water Res.*, 2015, **70**, 360–369.
86. S. Nouanthavong, D. Nacapricha, C. S. Henry and Y. Sameenoi, *Analyst*, 2016, **141**, 1837–1846.
87. M. Ornatska, E. Sharpe, D. Andreescu and S. Andreescu, *Anal. Chem.*, 2011, **83**, 4273–4280.
88. G. Sun, P. Wang, S. Ge, L. Ge, J. Yu and M. Yan, *Biosens. Bioelectron.*, 2014, **56**, 97–103.
89. S. Cho, J.-W. Jang, A. Jung, S.-H. Lee, J. Lee, J. S. Lee and K.-H. Lee, *Langmuir*, 2010, **27**, 371–378.
90. T. Colletes, P. Garcia, R. Campanha, P. Abdelnur, W. Romão, W. Coltro and B. Vaz, *Analyst*, 2016, **141**, 1707–1713.
91. M. Vaher and M. Kaljurand, *Anal. Bioanal. Chem.*, 2012, **404**, 627–633.
92. L. J. Dombrowski and E. J. Pratt, *Anal. Chem.*, 1972, **44**, 2268–2272.
93. J. Sanchez-Echaniz, J. Benito-Fernández and S. Mintegui-Raso, *Pediatrics*, 2001, **107**, 1024–1028.
94. H. Zhang, S. Qi, Y. Dong, X. Chen, Y. Xu, Y. Ma and X. Chen, *Food Chem.*, 2014, **151**, 429–434.
95. T. M. Cardoso, P. T. Garcia and W. K. Coltro, *Anal. Methods*, 2015, **7**, 7311–7317.
96. N. Lopez-Ruiz, V. F. Curto, M. M. Erenas, F. Benito-Lopez, D. Diamond, A. J. Palma and L. F. Capitan-Vallvey, *Anal. Chem.*, 2014, **86**, 9554–9562.
97. B. M. Jayawardane, S. Wei, I. D. McKelvie and S. D. Kolev, *Anal. Chem.*, 2014, **86**, 7274–7279.
98. P. Zuo, X. Li, D. C. Dominguez and B.-C. Ye, *Lab Chip*, 2013, **13**, 3921–3928.

CHAPTER 5

Colorimetric-based Sensing in Food Safety and Quality Analysis

AZADEH NILGHAZ, EVAN TROFIMCHUK AND XIAONAN LU*

Food, Nutrition and Health Program, Faculty of Land and Food Systems, The University of British Columbia, Vancouver, British Columbia V6T 1Z4, Canada
*Email: xiaonan.lu@ubc.ca

5.1 Introduction

Food safety is a scientific discipline regarding the understanding of handling, storage and preparation of food products to ensure their safety.[1–3] Safe food is a food product without contamination by biological contaminants as well as chemical and physical hazards, such as pathogens, toxins, allergens, chemical contaminations and physical components.[4] Exposure to high doses of food contaminants can pose immediate health risks.[5] Prolonged habitual exposure to low doses of the contaminants could cause long-term negative health effects such as cancer, diabetes and arthritic syndromes.[6] Some types of food contaminants are relatively stable, and therefore they are not affected by the many processing techniques taken by food manufacturers to prevent the distribution of biological and chemical contaminants.[7] Food manufacturers often implement Hazard Analysis and Critical Control Points (HACCP) to ensure that the final food product is free from biological, chemical and physical hazards. However, potential contaminant residue can still exist in the final processed food products after HACCP.

Food Chemistry, Function and Analysis No. 2
Sensing Techniques for Food Safety and Quality Control
Edited by Xiaonan Lu

Published by the Royal Society of Chemistry, www.rsc.org

Governmental health agencies have been working continuously to assess the risks that pathogens and chemical hazards pose to human health and estimate the average dietary exposure to these contaminants.[8] These agencies set strict regulations regarding the maximum levels for contaminants that can be identified in agri-food products. Some substances are classified as "zero tolerance", indicating that foods cannot even contain a small trace level.[9] In 1963, the Food and Agriculture Organization (FAO) and the World Health Organization (WHO) developed the Codex Alimentarius to enhance international food standards and safety,[10] and protect consumer health, as well as assure fair practices and competition in the agri-food industry.

Pathogenic bacteria, parasites, chemical hazards and physical elements can contaminate food products.[11] These substances may be accidentally present in food or, in some cases, have been intentionally added by the manufacturers during the food harvest, processing, storage, shipping and preparation of them. In either case, these unwanted substances can pose severe health risks. Frequent analysis of the targeted contaminants is one of the keys in ensuring the quality of food products and protecting consumer health.

According to the estimation of the global burden by the WHO, food contamination caused 600 million cases of foodborne diseases and 420 000 deaths in 2010.[12] The severity level of foodborne diseases depends upon age, the health status of the consumers, as well as the amount(s) and type(s) of contaminants.[13,14] The WHO Food Safety Group is most concerned about food contamination with pathogens, including bacteria, viruses and parasites. Foodborne pathogenic bacteria can cause severe illnesses at a very low concentration (*e.g.*, 10–100 colony forming units (CFU)). The five most common pathogens responsible for foodborne diseases, hospitalization and mortality are *Salmonella*, Shiga-toxin producing *Escherichia coli*, norovirus, *Campylobacter* and *Listeria monocytogenes*.[15] Some of these pathogens may exist in raw food products purchased by consumers.[16,17]

Pathogenic bacteria in foods can multiply quickly in the gastrointestinal system after food consumption.[18] Exposure to these pathogenic bacteria leads to various symptoms, such as nausea, vomiting, diarrhoea, abdominal pain, fever and other toxic shock syndromes.[19] For example, *Salmonella* is the leading cause of food poisoning globally. According to reports by Centres for Disease Control and Prevention, *Salmonella* infections resulted in ~20 000 hospitalizations and 400 deaths in a year in the United States.[20] Salmonellosis is derived from the consumption of contaminated eggs, raw dairy products, undercooked meat products and contaminated fresh produce.[21,22]

In addition to pathogenic bacteria, various fungi and moulds are also capable of producing secondary metabolite toxins (*e.g.*, mycotoxins) under suitable temperature, relative humidity and ecological conditions. Most of these metabolites are highly toxic to humans. Mycotoxins can remain in food products for a long period of time and some of these mycotoxins, such as aflatoxins and ochratoxins, are carcinogenic to humans.[23,24]

Apart from the aforementioned biological contaminations, food adulteration by adding chemical contaminants and allergens has grown dramatically in recent years. They can be added to food products either intentionally or non-intentionally. Generally, intentional food adulteration serves the purpose of augmenting a particularly desired sensory, chemical and/or physical property of food products desired by consumers. Alternatively, manufacturers may fully or partially substitute an ingredient with an inferior ingredient for economic gain. The substituted ingredients may be allergenic and pose health risks to consumers. For example, melamine was intentionally added to milk, infant formula and pet foods to enhance the detected protein level. This intentional adulteration contributed to around 300 000 cases of poisonings and hospitalizations globally during the past decade.[25,26]

Processing-induced contaminants are naturally absent in raw unprocessed food products. However, these contaminants can be generated through different processing steps, such as fermentation and heating.[27,28] In general, these contaminants are formed during chemical reactions between the raw food ingredients and the added food constituents during processing.

Besides the critical attribute of safety, external and internal properties of food products, such as sensory attributes, chemical compositions and shelf life, should be utilized to determine the quality of the food products. According to the Yearbook of Agriculture published in 1970, food quality "is the measure or expression of goodness".[29] This definition clearly displays that food quality is an arbitrary term and it is highly dependent upon changeable consumer preferences over time.[29–33] The measurement of sensory properties in food products requires advanced technology to result in a variety of sensor designs, which can be used to monitor food quality, such as freshness, leakage, carbon dioxide and oxygen levels, pH, shelf life and storage conditions.[34,35]

Food safety and quality control are global challenges. Therefore, robust, selective, sensitive, rapid and miniaturized methods are required to detect foodborne pathogens, chemical hazards, the level of nutrients and freshness in complicated food products. There are numerous analytical detection methods that have been developed to fulfil these requirements. For example, conventional plating, biochemical and DNA-based assays are widely used for the detection of pathogenic bacteria in foods.[36,37] However, most of these assays require large and expensive equipment and are not fully suitable for in-field detection of food contaminants in a straightforward manner.[38,39]

Colorimetric detection has a great potential to meet these requirements for food safety monitoring and quality analysis. In a colorimetric assay, the presence of particular pathogens and/or chemical contaminants can be directly observed by the naked eye without specific laboratory equipment and data analysis. Thus, the colorimetric method is suitable for in-field detection. Colorimetric assays are usually performed on simple and cheap substrates, such as plastic plates, glass slides, paper and textiles.[40–43] In colorimetric sensing assays, colour-based biomarkers are used to generate signals when they react with specific targets. The colorimetric reaction can be monitored and stopped when a desired level of signal is produced. In addition, users with minimum training can readily perform the testing. This chapter summarizes the key components of colorimetric detection methods, followed by selective examples of applying various colorimetric assays for the detection of pathogens and chemical contaminants in agri-food products. Finally, a summary and perspective on future research directions of using colorimetric assays in food safety and quality assurance are discussed.

5.2 Colorimetric Analysis

5.2.1 Overview

Colorimetric detection is a measurement method for determining the concentration of a target analyte in a sample with the aid of a colour reagent. This technique is suitable for either single or multiplex detection and for the analysis of organic and inorganic compounds and biomolecules. Colorimetric detection is commonly used in laboratories to determine different types of enzymes, antibodies, hormones, specific chemicals and reagents.

Various colorimetric methods on the basis of coloured pigments, fluorescent dyes and gold nanoparticles have been developed to sense food safety and quality. There are numerous advantages of this technique, such as specific and multiplex detection in a simultaneous manner.

5.2.2 Advantages and Limitations of Colorimetric Sensing in Food Safety and Quality Control

Colour reagents are widely used as analytical reagents in colorimetric detection methods because of their great optical properties that are highly advantageous for food safety and quality analysis. Colorimetric assays can be performed on substrates with low cost, such as a 96-well plate, plastic or glass slide/bottle, paper and textiles, using micro-millilitres of reagents and analyte samples. Colorimetric assays are also designed in simple formats and have the capability for real-time and in-field analysis of multiple pathogenic and chemical contaminants in foods. The results of the colorimetric assays can usually be observed by the naked eye in a short period of time and interpreted without either special equipment or trained laboratory personnel. Another advantage of using colour reagents is their ability to conjugate themselves to other compounds and molecules, such as antibodies, enzymes, proteins and nucleic acids, which enhance the sensitivity and specificity of the colorimetric reaction. However, the limitations of colorimetric detection are their relatively lower sensitivity in comparison to the techniques that can achieve single cell analysis and other laboratory-based instrumental analytical methods. In addition, colorimetric sensing is mainly a qualitative to semi-quantitative method, which makes it challenging to achieve absolute quantitative analysis.

5.3 Colorimetric Detection of Food Contaminants Using Gold Nanoparticles

5.3.1 General Overview

Different sizes of metallic nanoparticles have been extensively applied for detection purposes in the fields of biotechnology, biomedical engineering and food safety analysis.[44] Particularly, unique optical properties and low toxicity generated by gold nanoparticles are two key features that are suitable for highly sensitive and selective colorimetric detection, especially for the detection of pathogens and chemical hazards in agri-food products.[45] Gold nanoparticles can be chemically fabricated to different morphologies and sizes, such as nanospheres, nanorods, nanocubes and nanorings. Gold nanoparticles with different sizes and morphologies can be determined on the basis of different absorbance peaks. The majority of gold nanoparticle-based assays are dependent upon the changes of the surface plasmon resonance wavelength in different nanoparticle aggregation states by either particle aggregation or stabilization. The peak absorbance wavelength of gold nanoparticles is dependent upon the average particle distance, which is critical to lead to colour changes. In addition, the stability and sensitivity of gold nanoparticles are also dependent upon the particle size, in which stability decreases and sensitivity increases in large particles. Therefore, optimization of particle size is critical to achieve diagnostic purposes. Among different types of gold nanoparticles, nanospheres with size 13–20 nm and absorbance peak around 520 nm have a significant contribution to the development of various colorimetric detection methods.[46] In addition, gold nanorods generate two absorbance peaks at ~520 nm and in the infrared region, with the latter one more frequently employed for colorimetric sensing and detection.[47,48]

5.3.2 Applications of Using Gold Nanoparticles for Food Safety and Quality Analysis

A major research focus of applying gold nanoparticles for the detection of pathogens is to enhance the current genomic analysis methods in order to generate a quantitative detection method with extremely low detection limits. The application of non-functionalized gold nanoparticles is a suitable method to amplify nucleic acids. To avoid aggregation and the subsequent colour change of gold nanoparticles in the presence of salt, stabilization of the nucleic acid is required. In order to use gold nanoparticles for the detection of pathogenic bacteria, genomic DNA extracted from bacteria is required to be adsorbed by the surface of nanoparticles. The adsorption of DNA causes aggregation of the gold nanoparticles, which subsequently generates the colour signals (Figure 5.1).[46]

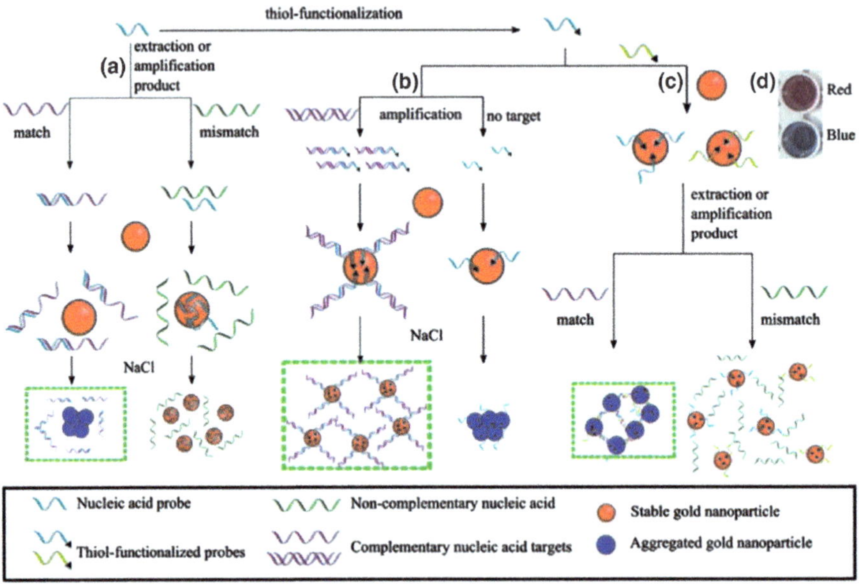

Figure 5.1 Schematic illustration of nucleic acid detection using gold nanoparticles. The probes can be designed and fabricated either without (a) or with thiolation (b) to recognize nucleic acids and cause a color change in the presence of non-functionalized gold nanoparticles. However, only thiolated probes are able to recognize nucleic acids and generate a color change in the presence of functionalized gold nanoparticles (c). Image of gold nanoparticle solutions exhibiting different colors (d).
Reprinted from *Biotechnology Advances*, Volume 33, Mohit S. Verma, Jacob L. Rogowski, Lyndon Jones, Frank X. Gu, Colorimetric biosensing of pathogens using gold nanoparticles. Copyright 2015, with permission from Elsevier.

Salmonella is a major pathogen responsible for ~16 million cases of foodborne diseases each year. Therefore, a very sensitive and rapid method for the detection of *Salmonella* in agri-foods is needed. The *stn* gene (Accession number L16014) of *Salmonella* was used as a target gene for the detection of *Salmonella* at low concentration in food commodities without false negative results. According to BLAST, the *stn* gene is specifically present in all of the *Salmonella* species and can be employed for the specific detection of *Salmonella*. Therefore, a primer was designed to amplify a single 116-bp product in the *Salmonella* family. The gene was targeted using gold nanoparticles with particle size of 22.6 nm and k_{max} of 520 nm for the colorimetric detection of *Salmonella*. The working principle of this method is based upon the fact that the ssDNA probe eliminates the aggregation of gold nanoparticles, while the dsDNA does not interact with gold nanoparticles and causes the aggregation of gold nanoparticles and a colour change from red to blue.[49]

Bacillus anthracis is another example of a pathogen that can be detected using gold nanoparticles. The nanoparticles could be employed to target the

DNA of *Bacillus* that is longer than 100 nt (ssDNA) or 100 bp (dsDNA). These types of DNA eliminate the aggregation of gold nanoparticles with a diameter of 15 nm, and subsequently produce colorimetric signals (Figure 5.1).[48,50]

Designing probes for the selective detection of different types of important molecules (*e.g.*, thiols, cysteine and homocysteine) in the environment and biological samples can play a critical role in improving the performance of the detection methods. The probes can be coupled with non-functionalized nanoparticles to produce a visible colour change upon reactions with other biomolecules. Among different probes used for diagnostic methods and assays, thiolated probes coupled with gold nanoparticles can target DNA in different types of pathogens. For example, thiolated probes are able to target the *hly* and *hut* genes in *Listeria monocytogenes* and *Salmonella enterica*, respectively, and produce a colour change to achieve diagnostic analysis.[48]

Gold nanoparticles are also able to functionalize with antibodies that target the specific pathogen. The antibody–antigen reaction induces the aggregation and subsequent colour change of gold nanoparticles. In this method, the surface properties of pathogens and gold nanoparticles change and subsequently cause an electrostatic interaction between the nanoparticles and pathogens before a colour change occurs. For the case of detection of *Salmonella*, *Salmonella* and anti-*Salmonella* antibody can be conjugated to gold nanoparticles and this complex can be used for label-free detection of *Salmonella* bacteria under near-infrared radiation with the detection limit of 10^3 CFU mL^{-1} (Figure 5.2).[51]

Besides the detection of pathogens, gold nanoparticles have also been employed for the detection of mycotoxins in agri-food products.[52] Aflatoxin is known as a poisonous chemical, which is a carcinogen to humans. This mycotoxin can exist in a wide variety of food products, such as cereal, coffee and fruit juice.[53] Thus, a highly sensitive method for the detection of aflatoxin is required. In order to achieve the detection of this mycotoxin, a method was developed to use aptamers to bind to the surface of gold nanoparticles. The interaction between the aptamers and aflatoxin resulted in the aggregation of gold nanoparticles, which subsequently produced a visible colour change. By using this colorimetric sensing method, a detection limit of 7 nM of aflatoxin was achieved.[54]

A rapid, robust and colorimetric detection of the chemical contaminants in dairy products was also developed using gold nanoparticles as probes.[55–57] As a representative contaminant in Chinese protein adulterations, melamine has been reported to be available in various types of products, such as dried infant formula powder. Specifically, gold nanoparticles can easily aggregate in the presence of melamine to result in a visible colour change from red to blue. Detection of melamine using gold nanoparticles with a diameter of 13 nm can be performed in raw milk after proteins are removed. The detection limit for this method was $\sim$0.4 mg L^{-1}.[58,59] A lower detection limit for melamine ($\sim$0.05 mg L^{-1}) was reported later on, which was achieved using gold nanoparticles with a diameter of 21 nm.[59] As shown in Figure 5.3, this method is based upon the

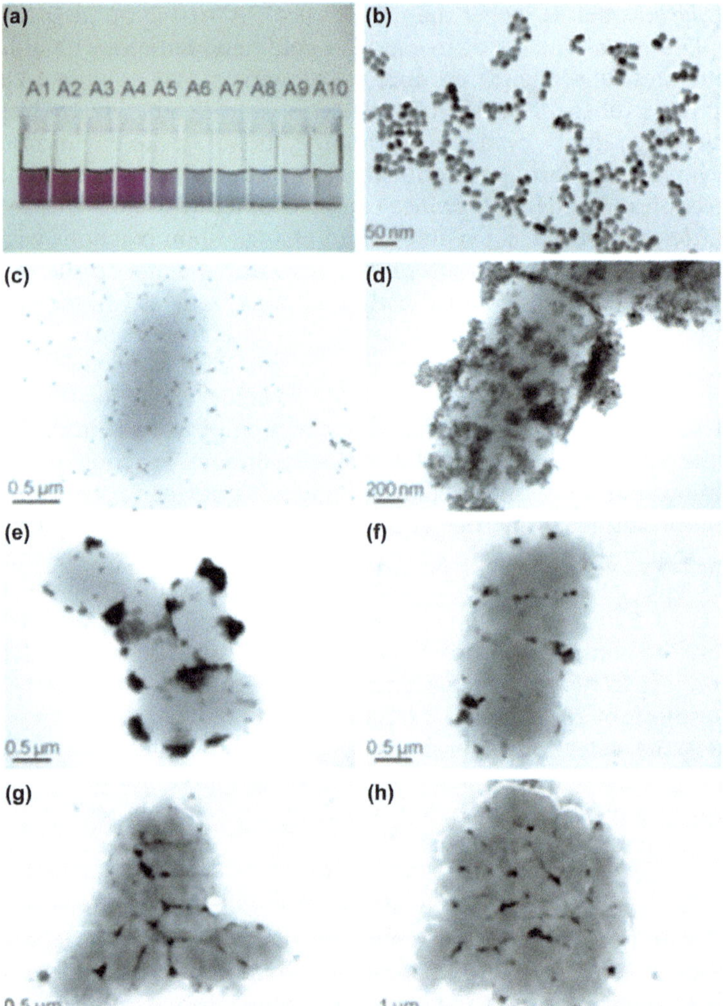

Figure 5.2 Image of colorimetric detection of *Salmonella* bacteria using oval-shaped gold nanoparticles conjugated with anti-*salmonella*-antibody. The color intensity of gold nanoparticles changes upon the addition of 10, 50, 100, 500, 1000, 5000, 10 000, 50 000, 100 000, and 500 000 *Salmonella* bacteria from left to right, respectively (a). TEM image of oval-shaped gold nanoparticles which are conjugated to anti-*salmonella* antibody before (b) and after (c), (d) addition of *salmonella.* TEM image after formation of different sizes of clusters (e)–(h).
Reproduced from Shuguang Wang, Anant K. Singh, Dulal Senapati, Adria Neely, Hongtao Yu, Paresh C. Ray, *Chemistry – A European Journal*, John Wiley and Sons, 2010, 16, 5600–5606, Copyright 2010 Wiley-VCH Verlag GmbH & Co. KGaA, Weinheim.

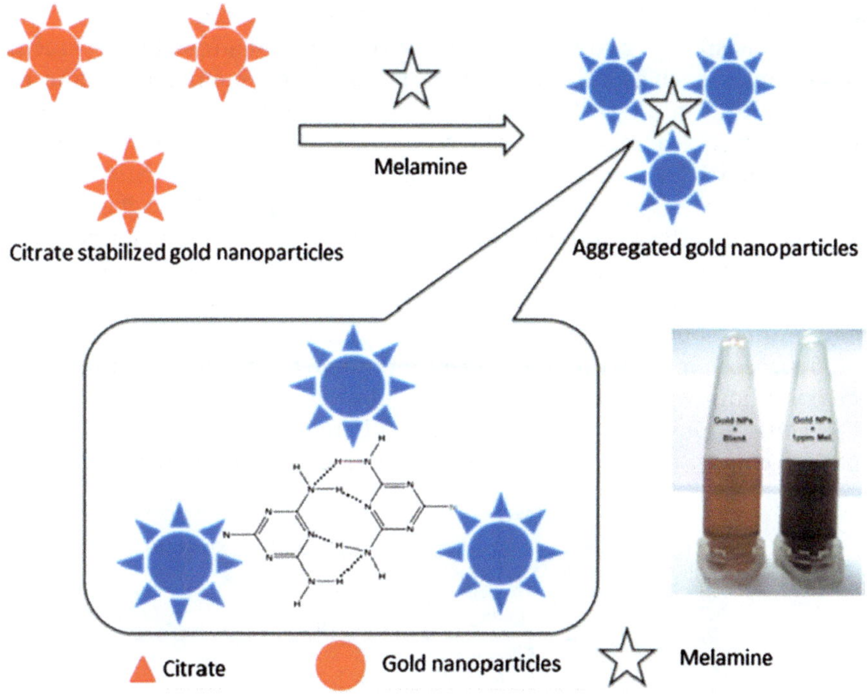

Figure 5.3 Schematic illustration of colorimetric detection mechanism of melamine using gold nanoparticles.
Reprinted from *Analytical Biochemistry*, Volume 456, N. Kumar, R. Seth and H. Kumar, Colorimetric detection of melamine in milk by citrate-stabilized gold nanoparticles. Copyright 2014, with permission from Elsevier.

distance-dependent properties of gold nanoparticles. As aforementioned, the absorbance wavelength in gold nanoparticles varies when the distance between nanoparticles changes. Once the aggregation occurred, the nanoparticles conjugated together and resulted in a colour change.[58,59]

Different types of antibiotics have been used in animal feed for over half a century to prevent diseases in animals, as well as to enhance growth performance. However, the use of antibiotics results in the accumulation of antibiotic residues in animal products,[50,60] such as meat, milk and eggs. There is a concern about the potential negative impact of antibiotic residue on human health due to either generating direct toxicity or facilitating the development of antibiotic resistant bacterial strains in the human body. Recently, a rapid and specific method was developed that applied gold nanoparticles to detect antibiotic residues in animal products. As a representative antibiotic, ampicillin residue could be successfully detected by using gold nanoparticles that produced visible fluorescence colour changes. The aptamer AMP17 had a specific interaction with ampicillin and it was conjugated onto the surface of gold nanoparticles.[61] In the presence of ampicillin, gold nanoparticles released from the aptamers and subsequently aggregated in the solution caused a colour change from red to purple.[62] This method was specific and sensitive for the detection of ampicillin residue in food samples, such as milk and animal tissues.[62] This method can achieve a detection limit of 2 ng mL^{-1} for ampicillin in milk.[62] As another representative antibiotic, tetracycline was also successfully detected in milk using gold nanoparticles conjugated to aptamers (Figure 5.4).[63] In this method, aptamers were used to stabilize the gold nanoparticles against high poly(diallyldimethylammonium) strength. Upon the addition of tetracycline, tetracycline-aptamer complex forms, which causes the aggregation of gold nanoparticles and a colour change from red to purple-blue in the sample solution in a short time period. The development of the purple-blue colour could be correlated with the concentration of tetracycline in milk. The detection limit was as low as 1 μM when observed by the naked eye.[63]

Biogenic amines are nitrogen compounds that can be correlated to the spoilage of both raw and processed food products.[64] Biogenic amines play a

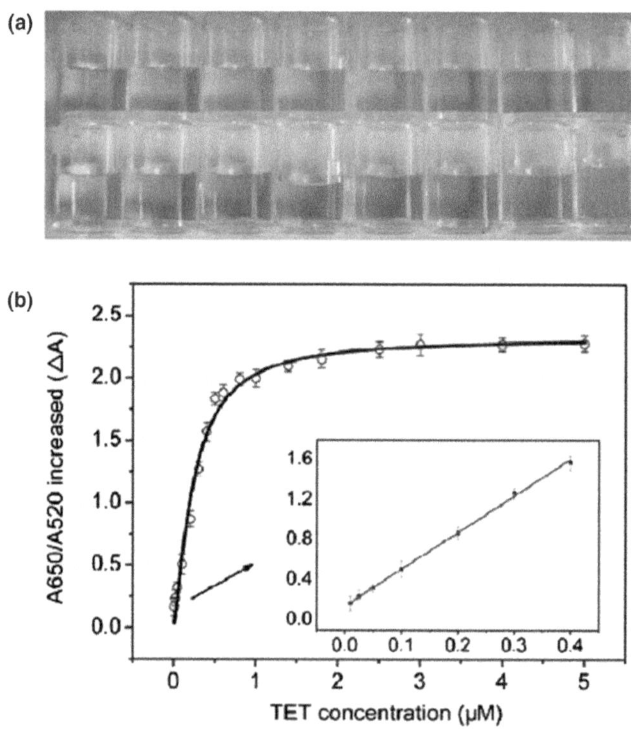

Figure 5.4 Colorimetric detection of serially diluted tetracycline (TET) solution using gold nanoparticles.
Reproduced from *Food Analytical Methods*, Colorimetric Sensing of Tetracyclines in Milk Based on the Assembly of Cationic Conjugated Polymer-Aggregated Gold Nanoparticles, Volume 6, 2013, 1704, L. He, Y. Luo, W. Zhi and P. Zhou, © Springer Science + Business Media New York 2013. With permission of Springer.

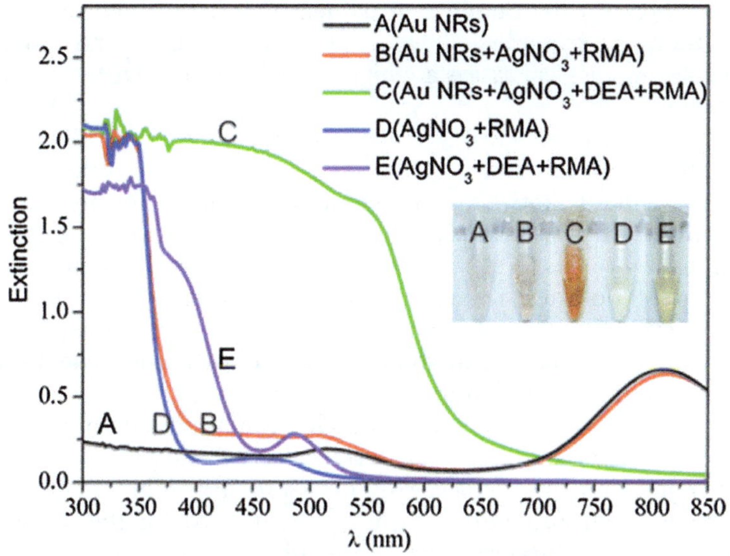

Figure 5.5 Colorimetric detection of biogenic amines using gold nanorods (Au NRs) to evaluate the level of spoilage in salmon. Biogenic amines were produced in hydrogel containing different amounts of $AgNO_3$, resorcinol monoacetate (RMA) and diethanolamine (DEA). The hydrolysis of RMA was reduced Ag+ to Ag0. The silver atoms were adsorbed on the surface of gold nanorods and caused a visible color change.
Reproduced with permission from T. Lin, Y. Wu, Z. Li, Z. Song, L. Guo, F. Fu, *Anal. Chem.*, 2016, **88**, 11022–11027. Copyright 2016 American Chemical Society.

role in evaluating the quality of food samples. A high-resolution colorimetric assay was developed to detect biogenic amines in fresh salmon with the aid of gold nanorods.[3] Biogenic amines were produced by storing salmon in the sensory hydrogels that contained gold nanorods (Au NRs), resorcinol monoacetate (RMA) and $AgNO_3$. The hydrolysis of RMA can affect and reduce the silver ions (Ag^+) to silver atoms (Ag^0). The produced silver atoms could get adsorbed onto the surface of gold nanorods, causing a visible colour change. The colour change and blue shift of longitudinal LSPR peak could be correlated with the concentration of biogenic amines in food products (Figure 5.5).[3]

5.4 Colorimetric Detection of Food Contaminants Using Immunological Methods

5.4.1 General Overview

Various rapid methods have been developed to detect food contaminants and analyze food quality. Each method has its own advantages and disadvantages. As a conventional method, polymerase chain reaction (PCR)-based techniques were used to detect the presence of pathogens in food products. However, these methods are relatively tedious and time consuming.[65] In contrast, colorimetric immunological assays are highly specific and convenient, and yield rapid results.[66]

Immunological assays are based upon extremely specific binding between an antigen and an antibody.[67] These assays have been developed for the quantitative and qualitative measurement of specific pathogens, antibiotics and proteins, in analyte sample solutions by means of the immunological reaction; therefore they are employed for food safety and quality analysis.[68] Hybridoma technologies are combined in immunoassays to provide a higher sensitivity over the conventional immunoassays. Immunological assays provide extremely sensitive methods to detect pathogens, chemical contaminations and adulteration in food products.

Either monoclonal or polyclonal antibodies can be used in immunological assays.[69] Polyclonal antibodies are a collection of antibodies with varying cellular origins and specificities. They have a higher affinity in comparison to the monoclonal antibodies and are usually used for the detection of denatured proteins. However, these antibodies generate a large amount of non-specific antibodies, and subsequently generate background signals. On the other hand, monoclonal antibodies that have higher specificity are more consistent. They are more reliable for immunological studies and are only able to detect specific antigens.[70]

There are two major immunological assays, which are homogenous immunoassays and heterogeneous assays.[71] In homogenous immunoassays, the antigens react with specific antibodies to form antigen–antibody complexes. In these assays, the bound and unbound antibodies do not require a separation and the result can be quantified visually by the naked eye after a short period of time. An example of the homogenous assays is the agglutination reaction. Heterogeneous assays require the bound and unbound antibodies to be separated. This category can be further segregated into competitive and non-competitive assays. The best-known example of the latter is the colorimetric enzyme linked immunosorbent assay (ELISA), which is commercially designed and available for laboratory-based studies. The colorimetric immunosorbent assays play an important role in food safety monitoring and include enzyme-linked immunosorbent assay (ELISA), radioimmunoassay, fluorescence immunoassay and luminescence immunoassay. Coupled with the advances in nanotechnology, sample handling is further simplified. Among the aforementioned methods, competitive and sandwich ELISA play the major roles for food safety and quality analysis.[72]

5.4.2 Applications of Colorimetric Immunological Methods and ELISA for Food Safety and Quality Analysis

Colorimetric ELISA has been commonly used in food safety monitoring to detect pathogens and chemical hazards using automated systems. These systems facilitate a high volume of routine laboratory tests where microplate readers or spectrometers also automatically quantify the results of the colorimetric tests. However, ELISA results can be qualitatively evaluated either by simple laboratory-based instruments or by the naked eye, which is suitable for use in field analysis or remote setting areas. Intensive research has employed colorimetric ELISA for food safety and quality analysis to detect pathogenic bacteria, parasites and chemical hazards.

Currently, there are numerous conventional colorimetric ELISA kits for the detection of different types of pathogenic bacteria used in different food products.[73] As an example, the *Salmonella* ELISA kit (TEK) was used to detect *Salmonella* in raw meat.[74] The meat was first minced and incubated in peptone water before enriching and collecting the liquid around the samples to perform a colorimetric ELISA. This method displayed a very low detection limit of 1–5 CFU 25 g^{-1} without a cross-reaction of the anti-*Salmonella* antibodies to other bacteria.[74]

Parasites are organisms that depend upon other living organisms to get their nutrients.[75] These species can be transmitted by consuming contaminated water and seafood as well as contact with infected individuals. Parasites can cause foodborne diseases.[76] In order to avoid the high cost of parasite infection treatment,[77] a rapid and sensitive method is required, such as colorimetric ELISA. The two most common foodborne parasites are protozoa and roundworms, such as *Cryptosporidium* spp. and *Anisakis* spp., respectively.[78] Parasites or parasite-related hazardous substances can be detected using colorimetric ELISA. As an example, *Anisakis* spp. was detected using sandwich ELISA in wild-caught marine fish. In this method, the polyclonal rabbit anti-*Anisakis* spp. antibody and its biotinylated conjugated antibodies were employed as the capture and labelled antibodies, respectively, to specifically capture *Anisakis* spp. This detection method had no cross-reaction with other species with the detection limit of 1.1 μg g^{-1}, which was sensitive enough to be used as a detection method in seafood products.[79]

In addition to pathogenic bacteria and parasites, the detection of allergens in foods is another application for colorimetric ELISA. The peanut (*Arachis hypogaea* L.) has been known as the most common cause of food allergies in babies and children, although it can cause allergies in adults as well.[80–82] Very low concentrations of peanut protein (*e.g.*, 100 μg) can cause allergenic reactions in individuals who are sensitive to peanuts.[83,84] A colorimetric ELISA test was developed to quantify the concentration of peanut allergens in foods containing peanuts, such as chocolate, butter, milk and cereal-based snacks.[84,85] In a study, mAb 6 and HRP-labelled mAb 4 were selected as capture and labelled antibodies, respectively, to specifically react with peanut allergens, Arah, in different types of food products.[85] The results displayed that these antibodies were able to specifically detect Arah with a detection limit of 0.34 ng mL^{-1}, which was low enough to be used for food safety monitoring.[85]

Ovalbumin in hen eggs is also considered as a frequent cause of allergenic reactions in babies.[86–88] In order to detect ovalbumin in solid foods, ELISA was applied with a detection limit of 0.51 $ng\,mL^{-1}$. The monoclonal antibodies 17 and 15 were employed as capture and labelled antibodies, respectively, to capture ovalbumin in solid food samples, such as ovalbumin-spiked wheat flour. The wheat sample was spiked by adding serial concentrations of ovalbumin (*e.g.*, 2.5, 5, 50 and 250 $ng\,g^{-1}$) diluted in 10 mL of 0.01 M PBS (pH 7.4) to 1 g of the sample. The spiked sample solution was homogenized and centrifuged before being tested by colorimetric ELISA.[86] The results verified that the monoclonal antibodies 17 and 15 were able to specifically detect trace levels of ovalbumin in food samples and this ELISA was suited to colorimetric detection for use in food safety analysis.[86]

Unlike most food allergies, fish allergies appear in adulthood and last for an individual's whole life. In order to manage this allergy, individuals are prevented from eating fish products.[89] Therefore, specific and sensitive detection methods of trace levels of fish allergens in food products are crucial. A study was developed to assess the colorimetric ELISA for the detection of trace levels of cooked and raw fish in food products by employing polyclonal antibodies against a 36-kDa thermal-stable fish muscle protein.[90] The sensitivity of the assays was high with a detection limit of 0.1 ppm for both raw and cooked fish. In addition, neither false-positive nor false-negative results were identified.[90]

Despite the high specificity and relatively simple designs of ELISA, this assay requires lengthy incubation and washing steps, which may limit its application array. In addition, directly detecting pathogens, parasites and allergens in food matrices is difficult as the detection limits range from 10^3–10^5 $CFU\,mL^{-1}$ for the whole bacterial cells and a few $ng\,mL^{-1}$ for toxins/proteins. Incorporation of magnetic nanoparticles is a classical modification method towards the traditional ELISA, which can be used to improve the performance of assays.[73,91] The separation and enrichment of magnetic nanoparticles can be easily achieved *via* the application of a magnetic field.

A magnetic particle-based ELISA was utilized to detect *Salmonella* in processed and non-processed food products, such as raw chicken neck skin. In this method, *Salmonella* cells were isolated by magnetic anti-*Salmonella* particles before adding the enzyme-labelled antibody to the analyte sample and achieving colorimetric results. The detection limit of this assay was 10^5 $CFU\,mL^{-1}$ of the sample solution.[92]

5.4.3 Colorimetric Detection of Food Contaminants on Paper as a Low-cost Substrate

In addition to magnetic nanoparticles, microfluidic devices have been developed to reduce the amount of required reagents and processing times of colorimetric detection, such as ELISA. Paper as a low-cost and disposable matrix has the capability to be used as immunological assay (*e.g.*, ELISA) substrate,[93,94] especially for food safety analysis that requires an inexpensive test substrate.[95–97] Paper-based microfluidic detection or "lab-on-paper" is an innovative technology, which was first introduced by Whiteside's group in 2007, and is still in its infancy stage.[98] They proposed integrating the function of micro total analytical systems onto a paper-based platform to minimize the need for reagents and analyte sample.[99,100] Paper-based microfluidic devices have exceptional promise for rapid colorimetric detection of food contamination that can be directly used by consumers and food manufacturers.[101]

Paper-based immunological assay was developed in a stacking manner, to detect *Escherichia coli* in order to evaluate the quality of drinking water and liquids. As shown in Figure 5.6, two different layers of paper were functionalized using HRP-labelled anti-*E. coli* antibodies as a signal antibody, and *E. coli* as a target bacterium. These layers were stacked using different layers of plain paper as separation layers to make a device for the detection of *E. coli*. In this method, samples contaminated with *E. coli* were conjugated to their specific HRP-labelled antibodies and they passed the layers from the top to the bottom layer of the device to generate the colorimetric signals.

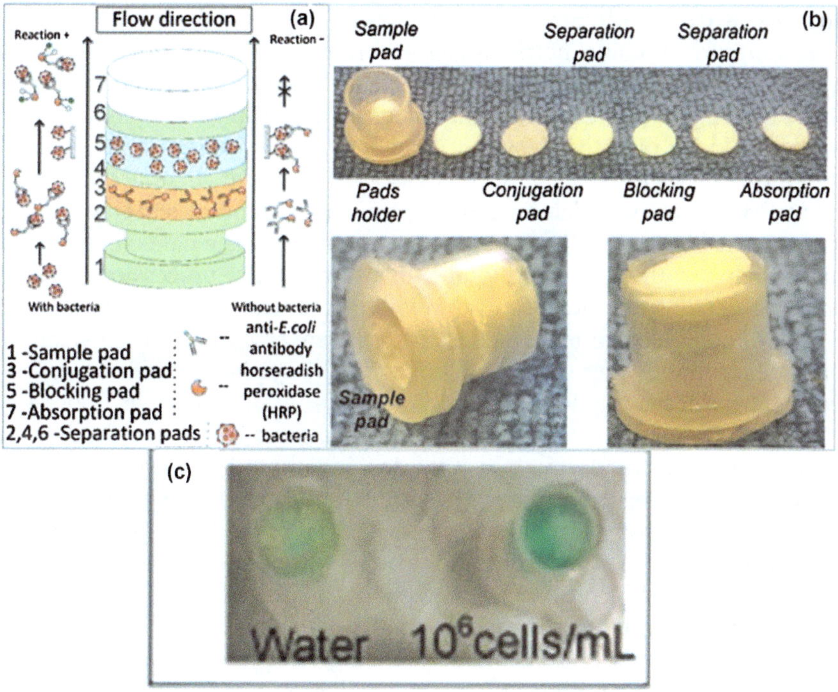

Figure 5.6 Schematic illustration of paper-based immunoassay for colorimetric detection of pathogenic bacteria in drinking water (A). The device was fabricated using different layers of papers functionalized with anti-*E. Coli* antibody-HRP and *E. Coli* before performing the test (B). Proof of concept of the optimized *E. Coli* detection (C).
Reprinted from *Biosensors and Bioelectronics*, Volume 87, E. Eltzov and R. S. Marks, Colorimetric stack pad immunoassay for bacterial identification, 572–578. Copyright 2017, with permission from Elsevier.

In samples without *E. coli* contamination, previously stored HRP-labelled antibodies were captured by *E. coli*, which had a functionalized layer and could not migrate from the upper layers to the bottom layer. The detection limit of this method was reported to be 1000-fold higher than performing ELISA in a 96-well plate with 10^2 CFU mL^{-1}.[96]

5.5 Summary and Perspective

Colorimetric sensing has proven to be useful as a simple, rapid, specific and low-cost research tool for the study of biological contaminants, chemical hazards and allergens in food products. This chapter provided an overview of colorimetric sensing for food safety and quality analysis. The working principle of colorimetric sensing is on the basis of the development of a visible colour change upon the reaction with target analytes in agri-food products. Besides the colour reagents, gold nanoparticles could also be used to specifically bind to different biomolecules, DNA and aptamers, and produce a colour change upon a reaction. The combination of gold nanoparticles with colorimetric sensing could improve the sensitivity and detection limit of colorimetric assays. There are many colorimetric sensing protocols and kits that have been successfully commercialized, including ELISA kits for food safety and environmental applications. The result of the assays could be quantified by automated laboratory-based instruments to determine the concentration of biological samples and chemical hazards in the analyte samples, which are suitable for use in laboratories. However, the colorimetric signals could also be either qualified by the naked eye or semi-quantified with the aid of a digital camera and image-processing software, which is ideally suited for use in developing countries and in field analysis.

Even though colorimetric sensing has been researched intensively for many years, there exist some challenges in the application of colorimetric methods for food safety and quality analysis. As an example, most food samples (*e.g.*, raw/cooked meat, fresh produce and processed foods) have a strong colour, rendering the samples incompatible with colorimetric sensing. Therefore, the separation of a colour from the food products is necessary, and this is highly dependent upon specific equipment and trained personnel. Future work should introduce easy sample preparation methods suitable for colorimetric detection. In addition, there is an absence of food testing commercially that is easy to operate and interpret and readily available to the public. Colorimetric sensing has shown great potential to fulfil this requirement, especially when it is designed to be tested on low-cost and portable testing substrates such as plastic plates, glass slides or paper.

References

1. J. C. Moore, J. Spink and M. Lipp, *J. Food Sci.*, 2012, **77**, 118–126.
2. Q. Chen, C. Zhang, J. Zhao and Q. Ouyang, *TrAC, Trends Anal. Chem.*, 2013, **52**, 261–274.

3. T. Lin, Y. Wu, Z. Li, Z. Song, L. Guo and F. Fu, *Anal. Chem.*, 2016, **88**, 11022–11027.
4. F. R. Shank and K. L. Carson, *Am. Chem. Soc.*, 1992, 492.
5. D. G. Barnes, M. Dourson and M. Dourson, *Regul. Toxicol. Pharmacol.*, 1988, **8**, 471–486.
6. K. R. Mahaffey, *Environ. Res.*, 2004, **95**, 414–428.
7. V. García-Cañas, C. Simó, M. Herrero, E. Ibáñez and A. Cifuentes, *Anal. Chem.*, 2012, **84**, 10150–10159.
8. L. Jacxsens, M. Uyttendaele and B. De Meulenaer, *Procedia Food Sci.*, 2016, **6**, 23–30.
9. N. L. W. Wilson and M. R. Worosz, *Food Policy*, 2014, **45**, 112–115.
10. W. Moon and S. K. Balasubramanian, *J. Consum. Aff.*, 2003, **37**, 317–339.
11. A. Reilly and F. Käferstein, *Aquacult. Res.*, 1997, **28**, 735–752.
12. F. Chaib and O. Lawe-Davies, http://www.who.int/mediacentre/news/releases/2015/foodborne-disease-estimates/en/, 2015, Accessed 02/20, 2017.
13. J. Trienekens and P. Zuurbier, *Int. J. Prod. Econ.*, 2008, **113**, 107–122.
14. J. D. Greig and A. Ravel, *Int. J. Food. Microbiol.*, 2009, **130**, 77–87.
15. H. L. DuPont, *Clin. Infect. Dis.*, 2007, **45**, 1353–1361.
16. S. Wang, H. Duan, W. Zhang and J. Li, *FEMS Immunol. Med. Microbiol.*, 2007, **51**, 8–13.
17. G. K. Adak, S. M. Long and S. J. Obrien, *Gut*, 2002, **51**, 832–841.
18. S. F. Altekruse, M. L. Cohen and D. L. Swerdlow, *Emerging Infect. Dis. J.*, 1997, **3**, 285–293.
19. C. R. Macaluso and R. M. McNamara, *Int. J. Gen. Med.*, 2012, **5**, 789–797.
20. B. K. Khajanchi, J. Han, K. Gokulan, S. Zhao, A. Gies and S. L. Foley, *Genome Announce.*, 2016, **4**, 01122.
21. R. K. Gast, R. Guraya, D. R. Jones and K. E. Anderson, *Poult. Sci.*, 2014, **93**, 728–733.
22. C. R. Braden, *Clin. Infect. Dis.*, 2006, **43**, 512–517.
23. T. Pusztahely, I. J. Holb and I. Pocsi, *Front. Plant Sci.*, 2015, **6**, 573.
24. M. Rohlfs, M. Albert, N. P. Keller and F. Kempken, *Biol. Lett.*, 2007, **3**, 523–525.
25. K. Ai, Y. Liu and L. Lu, *J Am. Chem. Soc.*, 2009, **131**, 9496–9497.
26. C. M. Gossner, J. Schlundt and P. B. Embarek, *Environ. Health Perspect.*, 2009, **117**, 1803–1808.
27. M. B. Marcó, S. Moineau and A. Quiberoni, *Bacteriophage*, 2012, **2**, 149–158.
28. M. B. Marco, S. Moineau and A. Quiberoni, *Bacteriophage*, 2012, **2**, 149–158.
29. M. Ferree, *J. Food Distrib. Res.*, 1973, **4**, 34–36.
30. M. Ghasemi-Varnamkhasti, S. S. Mohtasebi and M. Siadat, *J. Food Eng.*, 2010, **100**, 377–387.
31. W. D. Huang, S. Deb, Y. S. Seo, S. Rao, M. Chiao and J. C. Chiao, *IEEE Sens. J.*, 2012, **12**, 487–495.
32. K. G. Grunert, *Eur. Rev. Agric. Econ.*, 2005, **32**, 369–391.

33. P. J. Molnár, *Food Qual. Stand.*, 2009, **2**, 398–409.
34. H. S. Kwak, H. Kim and H. S. Kim, *Prev. Nutr. Food Sci.*, 2013, **18**, 67–75.
35. A. M. Paula and A. C. Conti-Silva, *J. Food Eng.*, 2014, **121**, 9–14.
36. J. L. McKillip and M. Drake, *J. Food Prot.*, 2004, **67**, 823–832.
37. P. M. Scheu, K. Berghof and U. Stahl, *Food Microbiol.*, 1998, **15**, 13–31.
38. E. De Boer and R. R. Beumer, *Int. J. Food Microbiol.*, 1999, **50**, 119–130.
39. B. Malorny, P. T. Tassios, P. Rådström, N. Cook, M. Wagner and J. Hoorfar, *Int. J. Food Microbiol.*, 2003, **83**, 39–48.
40. S. Choi, S. Kim, G. Lee and H. Park, *Sens. Actuators, B*, 2015, **219**, 245–250.
41. A. Nilghaz, D. H. B. Wicaksono, D. Gustiono, F. Abdul Majid, E. Supriyanto and M. Abdul Kadir, *Lab Chip*, 2012, **12**, 209–218.
42. A. Nilghaz, D. R. Ballerini, X. Fang and W. Shen, *Sens. Actuators, B*, 2014, **191**, 586–594.
43. A. Nilghaz and W. Shen, *RSC Adv.*, 2015, **5**, 53172–53179.
44. B. S. Inbaraj and B. H. Chen, *J. Food Drug Anal.*, 2016, **24**, 15–28.
45. G. Bulbul, A. Hayat and S. Andreescu, *Sensors*, 2015, **15**, 30736–30758.
46. M. S. Verma, J. L. Rogowski, L. Jones and F. X. Gu, *Biotechnol. Adv.*, 2015, **33**, 666–680.
47. X. Huang and M. A. El-Sayed, *J. Adv. Res.*, 2010, **1**, 13–28.
48. Z. Fu, X. Zhou and D. Xing, *Sens. Actuators, B*, 2013, **182**, 633–641.
49. H. Li and L. J. Rothberg, *J. Am. Chem. Soc.*, 2004, **126**, 10958–10961.
50. N. A. Mungroo and S. Neethirajan, *Biosensors*, 2014, **4**, 472–493.
51. S. Wang, A. Singh, D. Senapati, A. Neely, H. Yu and P. Ray, *Chem. – A Eur. J.*, 2010, **16**, 5600–5606.
52. N. W. Turner, S. Subrahmanyam and S. A. Piletsky, *Anal. Chim. Acta*, 2009, **632**, 168–180.
53. J. Varga, S. Kocsube, Z. Peteri, C. Vagvölgyi and B. Toth, *Toxins*, 2010, **2**, 1718–1750.
54. M. Hosseini, H. Khabbaz, M. Dadmehr, M. R. Ganjali and J. Mohamadnejad, *Acta Chim. Slov.*, 2015, **62**, 721–728.
55. Q. Zhou, N. Liu, Z. Qie, Y. Wang, B. Ning and Z. Gao, *J. Agric. Food Chem.*, 2011, **59**, 12006–12011.
56. F. Xu, K. Ren and Y. Yang, *J. Integr. Agric.*, 2015, **14**, 2282–2295.
57. K. Rovina, S. Siddiquee and N. K. Wong, *Sens. Biosens. Res.*, 2015, **4**, 16–22.
58. L. Li, B. Li, D. Cheng and L. Mao, *Food Chem.*, 2010, **122**, 895–900.
59. N. Kumar, R. Seth and H. Kumar, *Anal. Biochem.*, 2014, **456**, 43–49.
60. A. R. Nisha, *Vet. World*, 2008, **1**, 375–377.
61. S. K. Sonawane, S. S. Arya, J. G. LeBlanc and N. Jha, *Eur. J. Nutr. Food Saf.*, 2014, **4**, 2347–5641.
62. K. Song, E. Jeong, W. Jeon, M. Cho and C. Ban, *Anal. Bioanal. Chem.*, 2012, **402**, 2153–2161.
63. L. He, Y. Luo, W. Zhi and P. Zhou, *Food Anal. Methods*, 2013, **6**, 1704–1711.
64. M. H. S. Santos, *Int. J. Food Microbiol.*, 1996, **29**, 213–231.

65. A. Rasooly and K. E. Herold, *Foodborne Pathog. Dis.*, 2008, **5**, 531–550.
66. W. Chen, X. Fang, H. Li, H. Cao and J. Kong, *Sci. Rep.*, 2016, **6**, 31948.
67. M. E. Koivunen and R. L. Krogsrud, *Lab. Med.*, 2015, **37**, 490–497.
68. U. Samarajeewa, C. I. Wei, T. S. Huang and M. R. Marshall, *Crit. Rev. Food Sci. Nutr.*, 1991, **29**, 403–434.
69. N. S. Lipman, J. L. Fau, T. L. Fau and F. Weis-Garcia, *ILAR J.*, 2005, **46**, 258–268.
70. J. Descotes, *mAbs*, 2009, **1**, 104–111.
71. B. Mirhabibollahi, J. L. Brooks and R. G. Kroll, *Appl. Microbiol. Biotechnol.*, 1990, **34**, 242–247.
72. R. E. Poms, C. L. Klein and E. Anklam, *Food Addit. Contam.*, 2004, **21**, 1–31.
73. K. S. Cudjoe, T. Hagtvedt and R. Dainty, *Int. J. Food Microbiol.*, 1995, **27**, 11–25.
74. L. S. Van Poucke, *Appl. Environ. Microbiol.*, 1990, **56**, 924–927.
75. B. M. Allos, M. M. Fau, G. P. Fau and R. V. Tauxe, *Clin. Infect. Dis.*, 2004, **38**, 115–120.
76. Y. Nawa, C. F. Hatz and J. Blum, *Clin. Infect. Dis.*, 2005, **41**, 1297–1303.
77. T. Roberts, K. D. Murrell and S. Marks, *Parasitol. Today*, 1994, **10**, 419–423.
78. S. Farjallah, B. B. Slimane and M. Busi, *Parasitol Res.*, 2007, **102**, 371–379.
79. M. T. Werner, C. K. Faste, A. Levsen and E. A. Egaas, *Eur. Food Res. Technol.*, 2011, **232**, 157–166.
80. S. H. Sicherer, H. A. Sampson and S. M. Chan, *Expert Rev. Mol. Med.*, 2007, **120**, 491–503.
81. S. M. Chan, C. F. Dumitru and V. Turcanu, *Expert Rev. Mol. Med.*, 2012, **12**, 879–891.
82. M. P. de Leon, F. A. U. Rj and R. E. O'Hehir, *Expert Rev. Mol. Med.*, 2007, **9**, 1–18.
83. S. Al-Muhsen, A. E. Clarke and R. S. Kagan, *Can. Med. Assoc. J.*, 2003, **168**, 1279–1285.
84. M. Careri, L. Elviri, M. Maffini, A. Mangia, C. Mucchino and M. Terenghi, *Rapid Commun. Mass Spectrom.*, 2008, **22**, 807–811.
85. J. Peng, S. Song and L. Xu, *Int. J. Environ. Res. Public Health*, 2013, **10**, 2897–2905.
86. J. Peng, X. Meng, X. Deng, J. Zhu, H. Kuang and C. Xu, *Food Agric. Immunol.*, 2014, **25**, 1–8.
87. S. Lau, M. Thiemeier, R. Urbanek, M. Kemeny and U. Wahn, *Eur. J. Pediatr.*, 1988, **147**, 606–608.
88. J. Caubet and J. Wang, *Pediatr. Clin. North Am.*, 2011, **58**, 427–443.
89. H. Zhang, Y. Lu, H. Ushio and K. Shiomi, *Food Chem.*, 2014, **150**, 151–157.
90. Y. Chen and Y. P. Hsieh, *Food Control.*, 2014, **40**, 265–273.
91. O. Olsvik, T. F. Popovic, E. F. A. U. Skjerve and K. S. Cudjoe, *Clin. Microbiol. Rev.*, 1994, **7**, 43–54.

92. K. S. Cudjoe, T. Hagtvedt and R. Dainty, *Int. J. Food Microbiol.*, 1995, **27**, 11–25.
93. C. Cheng, A. Martinez and J. Gong, *Angew. Chem., Int, Ed.*, 2010, **49**, 4771–4774.
94. K. Zhu, R. Dietrich, A. Didier, D. Doyscher and E. Märtlbauer, *Toxins*, 2014, **6**, 1325–1348.
95. J. C. Jokerst, J. A. Adkins, B. Bisha, M. M. Mentele, L. D. Goodridge and C. S. Henry, *Anal. Chem.*, 2012, **6**, 2900–2907.
96. E. Eltzov and R. S. Marks, *Biosens. Bioelectron.*, 2017, **87**, 572–578.
97. C. Shih, C. Chang and M. Hsu, *Talanta*, 2015, **145**, 2–5.
98. A. W. Martinez, S. T. Phillips, M. J. Butte and J. M. Whitesides, *Angew. Chem.*, 2007, **46**, 1318–1320.
99. D. M. Cate, J. L. Adkins, J. Mettakoonpitak and C. S. Henry, *Anal. Chem.*, 2015, **87**, 19–41.
100. A. Nilghaz, L. Guan, W. Tan and W. Shen, *ACS Sens.*, 2016, **1**, 1382–1393.
101. L. S. A. Busa, S. Mohammadi, M. Maeki, A. Ishida, H. Tani and M. Tokeshi, *Micromachines*, 2016, **7**, 86.

CHAPTER 6

ELISA-based Sensing in Food Safety and Quality Analysis

YANG LU, WEI SHENG, BING LIU AND SHUO WANG*

Key Laboratory of Food Nutrition and Safety, Ministry of Education, School of Food Engineering and Biotechnology, Tianjin University of Science & Technology, Tianjin 300457, China
*Email: s.wang@tust.edu.cn

6.1 Introduction

An antibody is a protein molecule capable of binding with high specificity to an antigen. These antigens can be microbes, macromolecules, small molecules or even heavy atoms. For the exclusive properties of remarkable recognition breadth and high specificity, antibodies have been taken advantage of for the development of immunoassays to identify certain targets.

Enzyme-linked immunosorbent assay (ELISA) is an overwhelmingly influential immunoassay used in various areas. Since the 1970s, it has been used as a diagnostic tool in medicine and plant pathology, as well as a quality-control check in different industries.[1,2] In this chapter, we focus on the applications of ELISA in the food industry. Compared to instrumental techniques, ELISA offers a number of advantages, including but not limited to high sensitivity and specificity, small sample volume requirement, low costs, high-throughput sample processing and fewer sample-preparation steps.[3]

ELISA is based upon the principles of antibody–antigen interaction.[3] Normally, the target analyte (*i.e.*, the antigen) is used to immunize animals to generate specific antibodies. The general principle of ELISA is

Food Chemistry, Function and Analysis No. 2
Sensing Techniques for Food Safety and Quality Control
Edited by Xiaonan Lu

Published by the Royal Society of Chemistry, www.rsc.org

Figure 6.1 The general principle of ELISA. Enzyme-linked antibodies specifically bind to analytes, and thus the catalytic product of the enzyme generates a detectable signal in an analyte dose-dependent manner. Assay formats may vary depending on different purposes.

demonstrated in Figure 6.1. In simple terms, an unknown amount of antigen in the target sample is affixed to a surface, and then the specific antibody is applied over the surface so that it can bind to the antigen. This antibody is also covalently conjugated to an enzyme, which converts a certain substrate to some detectable signals, most commonly a colour change or fluorescence/luminescence signal stimulation that is facilitated by the catalysis of the linked enzyme in a dose-dependent manner. Assay formats may vary depending upon different purposes.

Identification of hazards (*e.g.*, allergens, microbes, illegal pesticides, veterinary drugs, *etc.*) is exceptionally difficult for foodstuffs. This is mainly because: (1) analyte extraction for most food samples is particularly difficult owing to the complexity of foods; (2) there is a high requirement of assay sensitivity for food hazards; (3) there is a requirement for rapid, low-cost detection. Due to the advantages of high affinity, specificity and variety of antibodies, together with the high catalytic efficiency of enzymes, ELISA is cheap and rapid, and thus applied for the detection of various samples, including complicated matrices such as agri-foods.

Despite all the advantages of ELISA applied to food safety studies, a major bottleneck in the development of ELISA is its sensitivity to certain targets, especially for small molecular analytes. For instance, glyphosate is one of the most widely used herbicides around the world.[4] A sensitive on-site detection technique is crucial. However, the sensitivity of glyphosate ELISA has long been limited by the low-affinity antibodies, owing to its low molecular-size and lack of immunogenicity. To overcome this issue, it has been widely considered as a vital issue from academia to industry to improve the sensitivity of ELISA, including efforts in rationalizing hapten design, improving tracer signal strength, *etc.*

In this chapter, we describe a whole range of steps in the development and research frontiers of ELISA, including hapten design (Section 6.2), antibody generation (Section 6.3), tracer labelling (Section 6.4), and development of sample preparation (Section 6.5), as well as different assay formats

(Section 6.6). In particular, a semi-quantitative on-site rapid assay derived from ELISA, the technique of immunochromatography, is introduced in Section 6.7. Examples of the applications of ELISA on the detection of food contaminants (*e.g.*, detection of pesticides, antibiotics and microbes, *etc.* in foods) are described in Section 6.8.

6.2 Principle and Practice of Hapten Design

Immunoassay is a method for trace analysis based upon specific and reversible recognition between the antigen and the antibody, which can specifically recognize a massive range of targets, from heavy atoms to microbes. Hence, ELISA is not only applicable to the detection of macromolecules (*e.g.*, proteins, nucleic acids and bacteria), but also for the determination of small molecules (with a molecular mass of less than 1 kDa, *e.g.*, pesticides, veterinary drugs, food additives, natural toxicants, hormones, *etc.*). The major concern of generating anti-small molecule antibodies is that most small molecular analytes lack immunogenicity to stimulate animals to produce specific antibodies.

To overcome this issue, haptens (*i.e.*, "artificial antigens") that mimic the structures of analytes are designed to conjugate to carrier proteins, which can facilitate the immune response of immunized animals. It can make use of T-cell epitopes to indirectly induce the proliferation and differentiation of B cells to produce the specific antibodies. General methods for covalently coupling functional groups of haptens and carrier proteins are summarized in Table 6.1. The hapten is the dominant epitope that can specially bind with antibodies. A good hapten should maximally preserve most of the chemical structure, electronic distribution and stereochemical conformation of the targeted molecule(s). Further, the special chemical structure of the target can be fully exposed to the greatest extent. However, the introduced conjugation sites of haptens may influence the chemical conservation of the antigen, and thereby decrease the specificity/affinity of the generated antibodies.[5] A poorly designed hapten may also be buried into its carrier protein, which decrease its exposure and thereby affect its specific immunogenicity. In contrast, the reasonable design of hapten can be greatly beneficial for producing high quality of antibodies. Hence, efforts have been made to rationalize hapten design to preserve the chemical conservation, whilst introducing conjugatable functional groups for immunogenicity.[6]

Using the small molecular target without modification for direct conjugation with carrier proteins is usually an easy approach for hapten design if the target itself has one of the functional groups (including carboxyl, hydroxyl, amino or carbonyl group). In this case, the carrier protein can be conjugated directly with the target by a suitable method as aforementioned. If there are two or more than two functional groups on the target, the other functional groups need to be protected by chemical methods. The only one remaining is used to conjugate with the carrier protein. The protective group needs to be removed after conjugation by chemical methods.

Table 6.1 Common coupling methods of hapten derivatives coupled with the carrier protein.

Coupling methods	Active groups	Reaction principle
Carbodiimide method	COOH	$\text{R-COOH} + \text{NH}_2\text{-protein} \xrightarrow{\text{R-N=C=N-R'}} \text{R-CO-NH-protein} + \text{R-NH-CO-NH-R'}$
Active ester method	COOH	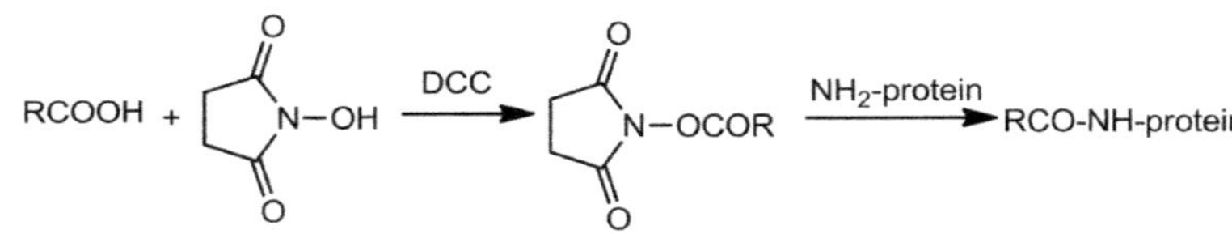
Mixed anhydride method	COOH	$\text{RCOOH} + \text{ClCOOCH}_2\text{CH(CH}_3)_2 \xrightarrow{(\text{C}_4\text{H}_9)_3\text{N}} \text{RCOOCOOCH}_2\text{CH(CH}_3)_2 \xrightarrow{\text{HO-protein}} \text{RCOO-protein}$
Azobenzoic acid method	COOH	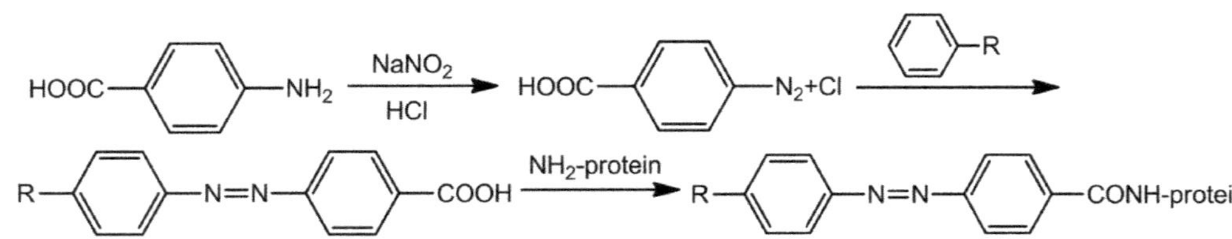
Glutaraldehyde method	NH_2	$\text{R-NH}_2 + \text{NH}_2\text{-protein} \xrightarrow{\text{OHC-(CH}_2)_3\text{-CHO}} \text{R-N=CH(CH}_2)_3\text{CH=N-protein}$
Diazotization method	NH_2	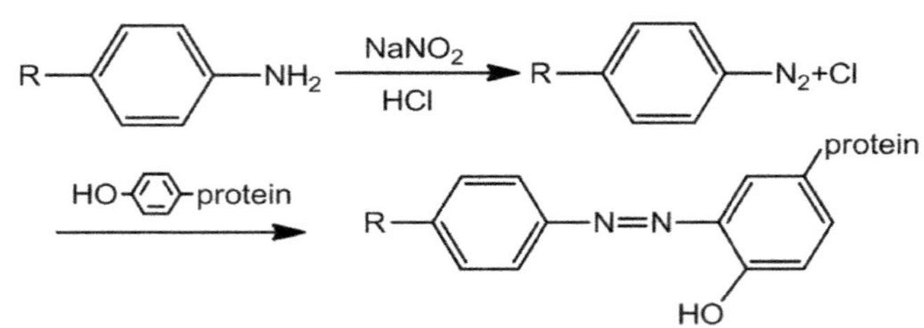

Succinic anhydride method	OH	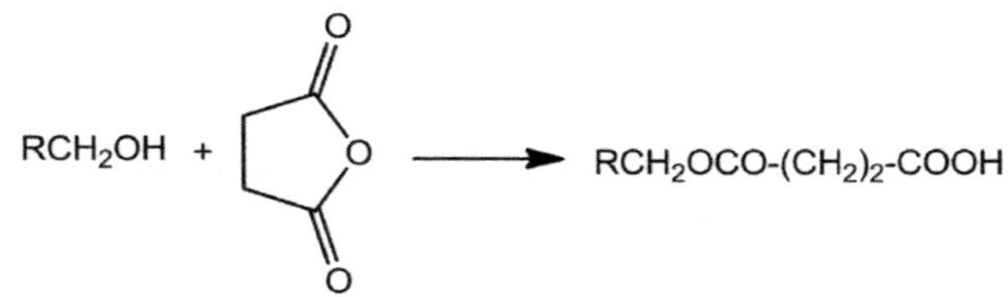
Chlorine acetate sodium	OH	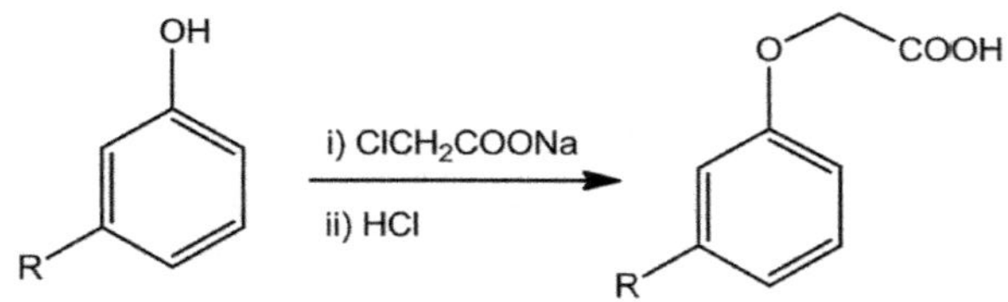
Cyanogen chloride method	OH	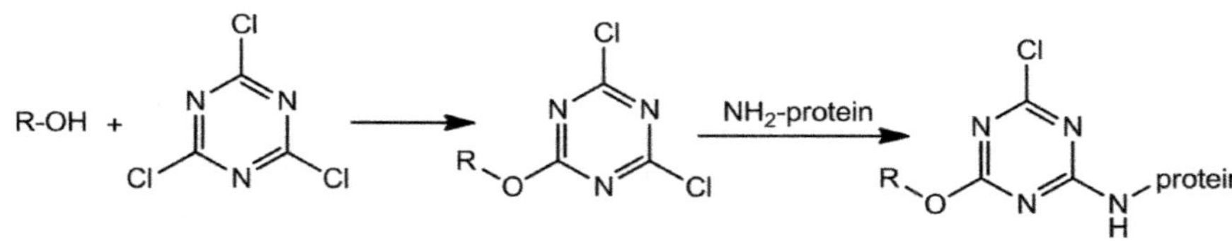
N,N'-Carbonyldiimidazole method	OH	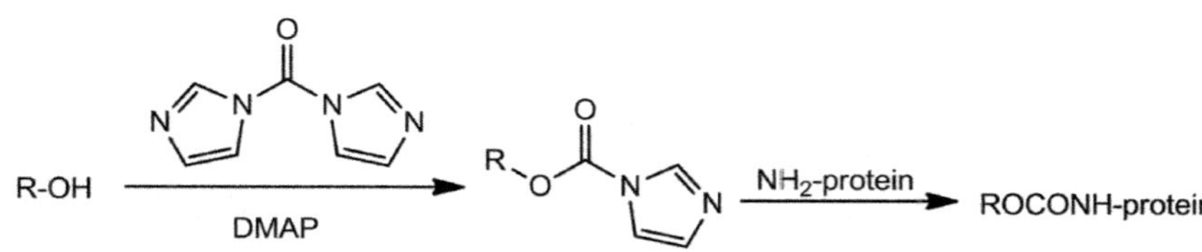
Azide method	$COOCH_3$	$RCOOCH_3 \xrightarrow{NH_2NH_2} RCONHNH_2 \longrightarrow RCON_3 \xrightarrow{NH_2\text{-protein}} RCONH\text{-protein}$

Table 6.1 (*Continued*)

Coupling methods	Active groups	Reaction principle
Carboxymethyl hydroxylamine	C=O	$R\text{-}CHO + H_2N\text{-}O\text{-}CH_2COOH \xrightarrow{-H_2O} R\text{-}CH{=}N\text{-}O\text{-}CH_2COOH$
O-Carboxy-methylhydroxylamine method	C=O	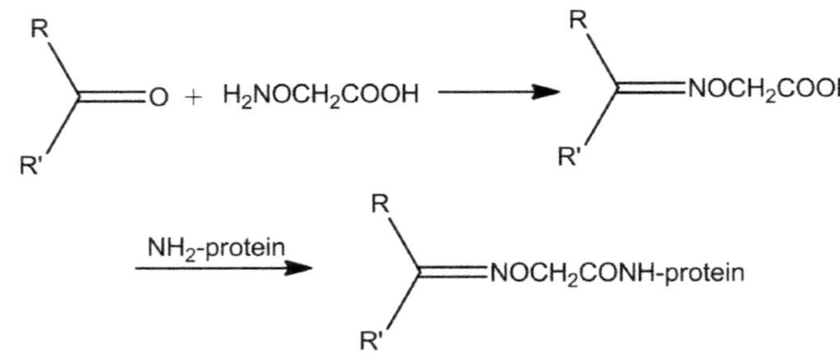
4-Hydrazin-obenzoic acid method	C=O	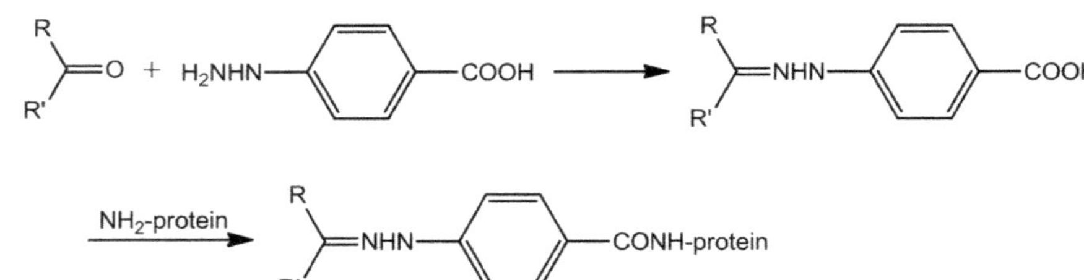

In many cases, the target has no functional group or the functional group is the characteristic portion of the molecule that is required to generate the antibody. Under these circumstances, structural analogues (which mimic the target molecule with functional groups introduced) of the target molecules are designed and used as haptens. Sometimes the molecular size of a hapten is too small, and thus it is likely to be buried into the carrier protein. In this case, an extra "spacer arm" (3–5 carbon atoms) that extends the length of the hapten is normally introduced to avoid embedding the hapten so that the epitope can be exposed. Although it usually requires more elaborate chemistry, derivatization through the carbon atoms of the analyte is preferred since these atoms do not contribute as much to the steric and electronic properties of the target molecule.

In addition to the chemical structures, many other factors of haptens have direct effects on generating antibodies with high specificity/affinity, including the design of the functional group, the length of the spacer arm and the molecular ratio between haptens and carrier proteins, *etc.* In principle, the haptens are supposed to be similar to the targets as much as possible in the aspects of chemical structures, electronic distributions and stereochemical properties.

Most designs of haptens and small-molecular immunoassays are based upon empirical methods due to the lack of understanding of the mechanism of antigen–antibody interaction. The introduced spacer arm, as well as the structural influence by the carrier protein, may lead to unpredictable structural changes and/or reduce the rationality of the hapten design. In particular, the development of broad spectrum ELISA (using an antibody that recognizes a range of structurally similar analytes) requires a rational structure-based hapten design to generate efficient antibodies that have both wide recognition breadth and high affinity. In recent years, computer-aided molecular modelling has been applied to aid in designing haptens, by which specific and broadly recognizing antibodies can be readily generated. For instance, in 2011, Yuan *et al.* studied the quantitative structure-activity relationship (QSAR) between a range of structurally similar phenylurea herbicides and their broad monoclonal antibody, providing insights into rationalizing hapten design.[7] In 2012, an improved computational methodology, comparative molecular similarity index analysis (COMSIA), was applied to investigate the antigen–antibody interactions of triazine herbicides.[8] Molecular models were also studied to improve hapten design for quinolone antibiotics,[9–13] organophosphorus pesticides[14,15] and sulfonamides.[16] Further, crystal structures of anti-ofloxacin provided more precise data for understanding the antigen–antibody interaction.[17]

6.3 Antibodies

The antibody, also known as immunoglobulin (Ig), is a class of Y-shaped glycoprotein produced by plasma cells, which are proliferated and differentiated by B cells after being subjected to antigenic stimulation. Antibodies circulate around the blood and lymph and can specifically bind to a large range of targets (antigens) to identify and neutralize foreign molecules, from microbes and macromolecules to small molecules and even heavy atoms.

Antibodies consist of heavy chains and light chains connected with disulfide bonds. Heavy chains are classified as α, γ, δ, ε and μ, and light chains are classified as κ and λ. According to the different types of heavy chains, antibody isotypes are classified as IgA, IgG, IgD, IgE and IgM, respectively. Along with the profound development of clinical and molecular biological studies of antibodies, various novel antibodies with different properties and applications have been developed. In addition to the classic types of antibodies (full-length polyclonal and monoclonal antibodies, shown in Figure 6.2), the fragment antigen-binding (Fab) fragment, single-chain variable fragment (scFv) retain native binding properties and are easily produced from recombinant DNA. Therefore, unlike full-length antibodies that are produced natively, recombinant antibodies can be easily manipulated and engineered with molecular biological techniques. For instance, phage display and yeast display have been extensively used for the screening and optimization of recombinant antibodies *in vitro*. Single-domain antibodies (mostly "nanobodies") have naturally extended CDR loops that improve their specificities and affinities to macromolecular antigens with deep cavities.[18] Bispecific and multi-specific antibodies are engineered to facilitate the multiple target binding property of a single antibody.[19] These novel techniques of antibody engineering have enormously improved the

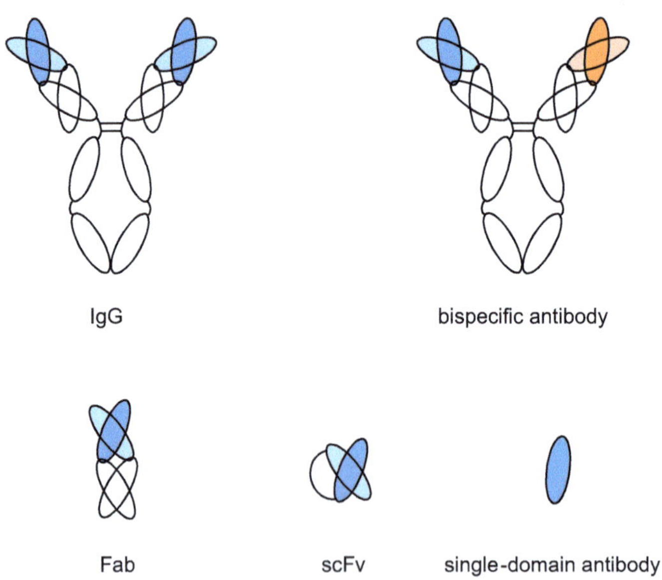

Figure 6.2 A panel of different types of antibodies. Blue and orange ellipses show variable domains with different targets. Ellipses with darker colours represent V_H domains, whilst light colours show V_L domains. White ellipses show constant domains.

applicability and selectivity of ELISA, and widened the future development of antibody-based assays.

6.3.1 Polyclonal Antibodies

Polyclonal antibodies are extracted from serum with various antibodies that are secreted by different B cell lineages within the body by immunizing animals. This includes some or all types of isotypes of immunoglobulin, which can recognize multifarious antigenic determinants. Polyclonal antibody is usually used in immunological detection of multivalent antigen and emergency prevention because of its important functions of neutralizing antigen, immune regulation and so on. Polyclonal antibody has the advantages of stability, high-yield and being relatively easy to generate. For these reasons, it has become a common means for the treatment of various diseases, as well as a powerful tool for diagnosis and immunoassays.

6.3.1.1 Preparation of Hapten and Immunogen

Among various target compounds, some small molecules cannot elicit immune response by themselves and some substances are either not stable or are easily metabolized *in vivo*. Due to these factors, appropriate hapten and immunogen derived from the target compound are necessary for immunization and production of antibody. Soluble immunogens are mixed with Freund's adjuvant (volume ratio of 1 : 1) and emulsified before the immunization.

6.3.1.2 Selection of Experimental Animals

Selecting the appropriate experimental animals is essential for the preparation of good quality antiserum. The following problems should be noticed as well. There should be a big taxonomical difference between the species of the antigen and the immunized animal. Besides, the selection of immunized animals is also made according to the requirement of antibody. Mammals, birds and amphibians can be selected to produce polyclonal antibodies. In the research facility, mice, rabbits and sheep are the most common experimental animals. For example, a female healthy New Zealand white rabbit with an age of three months and a weight of 1.5 kg is suitable for animal immunization. Large mammals, such as horses and goats, can be used to produce relatively high yields of antibodies. For the purpose of generating antibodies with long CDR loops, alpacas,[18] sharks[20] and bovines[21] are selected for immunization.

6.3.1.3 Immunization Procedure

In the immunization procedure, injecting the appropriate antigen has a critical effect on the immune efficacy. It should be accurately calculated before the initial immunization, according to the antigenic strength, molecular weight, animal size, time interval and some other factors. The dosage of immunogen is about 0.5–1 mg per injection for each large animal and 0.1–0.6 mg per injection for each small animal. Intraperitoneal injection, intra-muscular injection, intra-dermal injection and subcutaneous injection are the most commonly used methods for animal immunization. Normally, approximately 10 to 20 days of interval times between the first and second immunization can avoid immunosuppression. In addition, 7 to 10 days of interval time before booster immunization is a commonly used method to enhance the immune stimulation that increases the serum titre.

After the third injection, a small amount of blood from the marginal ear vein is collected a week after the injection. Subsequently, ELISA is used for the characterization of antiserum. When the antiserum titre and selectivity meet the requirements, blood is collected with carotid puncture of the animal approximately 5–7 days later after the boost immunization.

6.3.1.4 Determination of Titre and Affinity of Antiserum

Titre, binding capacity and selectivity of the antiserum are the most important indexes for evaluating the performance of antibodies. ELISA is commonly used in the determination and characterization of antiserum. It is noteworthy that the concentration, pH, ion strength and incubation time of antigen and antibody solutions should be optimized according to the characteristics of different experiments.

6.3.1.5 Purification of Polyclonal Antibodies

For polyclonal antibodies, miscellaneous substances exist in the antiserum, which may interfere with the antigen–antibody interaction. Therefore, the purification of antibody is necessary for obtaining specific antibodies for further experimental studies. Because of its ability to bind to the Fc region (and the Fab region for certain antibodies, too) of immunoglobulins from mammalian species and its good stability during the repeated cycles, Protein A and Protein G immunoaffinity chromatography is usually used to purify polyclonal antibodies. Since the elution buffer for Protein A/G purification normally has a low pH (pH 2.5–3.0), a subsequent step is applied to dialyze the purified antibodies to neutral pH.

6.3.1.6 Determination of Antibody Concentration

A UV spectrometer is commonly used for the measurement of the concentration of antibody based upon the Beer–Lambert law. A small fraction of antibody is usually diluted by 20 times into the dynamic range of a spectrometer. The dialysis buffer is used as the blank. An average extinction coefficient of antibodies (1.35) is used for the concentration calculation.

$$C_{\text{antibody}}(\text{mg mL}^{-1}) = \frac{A_1 - A_0}{1.35 \times L} \times 20$$

A_1: Absorbance value of antibody at 280 nm;
A_0: Absorbance value of blank control at 280 nm;
L: The pathlength of the cuvet in cm.

6.3.1.7 Storage of Polyclonal Antibody

Under normal conditions, antibodies are easy to preserve after the addition of a stabilizer, such as glycerol. To prevent the contamination of antibodies by bacteria or fungi, antimicrobial agents, such as 0.02% NaN_3 or 0.01% merthiolate, are usually added into the antibody solution. Under this condition, some antibodies can be stored for three to six months at 4 °C and stored at −20 °C to −40 °C for over five years. If an antibody is lyophilized into powder with less than 0.2% water, it can be stored for five to ten years without a significant titre-drop.

6.3.2 Monoclonal Antibodies

Unlike polyclonal antibodies, which are a collection of antibodies from different B cells recognizing multiple epitopes on the same antigen, a monoclonal antibody recognizes only one epitope. Monoclonal antibody is prepared with the cell hybridization technique. Monoclonal antibodies are produced by hybridoma cells, which are developed by fusing myeloma cells with mouse spleen cells immunized with the desired antigen. Hybridoma takes advantage of both types of cells, which facilitate both the abilities of stable replication and antibody production. Different from the polyclonal antibody, an isolated clone of hybridoma produces only one type of antibody, namely "monoclonal" antibody. Because of its characteristics of high purity and specificity, serological cross-reaction can be avoided. Another advantage of monoclonal antibodies is their high reproducibility, due to the natural "batch-to-batch homogeneity" of hybridoma.

6.3.2.1 Preparation of Hapten and Immunogen

Similar to the requirement for polyclonal antibodies, appropriate hapten and immunogen derived from the target compound are necessary for immunization and production of antibody. It is critical that the purity of the antigen, especially the antigen for the first immunization, should be as high as possible to improve the specificity of generated antibodies. The soluble immunogen is mixed with Freund's adjuvant (volume ratio of 1 : 1) and emulsified before immunization. If a cell immunogen is used, approximately 1.0×10^7 cells are suitable for intra-peritoneal injection.

6.3.2.2 Selection of Experimental Animals

For the preparation of monoclonal antibodies, experimental animals should be selected according to the type of myeloma cells used in the experiment. The most commonly used experimental animals in the laboratory are Balb/c mice and LOU/c rats. In particular, female healthy Balb/c mice with an age of 8–12 weeks are suitable for immunization. Similar to polyclonal antibodies, intra-peritoneal injection, intra-muscular injection, intra-dermal injection and subcutaneous injection are the most commonly used methods for animal immunization. The dosage of immunogen is about 0.1–0.6 mg per mouse. Normally, approximately two to three weeks of interval time between each immunization can enhance the immune stimulation and increase serum titre and antibody binding affinity. When the antiserum titre reaches the requirement, fresh spleen cells of the immunized mouse and pre-cultured myeloma cells are fused on the third day after the last booster injection.

6.3.2.3 Production of Hybridoma

Cell fusion is the core technique of producing hybridoma. Basically, spleen cells of the immunized mouse and myeloma cells are fused under a fusion condition. Chemical fusion with polyethylene glycol (PEG) and electrofusion are the most commonly used fusion methods. A brief introduction of producing monoclonal antibodies is described below and summarized in Figure 6.3.

Myeloma cells: Myeloma cells are cultured in Dulbecco's Modified Eagle Medium (DMEM) with foetal bovine serum supplemented. It is then passaged during the logarithmic growth phase of cells to ensure the stability of cell characteristics. The cells are centrifuged and washed for further experiment.

Spleen cells: The spleen of the immunized mouse is used for the B-cell resource under sterile condition. The spleen is ground in DMEM culture medium at a stainless-steel mesh and suspended. Erythrocyte is then lysed with erythrocyte lysis buffer. Afterwards, leukocytes are extracted by centrifugal separation, followed by wash steps with DMEM culture medium.

Cell fusion: Spleen and myeloma cells from the mice are fused at an appropriate ratio (1 : 2 to 1 : 10) in PEG. The reaction is terminated using DMEM without serum. The fused cells are washed with DMEM containing hypoxanthine-aminopterin-thymidine medium and serum. The fused cells are added into 96-well plates for a preliminary screening. The cells are cultured in 5% CO_2 at 37 °C.

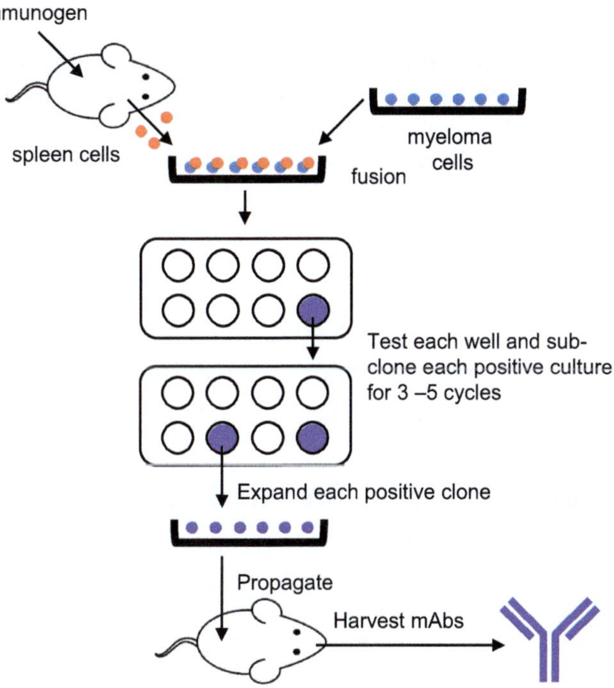

Figure 6.3 The process of producing monoclonal antibodies.

Screening of positive clones of hybridoma: Fused cells are allowed to grow in 96-well plates. ELISA is used to screen positive clones of hybridoma, so that the wells containing positive hybridoma can secrete antibodies that display higher signal. Each positive clone is further split into separate wells for subsequent culture. This "limited dilution" process is carried out a few times (usually three to five times) until positive, single and stable hybridoma strains are separated.

Cryopreservation: It is very important to cryopreserve the primitive positive hybridoma and sub-cloned hybridoma in a timely manner. Hybridoma cells are normally cryopreserved in liquid nitrogen. Dimethyl sulfoxide (DMSO) is a commonly used cryoprotectant.

6.3.2.4 *Amplification of Hybridoma*

Inoculation *in vivo*: Ascites obtained by inoculation *in vivo* can have a high level of amplified monoclonal antibody and less miscellaneous protein. A total of 0.5 ml of pristine or liquid paraffin is injected into a mouse peritoneal cavity one week prior to the hybridoma injection. Positive clones of hybridoma cells are re-suspended in sterile phosphate buffered saline to a final concentration of 2×10^6 cells ml^{-1} and injected with 0.5 ml suspension into the peritoneal cavity of each mouse. After 7–10 days, ascites fluids are collected, centrifuged and stored at -20 °C until use.

6.3.2.5 Purification of Monoclonal Antibodies

Ammonium sulfate precipitation is used for preliminary purification of antibodies from serum, ascites fluid or cell culture supernatant. Immunoglobulins precipitate while other proteins remain in the solution at 40% to 50% ammonium sulfate saturation (100% saturation equals 4.32 M).[22] Alternatively, Protein A/Protein G/Protein L immunoaffinity methods are also selected for purification according to the species, subtypes and types of light chains of the target antibody.

6.3.2.6 Determination of Antibody Concentration

The method of determining the concentration of monoclonal antibodies is the same as for polyclonal antibodies, as shown in Section 6.3.1.6.

6.4 Tracers for ELISA: Enzymes and Beyond

A single molecule of enzyme can catalyze 10^7 molecules of substrate per minute for signal enhancement. Because of the high efficiency and stability of some enzymes, they are applied as labels conjugated on antibodies to display the signal of the detected trace amount of analyte.

Horseradish peroxidase (HRP) and alkaline phosphatase (AP) are mainly used as labels for ELISA. HRP is a 44 kDa glycosylated haemoprotein that consists of a protoporphyrin IX hemin prosthetic group. It is an oxidoreductase that can be used with a wide variety of hydrogen donors to reduce the level of hydrogen peroxide. The working range of HRP is pH 4.0–8.0. It can be used to display coloured, fluorescent or luminescent signals with different substrates. AP is a 140 kDa dimeric glycoprotein with a working range of pH 9.5–10.5. It catalyzes the hydrolysis of phosphate esters of primary alcohols, phenols and amines.[23]

Efforts have been made to improve the sensitivity of the labels and to overcome the disadvantages of enzyme labels (*e.g.*, its relatively short shelf-life). In 2008, a nanoparticle label made of CSe@ZnS facilitated a detection of prostate-specific antigen as sensitive as 0.02 ng ml^{-1}.[24] In 2012, an Envision signal antibody copolymer composed of a dextran amine skeleton anchoring more than 100 molecules of HRP and 15 molecules of secondary antibodies was developed, which facilitated the detection sensitivity down to 2 pg ml^{-1} for alfa-fetoprotein.[25] In 2016, Zhan and others showed that a quantum-dot based fluorescent ELISA could improve the LOD 17-fold.[26]

6.5 Sample Preparation

Components in food matrices, such as fats, pigments, proteins and minerals can interfere with the detection of analytes and thus cause false positive results.[3] Therefore, the reduction of matrix effects is vital in sample preparation to improve the precision and accuracy. For example, milk, soya powder, biscuits and pet foods normally contain fat globules, casein micelles, solubilized lactose, whey proteins and some minerals.[27,28] During the preparation of these samples, proteins and fats are removed prior to the ELISA experiment. For this purpose, addition of trichloroacetic acid, methylene chloride, dichloromethane and/or hexane is normally applied.[27–29] Thereafter, solvent extraction is usually applied, followed by solid phase extraction for further purification.

Solvent extraction is an efficient and highly selective separation and purification method based upon the relative solubility of substances in two different immiscible liquids. The selection of mobile phase is based upon the polarity of the analyte. For instance, melamine is a small polar compound that can be hydrolyzed under both acidic and alkaline conditions.[30] Therefore, a polar solution, such as methanol/water, acetonitrile/water, diethylamine/water, cyanuric acid, acetic acid, hydrochloric acid and carboxylic acid, is generally used for the extraction of melamine.[30–32] Solid phase extraction (SPE) is another frequently used method for on-line and off-line separation and pre-concentration for trace analytes in food products.[33,34] This technique has the advantages of rapid response, high efficiency, high reproducibility and low or zero organic-solvent consumption.[35] As melamine is positively charged under acidic conditions, cation exchange SPE (*e.g.*, strong cation exchange sorbent, polystyrene cation exchange, mixed-mode anion-exchange, *etc.*) is usually used to extract melamine mainly from dairy products.

6.6 Assay Format

Here we describe a variety of different ELISA assay formats, normally used for different purposes and target types (Figure 6.4).

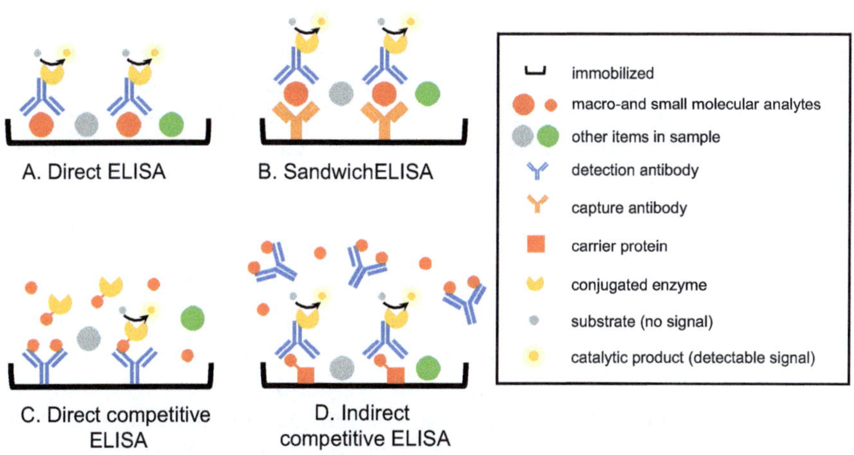

Figure 6.4 Principles of different formats of ELISA.

6.6.1 Direct and Sandwich ELISAs

A simple ELISA (direct ELISA) is shown in Figure 6.4A. Briefly, the sample is immobilized to a solid phase (normally a 96-well plate), followed by washing steps to remove the non-immobilized components. Under this condition, the target analyte together with some other components are immobilized. In this case, enzyme-linked specific antibodies bind to the analytes in a dose-dependent manner, which is displayed by the colour/fluorescence-producing catalysis.

Sandwich ELISA (Figure 6.4B) is designed to improve the sensitivity of the assay. Analytes in the sample bind to pre-immobilized capture antibodies, so that non-bound components in the sample can be removed by washing steps. Afterwards, an enzyme-linked specific detection antibody (normally different from the capture antibody) is applied to bind to the analytes. Sandwich ELISA does not apply to the determination of small molecular targets because their small epitopes are not able to accommodate more than one antibody.

6.6.2 Indirect and Direct Competitive ELISAs

Competitive ELISAs are used for identification of small molecular analytes. A competitive assay format is that the analyte from the sample and the hapten-protein conjugates compete to bind to a limited number of antibody binding sites (Figure 6.4C and D). This assay can be performed with either immobilizing antibodies (Figure 6.4C, *i.e.*, direct competitive ELISA) or immobilizing analyte conjugated with a carrier protein (Figure 6.4D, *i.e.*, indirect competitive ELISA).

In the direct competitive ELISA format (Figure 6.4C) with the solid phase attached with antibodies, enzyme-labelled hapten (structurally similar to the analyte) competes with analytes in a sample for binding to a limited amount of antibody. In the indirect competitive ELISA format (Figure 6.4D) with the solid phase attached with hapten-protein conjugates, sample analytes compete with hapten-protein conjugates for binding to a limited amount of the labelled antibody. Quantitative results can be obtained with the reference of a standard curve (*i.e.*, a dose-dependent curve generated by a series of known amounts of analyte).

6.6.3 Homogeneous and Heterogeneous ELISAs

In homogeneous ELISAs the haptens used for protein conjugates are the same as the analytes. In contrast, molecular structures for the analyte and the hapten used for protein conjugation are deliberately designed slightly differently. Therefore, the analyte-like hapten generated antibodies have higher binding affinity to the analytes over the competitor, which increase the sensitivity of the assay.

6.7 Lateral-flow Immunochromatographic Assays

Lateral-flow immunochromatographic assay (LFIA) is an on-site semi-quantitative method for the determination of various components. LFIA is rapid, equipment-free and very easy to use. Complicated sample preparation is not required for LFIA. Despite being relatively less quantitative and sensitive, this technique is particularly suitable for the on-site detection of food contaminants and adulterants. Indeed, commercial LFIA strips are widely used in the determination of many targets in various foods. The advantages and disadvantages of LFIA in comparison with other techniques normally used for food safety detection are summarized in Table 6.2.

The basic principle of LFIA is very similar to that of ELISA. Briefly, a liquid sample containing the analyte moves with capillary action towards the absorbing pad through various zones of polymeric strips, on which antibodies that can interact with the analyte are attached. As shown in Figure 6.5, the sample is applied onto a sample pad. The sample then migrates through the conjugate release pad, which contains antibodies that are specific to the target analyte and are conjugated to coloured or fluorescent particles – most commonly colloidal gold. In this case, the analyte is able to bind to the antibodies. The sample, together with the conjugated antibody bound to the target analyte, migrates along the strip into the detection zone. The specific antibodies pre-immobilized onto the test line of the nitrocellulose strips were able to bind present analytes, and thereby display the colour of colloidal gold on the test line. An unspecific secondary antibody is pre-immobilized onto the control line for the quality control of the assay.

6.8 Application of ELISA on Food Safety Detection

6.8.1 Pesticides

Some pesticides are difficult to metabolize. They can accumulate on the crop culture and cause food safety problems. To ensure that the residue level is lower than the legal limit, many ELISA methods have been developed to determine the level of pesticide residues in foods.

Table 6.2 Advantages and pitfalls of lateral-flow immunochromatographic assays.

Advantages	Disadvantages
• Less time consuming • No equipment required • Low cost • No trained personnel required • No complicated sample preparation required • Long shelf-life • Requirement of small sample volume	• Semi-quantitative • Variable reproducibility • Not high-throughput

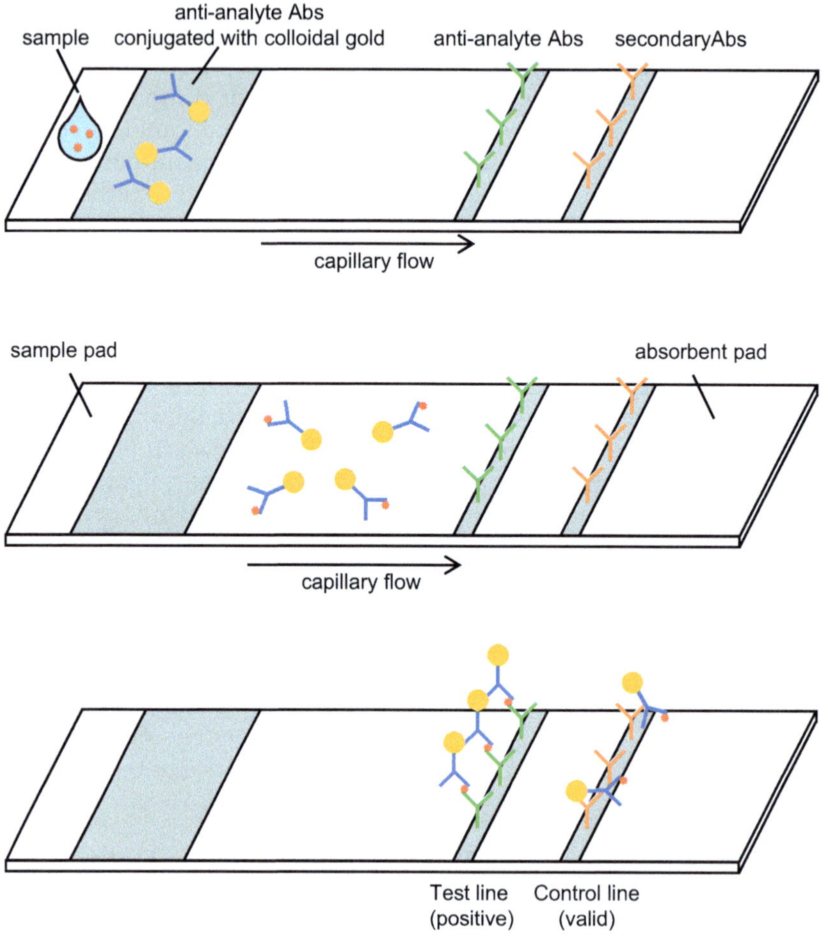

Figure 6.5 Principle of lateral-flow immunochromatographic assays. A sandwich format is shown here as an example. The sample containing analytes flow through the test line and the control line pre-immobilized with specific anti-analyte antibodies and secondary antibodies, respectively. Colloidal gold displays colour signal on test line in correspondence with the presence of analytes.

Abad and Montoya developed an ELISA with monoclonal antibody for the detection of carbaryl (1-naphthyl *N*-methylcarbamate).[36] The detection limit was 6.5×10^{-5} mg L^{-1} and the IC_{50} (measured as the 50% inhibition of antibody binding value) value was 7.06×10^{-4} mg L^{-1}. The ELISA was applied for the determination of carbaryl in apple and grape juices. It determines carbaryl in fruit juices at the low 10^{-3} mg L^{-1} level by a simple step of diluting the sample, without any clean-up or concentration step. Recovery and precision of the method were evaluated by spiking apple and grape juices with carbaryl in the $0.5–5\times10^{-4}$ mg L^{-1} range. Coefficients of variation were lower than 13.0% in all of the cases, and the mean recoveries were 100.0% and 92.2% for apple and grape juices, respectively. The method was validated to be simple and sensitive, with a quantification limit well below the maximum residue limits for carbaryl in apple and grape juices.

Lu and coworkers described a general and broad ELISA approach for the detection of pyrethroids.[37] Polyclonal antibodies were generated against the preparation of hapten by attaching a 3-carbon spacer arm to the pyrethroids metabolite analogue, 3-phenoxybenzoic acid. The IC_{50} for different pyrethroids was 20 $mg L^{-1}$ for deltamethrin, 16 $mg L^{-1}$ for cypermethrin, 11 $mg L^{-1}$ for fluvalinate, 15 $mg L^{-1}$ for fenvalerate and 20 $mg L^{-1}$ for fenpropathrin, respectively. This method was also optimized for the detection of pyrethroids in *Brassica campestris* L., peach and green tea by using a simple and rapid sample preparation.

Kido and coworkers reported an ELISA in various formats for the detection of hydroxypropazine, hydroxyatrazine and hydroxysimazine.[38] Aminohexanoic acid was used as the immunizing hapten. One of the resulting ELISAs had IC_{50} values of 0.4×10^{-3}, 0.8×10^{-3} and 3.1×10^{-3} $mg L^{-1}$ for hydroxypropazine, hydroxyatrazine and hydroxysimazine, respectively. This heterogeneous indirect competitive ELISA utilized a coated antigen with a simazine–mercaptopropanoic acid hapten conjugated to conalbumin. Groundwater fortified with hydroxyatrazine was analyzed using this method. Good correlations with different spiking levels were observed, making this assay potentially valuable to screen groundwater samples that contain parts per trillion levels of hydroxytriazine contamination.

Nakata and coworkers reported a competitive ELISA for the detection of the insecticide flucythrinate in environmental and food samples.[39] Monoclonal antibodies were generated against two types of haptens, namely the acid moiety that is the hydrolyzed product of flucythrinate and the carboxylated propyl derivative of the alcohol moiety. One format of the monoclonal antibodies showed the highest activity towards flucythrinate and did not cross-react with other pyrethroids, such as cycloprothrin, fenvalerate, fluvalinate, etofenprox and silafluofen. The assay conditions of indirect competitive ELISA were optimized for the detection of flucythrinate in the environmental and food samples. Under the optimal condition, the limits of detection of flucythrinate spiked in water, soil and extracts of apple and tea samples were 10 $mg L^{-1}$, 0.2 $mg L^{-1}$, 0.3 $mg L^{-1}$ and 0.3 $mg L^{-1}$, respectively. The mean recovery and coefficient of variation ranged from 91% to 120% and from 5% to 12%, respectively. The ELISA results for apple samples correlated well with those from LC–MS analysis ($R^2 = 0.99$, $n = 12$).

6.8.2 Veterinary Drugs

He and coworkers developed an indirect competitive ELISA with anti-sulfamethazine monoclonal antibody generated against sulfamethazine-human serum albumin.[40] This method was utilized to detect sulfamethazine and its metabolites in chicken breast muscle tissue. The IC_{50} was

9.3×10^{-3} mg L^{-1} and the recoveries ranged from 81.3% to 104.2% with coefficients of variation of 4.3% to 19.3%. The metabolite N^4-acetyl sulfamethazine was also evaluated using the same assay. In another study, Schnappinger and coworkers generated polyclonal antibodies against streptomycin and used these antibodies to develop a competitive ELISA.[41] The detection limit in buffer solution was 6×10^{-4} mg L^{-1} for streptomycin and 0.4×10^{-3} mg L^{-1} for dihydrostreptomycin. Employing rapid sample preparation procedure, streptomycin and dihydrostreptomycin were detected in milk, with the limit of detection of 6 and 0.8 ng mL^{-1}, respectively.

Wang and coworkers developed a direct competitive ELISA with a chemiluminescent method for the detection of 1-amino-hydantoin.[42] Monoclonal antibodies against 1-amino-hydantoin were prepared by immunization of BALB/c mice with the synthesized 1-[(4-carbo-benzylidene)-amino]-imidazolidin-2,4-dione (CPAHD)–Jeffamine–bovine serum albumin (BSA) as the antigen. Luminol, *p*-iodophenol and urea peroxide mixture solution served as the substrate. The cross-reactivities of the monoclonal antibody, for 1-[(4-nitro-benzylidene)-amino]-imidazolidin-2,4-dione (NPAHD) and CPAHD were 100% and 39.67%, respectively. All other compounds showed less than 0.01% cross-reactivity. The sensitivity of the antibody was determined to be 6×10^{-4} mg L^{-1}. The limits of detection in fish and honey samples were 1×10^{-4} and 2.8×10^{-4} mg kg^{-1}, respectively, and the mean recovery values ranged from 83.6% to 94.7% with coefficient of variation values less than 15%.

Wang and coworkers reported a colorimetric competitive direct ELISA using polyclonal antibody to determine neomycin residues in food of animal origins.[43] No cross-reactivity of antibody was observed by using other aminoglycosides. The limit of detection of this method was 1×10^{-4} mg kg^{-1}. A simple and efficient sample extraction method was developed with the recoveries of neomycin ranging from 75% to 105%. The limit of detection was 5×10^{-3} mg kg^{-1} in pig muscle, chicken muscle, fish and milk, 1×10^{-4} mg kg^{-1} in kidney and 2×10^{-4} mg kg^{-1} in egg, respectively. A chemiluminescence assay was developed for the detection of neomycin residues in pig muscle and chicken muscle. The limit of detection of this method for the standard solution was 1.5×10^{-5} mg kg^{-1}, while the limit of detection was 1.5×10^{-3} mg kg^{-1} in pig muscle and 6×10^{-3} mg kg^{-1} in chicken muscle, respectively. The ELISA tests were validated by HPLC, and the results showed a good correlation with R^2 value higher than 0.9.

6.8.3 Plasticizer

Zhao and coworkers generated a specific polyclonal antibody against bisphenols by immunizing rabbits with a conjugate of 4,4-bis (4-hydroxyphenyl) valeric acid and bovine serum albumin.[44] The polyclonal antibody showed specific recognition to the bisphenol structure, while the cross reaction of other common phenolic compounds, such as phenol, hydroquinol and *p*-hydroxybenzoic acid, was lower than 1%. The linear range of the calibration curve for bisphenol A was 1×10^{-3}–10 $mg\,kg^{-1}$. Real water samples and mouse serum samples were spiked with known amounts of bisphenol A and measured by competitive ELISA. Recoveries were between 92% and 105%. The limit of detection was determined to be 1×10^{-4} $mg\,kg^{-1}$ for real water samples and 2×10^{-4} $mg\,kg^{-1}$ for serum samples, respectively.

Kuang and coworkers generated a polyclonal antibody against di-*n*-octyl 4-aminophthalate coupled with bovine serum albumin.[45] After optimization, an indirect competitive ELISA was developed. The quantitative working range for DOP was 0.5–7.5×10^{-4} $mg\,kg^{-1}$ with the limit of detection being $1.9\pm0.1\times10^{-3}$ $mg\,kg^{-1}$ and IC_{50} being $1.92\pm0.11\times10^{-3}$ $mg\,kg^{-1}$. The optimized ELISA had cross-reactivity of 22.6%, 17.6% and 21.2% with di-iso-octyl phthalate, di-n-butyl phthalate and di-hexyl phthalate, respectively.

Lu and coworkers generated a series of polyclonal antibodies *in vivo* using bisphenol-A (BPA)-butyrate-protein conjugate and BPA valerate-protein conjugate.[46] Two indirect and one direct ELISAs were further developed for the detection of BPA. The assays exhibited high sensitivity and specificity for BPA. IC_{50} values were 0.78 ± 0.01–$1.20\pm0.26\times10^{-3}$ $mg\,L^{-1}$, and the limit of detection was measured by IC_{20} values as 1.0 ± 0.3–$2.0\pm0.4\times10^{-4}$ $mg\,L^{-1}$. The assays were highly specific to BPA, only displaying low cross-reactivity for 4-cumylphenol at pH 7.2. The degree of cross-reaction of 4-cumylphenol was influenced by hapten conjugates, assay condition and assay format. Based upon the spiking and recovery data, the limit of quantification for canned vegetables, bottled water and carbonated drinks was 5×10^{-4}, 2.5×10^{-3} and 0.1 $mg\,L^{-1}$, respectively.

In another study, Zhang and coworkers developed a direct competitive ELISA for the detection of diethyl phthalate (DEP).[47] The limit of detection was 9.6×10^{-5} $mg\,L^{-1}$, and the linear range was 1×10^{-4}–3.5 $mg\,L^{-1}$ with R^2 of 0.9957. The cross-reactivity of the anti-DEP antibody to six structurally related phthalate esters was less than 9%. Recoveries were between 96.4 and 106.2%. The method was successfully applied to determinate DEP in tap water, Yangtze River water and leachate from plastic drinking bottles.

6.9 Concluding Remarks

ELISA is a technique for determination of trace level contaminations in environmental and agri-food products. The working principle is based upon specific interaction between antigen and antibody. Since it is specific, sensitive, simple, rapid, cheap and suitable for on-site screening of a large number of samples, the application of ELISA is widely available in food analysis, such as the detection of pesticides, veterinary drugs, food additives, contaminants, natural toxicants and adulterants. The critical steps mainly include the generation of highly specific antibodies and the design of hapten for small size analytes.

References

1. E. Engvall and P. Perlmann, *Immunochemistry*, 1971, **8**, 871–874.
2. J. K. Lee, K. C. Ahn, O. S. Park, S. Y. Kang and B. D. Hammock, *J. Agric. Food Chem.*, 2001, **49**, 2159–2167.
3. N. A. Lee and I. R. Kennedy, Immunoassays, in ed. Y. Picó, *Food Toxicants Analysis: Techniques, Strategies and Developments*, Elsevier Science, 2007, ch. 5.
4. M. Krüger, P. Schledorn, W. Schrödl, H.-W. Hoppe, W. Lutz and A. A. Shehata, *J. Environ. Anal. Toxicol.*, 2014, **4**, 1.
5. Z. Xu, Y. Shen, H. Wang, H. Lei, J. Yang, Z. Xiao and Y. Sun, *Sci. Sin.: Chim.*, 2015, **45**, 168–176.
6. G. T. Hermanson, *Bioconjugate Techniques*, Academic Press, 2013.
7. M. Yuan, B. Liu, E. Liu, W. Sheng, Y. Zhang, A. Crossan, I. Kennedy and S. Wang, *Anal. Chem.*, 2011, **83**, 4767–4774.
8. M. Yuan, Y. Na, L. Li, B. Liu, W. Sheng, X. Lu, I. Kennedy, A. Crossan and S. Wang, *J. Agric. Food Chem.*, 2012, **60**, 10486–10493.
9. D. G. Pinacho, F. Sanchez-Baeza and M. P. Marco, *Anal. Chem.*, 2012, **84**, 4527–4534.
10. L. Cao, D. Kong, J. Sui, T. Jiang, Z. Li, L. Ma and H. Lin, *Anal. Chem.*, 2009, **81**, 3246–3251.
11. H. Mu, B. Wang, Z. Xu, Y. Sun, X. Huang, Y. Shen, S. A. Eremin, A. V. Zherdev, B. B. Dzantiev and H. Lei, *Analyst*, 2015, **140**, 1037–1045.
12. H. Mu, H. Lei, B. Wang, Z. Xu, C. Zhang, L. Ling, Y. Tian, J. Hu and Y. Sun, *J. Agric. Food Chem.*, 2014, **62**, 7804–7812.
13. J. Chen, N. Lu, X. Shen, Q. Tang, C. Zhang, J. Xu, Y. Sun, X. A. Huang, Z. Xu and H. Lei, *J. Agric. Food Chem.*, 2016, **64**, 2772–2779.
14. Z. L. Xu, Y. D. Shen, W. X. Zheng, R. C. Beier, G. M. Xie, J. X. Dong, J. Y. Yang, H. Wang, H. T. Lei, Z. G. She and Y. M. Sun, *Anal. Chem.*, 2010, **82**, 9314–9321.
15. Z. L. Xu, G. M. Xie, Y. X. Li, B. F. Wang, R. C. Beier, H. T. Lei, H. Wang, Y. D. Shen and Y. M. Sun, *Anal. Chim. Acta*, 2009, **647**, 90–96.
16. Z. Wang, Z. Kai, R. C. Beier, J. Shen and X. Yang, *Int. J. Mol. Sci.*, 2012, **13**, 6334–6351.
17. K. He, X. Du, W. Sheng, X. Zhou, J. Wang and S. Wang, *J. Agric. Food Chem.*, 2016, **64**, 2627–2634.
18. S. Muyldermans, *Annu. Rev. Biochem.*, 2013, **82**, 775–797.
19. R. E. Kontermann, *Acta Pharmacol. Sin.*, 2005, **26**, 1–9.
20. R. L. Stanfield, H. Dooley, M. F. Flajnik and I. A. Wilson, *Science*, 2004, **305**, 1770–1773.
21. R. L. Stanfield, I. A. Wilson and V. V. Smider, *Sci. Immunol.*, 2016, **1**, aaf7962.
22. C. Boi, *J. Chromatogr. B: Anal. Technol. Biomed. Life Sci.*, 2007, **848**, 19–27.
23. I. Weeks, L. J. Kricka and D. Wild, Signal generation and detection systems (excluding homogeneous assays), in ed. D. Wild, *The Immunoassay Handbook*, 2013.

24. Y. Y. Lin, J. Wang, G. Liu, H. Wu, C. M. Wai and Y. Lin, *Biosens. Bioelectron.*, 2008, **23**, 1659–1665.
25. P. Xiong, N. Gan, Y. Cao, F. Hu, T. Li and L. Zheng, *Materials*, 2012, **5**, 2757.
26. S. Zhan, X. Huang, R. Chen, J. Li and Y. Xiong, *Talanta*, 2016, **158**, 51–56.
27. C. M. Karbiwnyk, W. C. Andersen, S. B. Turnipseed, J. M. Storey, M. R. Madson, K. E. Miller, C. M. Gieseker, R. A. Miller, N. G. Rummel and R. Reimschuessel, *Anal. Chim. Acta*, 2009, **637**, 101–111.
28. S. A. Tittlemier, B. P. Y. Lau, C. Ménard, C. Corrigan, M. Sparling, D. Gaertner, K. Pepper and M. Feeley, *J. Agric. Food Chem.*, 2009, **57**, 5340–5344.
29. W. C. Andersen, S. B. Turnipseed, C. M. Karbiwnyk, S. B. Clark, M. R. Madson, C. M. Gieseker, R. A. Miller, N. G. Rummel and R. Reimschuessel, *J. Agric. Food Chem.*, 2008, **56**, 4340–4347.
30. H. Sun, L. Wang, L. Ai, S. Liang and H. Wu, *Food Control*, 2010, **21**, 686–691.
31. H. Yu, Y. Tao, D. Chen, Y. Wang, Z. Liu, Y. Pan, L. Huang, D. Peng, M. Dai, Z. Liu and Z. Yuan, *Anal. Chim. Acta*, 2010, **682**, 48–58.
32. R. Araújo, J. L. Moreira, N. Ratola, L. Santos and A. Alves, *Anal. Lett.*, 2012, **45**, 613–624.
33. M. H. Mashhadizadeh, M. Amoli-Diva, M. R. Shapouri and H. Afruzi, *Food Chem.*, 2014, **151**, 300–305.
34. M. Rezaee, F. Khalilian, H. A. Mashayekhi and N. Fattahi, *Anal. Methods*, 2014, **6**, 3456–3461.
35. R. D. Oleschuk, L. L. Shultz-Lockyear, Y. Ning and D. J. Harrison, *Anal. Chem.*, 2000, **72**, 585–590.
36. A. Abad and A. Montoya, *Anal. Chim. Acta*, 1995, **311**, 365–370.
37. Y. Lu, N. Xu, Y. Zhang, B. Liu, Y. Song and S. Wang, *Food Agric. Immunol.*, 2010, **21**, 27–45.
38. H. Kido, M. H. Goodrow, V. Griffeth, A. D. Lucas, S. J. Gee and B. D. Hammock, *J. Agric. Food Chem.*, 1997, **45**, 414–424.
39. M. Nakata, A. Fukushima and H. Ohkawa, *Pest Manage. Sci.*, 2001, **57**, 269–277.
40. J. He, J. Shen, X. Suo, H. Jiang and X. Hou, *J. Food Sci.*, 2005, **70**, C113–C117.
41. P. Schnappinger, E. Usleber, E. Märtlbauer and G. Terplan, *Food Agric. Immunol.*, 1993, **5**, 67–73.
42. Q. Wang, Y.-C. Liu, Y.-J. Chen, W. Jiang, J.-L. Shi, Y. Xiao and M. Zhang, *Anal. Methods*, 2014, **6**, 4414–4420.
43. S. Wang, B. Xu, Y. Zhang and J. X. He, *Meat Sci.*, 2009, **82**, 53–58.
44. M.-P. Zhao, Y.-Z. Li, Z.-Q. Guo, X.-X. Zhang and W.-B. Chang, *Talanta*, 2002, **57**, 1205–1210.
45. H. Kuang, L. Xu, G. Cui, W. Ma and C. Xu, *Food Agric. Immunol.*, 2010, **21**, 265–277.
46. Y. Lu, J. R. Peterson, J. J. Gooding and N. A. Lee, *Anal. Bioanal. Chem.*, 2012, **403**, 1607–1618.
47. M. Zhang, Y. Cong, Y. Sheng and B. Liu, *Anal. Biochem.*, 2010, **406**, 24–28.

CHAPTER 7

Molecularly Imprinted Polymers-based Sensing in Food Safety and Quality Analysis

YIWEI TANG,* HONG ZHANG AND YUCHEN ZHANG

College of Food Science and Engineering, Bohai University, 19 Keji Road, Jinzhou 121013, China
*Email: tangyiwei81@163.com

7.1 Introduction

The molecular imprinting technique is now an ideal way to prepare functional polymer materials, which are called "molecularly imprinted polymers (MIPs)", with specific recognition capability towards analytes. During the process of molecular imprinting, functional monomers and cross-linkers around an imprinting molecule, also called the "molecular template" (the target molecule or a derivative of the target molecule), are copolymerized in a suitable solvent. After removal of the template, three-dimensional cavities complementary in size, shape and positioning of chemical groups to the template remain in the polymer matrices, as illustrated in Figure 7.1.[1,2] These cavities in the polymers and the interacting forces between the analyte and the binding sites include ionic interactions, hydrogen bonds, π–π interactions and hydrophobic interactions,[3] all of which are vital to guarantee the excellent selectivity of the MIPs. Depending upon the molecular

Food Chemistry, Function and Analysis No. 2
Sensing Techniques for Food Safety and Quality Control
Edited by Xiaonan Lu

Published by the Royal Society of Chemistry, www.rsc.org

Figure 7.1 The principle of molecular imprinting. (a) functional monomers, (b) cross-linker, (c) template molecule, (1) assembly of the pre-polymerization complex, (2) polymerization, (3) extraction, (4) rebinding.
Reproduced with permission from K. Haupt, *Analytical Chemistry*, 2003, 75, 376A–383A. Copyright 2003 American Chemical Society.

interactions of the template-functional monomer complex prior to polymerization, the approaches of synthesis of MIPs can be divided into covalent molecular imprinting,[4] non-covalent molecular imprinting[5] and semi-covalent molecular imprinting.[6,7] Currently, MIPs have been regarded as powerful molecular recognition materials that are widely used in the detection of pesticides[8] and biological sensing.[9]

Compared with biomaterials (*e.g.*, antibody and aptamer), MIPs have unique features, such as high stability, ease of preparation, low cost and high reproducibility, all of which make them an ideal molecular recognition element.[10,11] Combination of MIPs and various types of signal elements to construct sensors has attracted extensive interest. Currently, different types of MIPs-based sensors have been constructed, including electrochemical sensors,[12] acoustic sensors,[13] calorimetric sensors[14] and optical sensors.[15]

Food is one of the most important pathways of transmission for global disease derived from microbial contaminations.[16] It was reported that over 2 million deaths occur per year due to foodborne diseases in developing countries.[17] Rapid and early detection of the potential contaminants in agri-food products is critical to reduce cases of mortality and morbidity. These contaminants mainly include pathogenic microorganisms and chemical pollutants (*e.g.*, pesticides, veterinary drugs and heavy metals).[18] One attractive feature of MIPs is that they can be prepared using a wide range of target molecules as molecular template. The imprinting of chemical molecules is well established. As large pathogenic microorganism templates (analytes), two-dimensional surface imprinting and/or three-dimensional imprinting are usually used to prepare the MIPs, leading to an effective binding and extraction of the large analytes from a solution. As a new molecular recognition material and biomimetic antibodies, MIPs have been widely used in sensing the safety and quality of food products.

This chapter will review MIPs-based sensors, the performance of the developed methods and specific applications. The working mechanism and categories of MIPs and their characteristics will be summarized. The principles and design schemes of the sensors will be described and explained.

7.2 Materials

7.2.1 Molecularly Imprinted Polymers

In brief, the working mechanism of MIPs can be described as follows: the complexes of functional monomers and the template are generated in a solvent. Then, copolymerization is carried out between functional monomers and cross-linkers in the presence of an initiator. After polymerization, rigid MIPs are obtained. Recognition binding sites in MIPs matrices are left, followed by the removal of the template using chemical and/or physical methods.

Three types of approach have been reported for the preparation of MIPs, namely covalent, non-covalent and semi-covalent imprinting approaches. For covalent imprinting, an independent chemical step is usually required to attach the template with polymerizable units by covalent bonds. Then, copolymerization of this covalent chemical compound with additional cross-linker is carried out in a porogenic solvent. Specially, removal of the template and rebinding of the target are both chemical reactions. A typical example based upon boronate esters is shown in Figure 7.2.[19] Additionally, covalent imprinting polymers are also prepared based upon Schiff's bases and ketals (or acetal). The covalent approach gives rise to a higher stability and selectivity of the MIPs than the non-covalent approach.[20] However, it is relatively challenging to remove imprinting molecules by using this

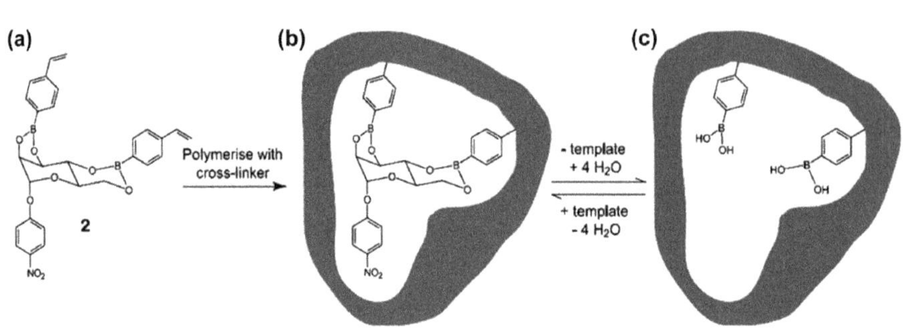

Figure 7.2 Imprinting of 4-nitrophenyl-a-d-mannopyranoside-2,3:4,6-di-*O*-(4-vinylphenylboronate) (2): the covalent template monomer prepared by condensation of 4-vinylbenzeneboronic acid with 4-nitrophenyl-a-d-mannopyranoside (a) is copolymerized with cross-linker (divinylbenzene or ethyleneglycol dimethacrylate) to generate the imprinted polymer (b). Template removal involves the hydrolysis of the boronate ester groups and the addition of 4 equivalents of water to reveal the imprinted site (c). Template uptake and release are reversible and rapid enough for the polymers to be used as the HPLC stationary phase.
Reproduced from *Advanced Drug Delivery Reviews*, Volume 57, A. G. Mayes and M. J. Whitcombe, Synthetic strategies for the generation of molecularly imprinted organic polymers, 1742–1778, Copyright 2005, with permission from Elsevier.

approach and a long rebinding time is often reported, both of which limit its application as the molecule recognition element in a sensor.

For non-covalent imprinting, the template and functional monomer form a complex by self-assembly in the porogenic solvent. Then, a cross-linker is added to copolymerize with the complex in the presence of an initiator under certain conditions. After removal of the template, the template molecules can be rebound through non-covalent bonds, such as hydrogen-bonding, electrostatic force, dipole–dipole and van der Waals interaction. MIPs prepared by the non-covalent imprinting method are recognized as more similar to natural interactions (*e.g.*, enzyme-substrates and antibody–antigen) than covalent imprinting MIPs.[21] To obtain the candidate of functional monomers and the stoichiometry between template and functional monomers, UV-Vis spectroscopy, nuclear magnetic resonance spectroscopy and the computational method are commonly employed during the non-covalent imprinting process.[22,23] MIPs prepared by the non-covalent imprinting method are by far the most frequently used as the molecular recognition element of sensors due to the simplicity of their polymerization process.

For semi-covalent imprinting polymers, the imprinting process is the same as for covalent imprinting, a covalent chemical compound attaching the template with polymerizable units, but the rebinding process is the same as for non-covalent imprinting. A typical example is shown in Figure 7.3,[24] in which clenbuterol imprinting polymers were prepared by the semi-covalent method. Firstly, the template clenbuterol has been linked with methacryloyl

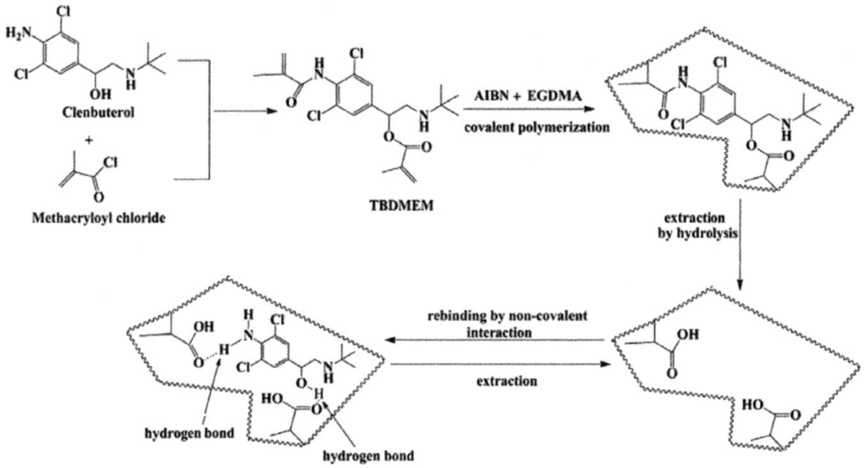

Figure 7.3 Imprinting of clenbuterol by semi-covalent imprinting method. Reproduced from Food Chemistry, Volume 190, Y. Tang, J. Lan, X. Gao, X. Liu, D. Zhang, L. Wei, Z. Gao and J. Li, Determination of clenbuterol in pork and potable water samples by molecularly imprinted polymer through the use of covalent imprinting method, 952–959. Copyright 2016, with permission from Elsevier.

chloride by covalent bonding before polymerization, improving the selectivity of the MIPs. The template is then extracted by hydrolysis. Typically, the template can be rebound by non-covalent interaction, minimizing the existence of non-specific sites along with a fast binding rate.

7.2.2 Polymerization Techniques

Various methods of polymerization have been used to obtain molecular recognition element MIPs.[25] Some of the MIPs used in sensors were prepared by the bulk polymerization method and then grafted onto the surface of the sensors.[26,27] Others were produced by *in situ* methods directly onto the sensor surface.[28,29] The advantage of the *in situ* MIPs preparation method is that the thickness of the MIPs layer can be precisely controlled.[30] This is very important to some sensors that require a specific thickness of the MIPs layer, for example, surface plasma resonance.[31]

Currently, numerous examples of MIPs have been successfully synthesized, almost 10% of which are related to the development of the MIPs-based sensor.[32] The MIPs have excellent selective recognition capability towards the analytes, which can be comparable to antibodies and other biological receptors. In addition, MIPs are highly stable and can be regenerated many times. Last but not least, MIPs are much cheaper than antibodies due to their easy synthesis in a relatively large amount per batch of synthesis.

7.3 Molecularly Imprinted Polymers-based Sensors in Food Safety and Quality Analysis

MIPs-based sensors used in food safety and quality analysis can be categorized into different groups depending on the basic mechanism and principles of signal transduction. The major types of MIPs-based sensors include electrochemical sensors, quartz crystal microbalance sensors, fluorescence sensors, surface enhanced Raman scattering sensors, surface plasmon resonance sensors and MIPs-based enzyme-linked immunoassays. For MIPs-based sensors, when the target compounds bind to the MIPs recognition element, the signals change correspondingly. There is a relationship between the concentration of the target compound and the intensity of the signals.

7.3.1 Electrochemical Sensors

Electrochemical sensor assay is one of the most popular analytical methods in food analysis due to its ease of measurement, availability of devices and instruments and simplicity. Current, voltage, conductivity or impedance is usually used as the monitoring signal of electrochemical sensors.[33] In these sensors, the transducer is a very important component, which can translate the signal into a quantifiable electrical output. Next to the transducer, another key component is the molecular recognition element, which can recognize and bind to the target analyte in a specific manner. When the binding between a target molecule and the MIPs recognition element takes place, the changes of signals can be monitored. The function of the transducer is critical to MIPs-based sensing technology. The molecular recognition element, MIPs, must be in close proximity to the surface of the transducer,[33] which is related to the response time of the developed sensor. There are three ways of integrating the MIPs-transducer: the deposition of the synthesized MIPs particles onto or into the electrode, grafting of thin MIPs films onto an electrode and *in situ* polymerization method.[34–36]

Erythromycin is widely used to prevent and treat numerous diseases in agriculture and the cattle industry due to its great antimicrobial effect against Gram-positive and Gram-negative bacteria. The widespread use of erythromycin leads to its residues being present in different agri-food products. Song and coworkers developed a MIPs-based electrochemical sensor for the determination of erythromycin.[37] The molecular recognition element MIPs were prepared using erythromycin as the template, methacrylic acid (MAA) as the functional monomer, ethylene glycol dimethacrylate (EGDMA) as the cross-linker and 2,2-azobisisobutyronitrile (AIBN) as the initiator. Then, the imprinted polymers were extracted in a Soxhlet apparatus. For the preparation of MIPs-carbon paste electrode, graphite powder was homogenized with MIPs and paraffin oil in an agate mortar, and the final homogenous carbon paste was tightly pressed into the cavity (4 mm in diameter) at the end of an electrode body. Then, the surface was further polished. The prepared electrode was used for quantitative determination of erythromycin by differential pulse voltammetry methods with a low limit of detection and high selectivity. On the basis of the same principle, Motaharian and coworkers developed an electrochemical sensor based upon a carbon paste electrode modified with MIP nanoparticles for selective and sensitive determination of diazinon pesticide residues in well water and apple fruit samples using cyclic voltammetry and square wave voltammetry methods.[38]

In comparison to the conventional method, grafting thin MIP film onto the electrode surface has good site accessibility and fast mass transfer characteristics.[39,40] Hedborg and colleagues investigated the application of thin MIPs membranes in field-effect capacitors sensors on the basis of the principle of the plate capacitor with double layer phenomenon.[41] Direct electro-polymerization of the MIPs onto the surface of the electrode is a promising approach for the integration of MIPs-transducers, also known as *in situ* polymerization.[35] Kong and others reported a ractopamine-imprinted film on a gold electrode surface, prepared through an electrochemical method using *o*-aminothiophenol as the functional monomer.[42] Firstly, the surface of gold electrodes was polished using metallographic abrasive paper and alumina slurry and they were then immersed in "Piranha" solution (30% H_2O_2/98% $H_2SO_4 = 1:3$, v/v) for 5 min. After ultrasonic cleaning with ethanol and deionized water, the electrode was electrochemically activated

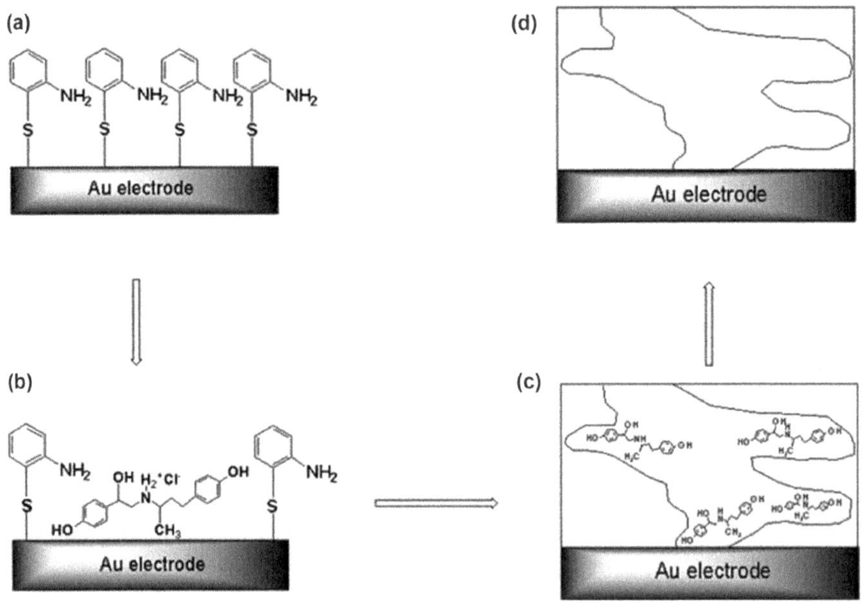

Figure 7.4 The process for the preparation of ractopamine-imprinted films on gold electrode surfaces. (a): Monomer self-assembly. The pretreated electrode was immersed in an electro-polymerization solution (20 mL) containing methanol, *o*-aminothiophenol (30 mmol L^{-1}), ractopamine (10 mmol L^{-1}), tetrabutylammonium perchlorate (5 mmol L^{-1}) and HCl (10 mmol L^{-1}) overnight to allow *o*-aminothiophenol self-assembly. (b): Electrostatic gravitation. After self-assembly, the electrode was treated at −0.8 V for 10 min in the electro-polymerization solution. Prior to use, nitrogen gas (99%) was bubbled through the solution for about 5 min to remove oxygen. (c): Electro-polymerization. Twenty cyclic voltammetry scans were performed on the gold electrode to electrosynthesize the ractopamine-imprinted film over the potential range −0.4 to 1.4 V (*versus* the saturated calomel electrode) with a scanning rate of 50 mV s^{-1}. (d): Template removal. The modified electrode was treated at a potential of 0.8 V for 10 min in HCl (1.0 mol L^{-1}) and ultrasonically cleaned in methanol and doubly deionized water twice to remove the template molecule entrapped within the polymeric matrix.
Reproduced from *Analytical and Bioanalytical Chemistry*, An electrochemical sensor for rapid determination of ractopamine based on a molecularly imprinted electrosynthesized o-aminothiophenol film, Volume 404, 2012, L. J. Kong, M. F. Pan, G. Z. Fang, K. Qian and S. Wang, © Springer-Verlag 2012. With permission of Springer.

in 1.0 mol L^{-1} H_2SO_4 by cyclic voltammetry scanning from −0.2 to 1.0 V until a stable electrochemical response was received. The process of preparation of ractopamine-imprinted film on the gold electrode is shown in Figure 7.4. In brief, there are four major steps: monomer self-assembly, electrostatic gravitation, electro-polymerization and template removal.[42]

To improve the sensitivity of MIPs-based electrochemical sensors, a MIPs@Prussian blue film was electrochemically polymerized on the surface

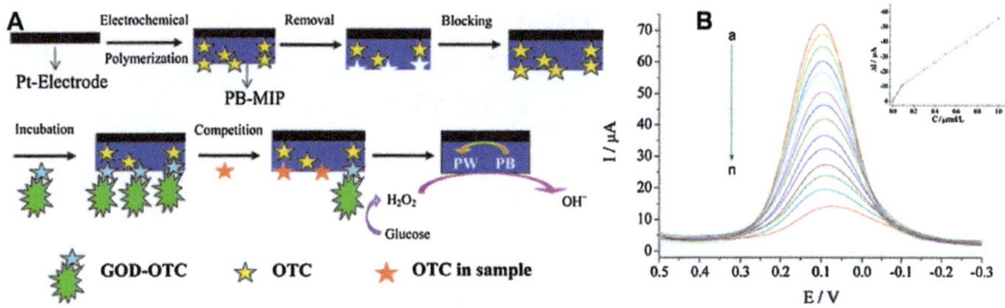

Figure 7.5 (A) Schematic diagram of the principle of the preparation and competition assay of the Prussian blue (PB)-based oxytetracycline (OTC) molecularly imprinted electrochemical sensors. (B) Differential pulse voltammetries of the Prussian blue-MIP sensor after incubation in samples with varying concentrations of oxytetracycline: (a–n) 0, 0.005, 0.01, 0.05, 0.08, 0.1, 0.3, 0.4, 0.5, 0.6, 0.7, 0.8, 0.9, and 1.0 μmol L^{-1} oxytetracycline, respectively. The inset shows the resulting calibration curve.
Reproduced from J. Li, Y. Li, Y. Zhang and G. Wei, *Analytical Chemistry*, 2012, **84**, 1888–1893. Copyright 2012 American Chemical Society.

of an electrode using oxytetracycline as the template.[43] The determination of oxytetracycline relied upon a competition reaction between oxytetracycline and glucose-oxidase-labelled oxytetracycline and the catalytic reduction of hydrogen peroxide by the MIPs@Prussian blue film. The principle of the preparation and competition assays of the Prussian blue-based oxytetracycline molecularly imprinted electrochemical sensors is shown in Figure 7.5A.[43] The differential pulse voltammetries were recorded with various concentrations of oxytetracycline under the optimal experimental conditions and are shown in Figure 7.5B.[43] The calibration curve consists of two linear sections (Figure 7.5B, inset).[43] The results showed that the double amplification based upon the Prussian blue film and the enzymatic effect of glucose oxidase (GOD) improved the sensitivity and demonstrated the feasibility of its practical applications in the analysis of real milk samples.

In addition, a variety of MIPs-based electrochemical sensors were prepared and used in food safety and quality analysis, such as the detection of tetracycline in honey,[44] detection of 2,4-dichlorophenol in water,[45] detection of glyphosate in tap water[46] and detection of diphenylamine in apple juice.[47]

7.3.2 Quartz Crystal Microbalance Sensors

A quartz crystal microbalance (QCM) consisting of a thin quartz disk with coated electrodes is a mass sensor based upon the piezoelectric properties of quartz crystals. Typically, quartz crystals are cut in the AT form, at a 35°10′ angle from the *z*-axis of the crystal (Figure 7.6[48]), providing a stable oscillation with almost no temperature fluctuation in *f* at room temperature.[49] The mass sensitivity of the QCM is dependent upon the thickness of the crystal. Most commercial QCM crystals are 0.5 inches in diameter.

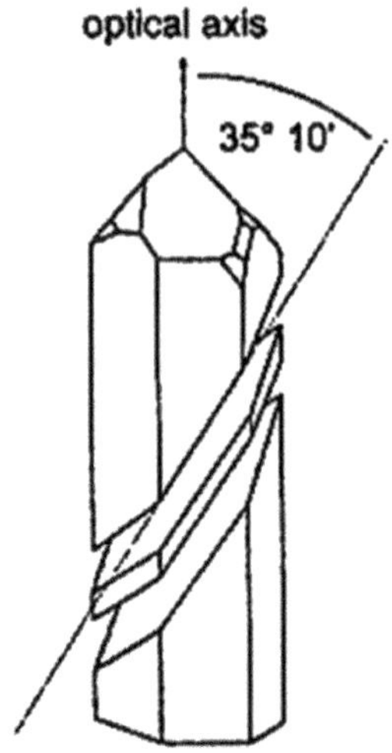

Figure 7.6 AT-cut of a quartz crystal.
Reproduced with permission from A. Janshoff, H. J. Galla and C. Steinem, *Angewandte Chemie International Edition*, 2000, **39**, 4004–4032. © WILEY-VCH Verlag GmbH, D-69451 Weinheim, 2000.

When voltage is applied to a quartz crystal to induce oscillation at a specific frequency, the change in mass on the surface of quartz is directly related to the change in frequency of the oscillating crystal,[50] as shown by the Sauerbrey Equation, $\Delta f = -\frac{2f^2}{\sqrt{\rho_q u_q}}\Delta m = -C_f \Delta m$, where Δf is the frequency shift, Δm is the mass change, f is the intrinsic crystal frequency, ρ_q is the density of the quartz and u_q is the shear modulus of the quartz plate. Recently, QCM with dissipation monitoring technique was developed to measure the changes in the induced energy dissipation and the frequency, enabling a label-free detection and analysis of bio-recognition events in a real-time manner.[51] QCM-based sensors have been developed for the detection of chemicals,[52] biomolecules[53] and bacteria.[54] These sensors have the features of low cost, rapid response, portability and high sensitivity.

Eren and coauthors developed a sensitive MIPs-based QCM sensor by fabricating a self-assembling monolayer formation of allylmercaptane on the QCM chip surface. This sensor was applied to determine lovastatin in red yeast rice with high selectivity.[55] The preparation of a lovastatin-imprinted QCM nanosensor is shown in Figure 7.7.[55] Firstly, the QCM chip was treated with acidic piranha solution to modify the surface by using CH_2CHCH_2SH. Then, the QCM chip was washed and dried at 35 °C for 2 h. Afterwards, the QCM chip was sinked in an ethanol/water (4 : 1, v/v) solution containing 3.0 mol L^{-1} of CH_2CHCH_2SH to allow vinyl groups to form a self-assembled monolayer onto the gold surface for about 1 day. Then, the chip was washed with ethanol and dried under nitrogen flow. Lovastatin-imprinted film on a CH_2CHCH_2SH-modified QCM chip was prepared as follows: firstly,

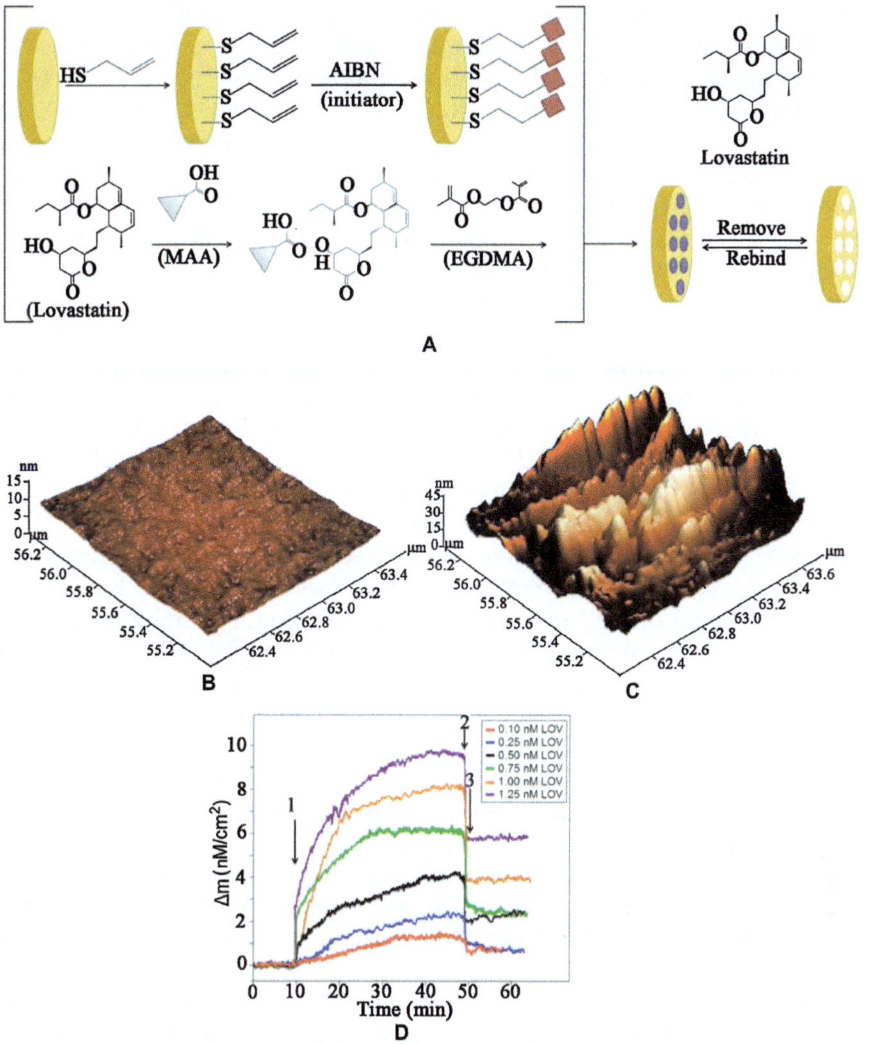

Figure 7.7 Schematic representation of the preparation of QCM chip by surface polymerization (A). AFM images of non-modified QCM chip (B). AFM images of lovastatin (LOV)-imprinted film on QCM chip (C). Effect of the concentration on QCM response of LOV-imprinted nanosensor (1) adsorption, (2) desorption, (3) regeneration (D).
Reproduced from *Food Chemistry*, Volume 185, T. Eren, N. Atar, M. L. Yola and H. Karimi-Maleh, A sensitive molecularly imprinted polymer based quartz crystal microbalance nanosensor for selective determination of lovastatin in red yeast rice, 430–436. Copyright 2015, with permission from Elsevier.

lovastatin was mixed with MAA monomer with a molar ratio of 2 : 10000 in 500 μL Gly-NaOH buffer (pH 10.0) at 20 °C for 3 h. Then, a complex mixture of AIBN (5.0 mg), MAA-lovastatin (200 μL), 2-hydroxyethyl methacrylate

(HEMA, 1250 μL) and EGDMA (500 μL) was prepared and sparged with a gentle flow of nitrogen for 15 min, as the stock monomer solution. After treatment with 20 μL of stock monomer solution on the surface of the QCM chip for 10–15 s, the polymerization was executed with UV light (100 W, 365 nm) for 1 h. Last, the QCM chip coating lovastatin MIPs was washed with ethanol several times, and then dried in a vacuum oven. Atomic force microscopy (AFM) images were obtained to characterize the morphology of the surface of the QCM chips. The amount of the surface depth for bare (Figue 7.7B) and lovastatin imprinted film modified QCM chips was 3.64 ± 0.21 and 24.13 ± 1.62 nm, respectively, indicating that the polymerization was successfully completed. Figure 7.7D shows the sensorgrams of concentration effect on QCM response of lovastatin imprinted nanosensor, and the increase of the concentration of lovastatin indicated that the mass shift increased linearly along with the increase in concentration.

Similar to the work reported by Eren and coworkers,[55] the QCM chips were modified with 11-mercaptoundecanoic acid (MUA) by forming a self-assembled monolayer (SAM) to fabricate MIPs-based QCM sensor.[56] Then, the carboxyl groups on the SAM were activated by 1-(3-dimethylaminopropyl)-3-ethylcarbodiimide hydrochloride and *N*-hydroxysuccinimide. The chip was then transferred into 2, 2¢-azobis(2-amidinopropane) hydrochloride aqueous solution at room temperature for 3 h. After being dried by nitrogen, the initiator-covered chip was immediately dipped into the pre-polymerization solution that contained profenofos (template), MAA (function monomer), trimethylopropane trimethacrylate (cross-linker) and dimethyl sulfoxide (porogen). The oxygen dissolved in the solution was removed by purging with nitrogen for 10 min. After that, the container was sealed and the polymerization was carried out at 60 °C for 18 h. The template was extracted by immersing the QCM into the methanol-acetic acid solution with gentle shaking. Finally, the chip was dried by nitrogen. This MIPs-based QCM sensor combined the advantages of the high selectivity of MIPs and the high sensitivity of QCM sensors.

Apart from directly polymerizing the thin MIPs films onto the QCM chip, grafting MIPs particles onto the QCM surface is another important method, such as the method reported by Liu and coworkers.[57] Firstly, endosulfan MIPs particles were synthesized by precipitation polymerization. Then, endosulfan-MIPs (6.0 mg) were mixed with polyvinyl chloride (4.0 mg) dissolved in tetrahydrofuran (10.0 mL). The mixture (100 μL) was dispensed onto the centre of the QCM chips with Au electrode for spin coating at 1000 rpm for 30 s by a spin coater. Then, the mixture was coated uniformly onto the surface of the QCM chip. After that, the template endosulfan was extracted with a methanol-acetic acid solution (9 : 1, v/v). Finally, the surface of the QCM chips was dried by nitrogen blowing for 5 min. Figure 7.8 demonstrates the microstructure of the bare QCM chip (left) and the MIPs@QCM chip (right), indicating that a monolayer of MIPs was uniformly and orderly deposited onto the surface of the chip under the cohering contribution of polyvinyl chloride. Lastly, a MIPs-based QCM

Figure 7.8 Scanning electron microscope images of the surface of a bare QCM chip (left) and the MIPs coated QCM chip (right).
Reproduced from ref. 57 with permission from the Royal Society of Chemistry.

sensor method was established for rapid detection of endosulfan in drinking water and milk samples.

In order to improve the sensitivity of the MIPs-based QCM sensor, Fang and others firstly introduced the composite (AuNPs@CMK-3) of Au nanoparticles (AuNPs) and ordered mesoporous carbon CMK-3 acting as a signal amplifier because the larger specific surface area of AuNPs@CMK-3 was beneficial to increase the amount of effective imprinted sites and subsequently improve the sensitivity of the sensor.[58] Herein, the AuNPs were synthesized *via* the reduction of $HAuCl_4$. Then, 30 mg CMK-3 was added into 100 mL AuNPs solution with stirring for 4 h to fabricate AuNPs@CMK-3. The preparation of MIPs-based QCM sensor included four steps: (1) preparation of AuNPs@CMK-3/AuE (QCM electrode with Au coating), (2) *o*-aminothiophenol and 1-hydroxy-2-naphthoic acid (as dummy template) self-assembly onto the AuNPs@CMK-3/AuE, (3) electropolymerization and (4) removal of the template. Under optimal conditions, the proposed QCM sensor using MIPs-based AuNPs@CMK-3/AuE as working electrode could selectively recognize trace citrinin under the interference of structural analogues. Moreover, the proposed sensor was successfully applied to detect trace citrinin in cereals with satisfactory recoveries, making it possible to detect trace citrinin in foods. Electropolymerization used in this protocol is a simple and effective method to prepare MIPs membrane directly onto the surface of QCM since the thickness of MIPs films can be easily controlled by varying the number of cyclic voltammetry scanning cycles.

In addition, many other MIPs-based QCM sensors were developed and used in food safety and quality analysis, such as detection of domoic acid in mussel,[59] detection of patulin in apple juice, pear juice and haw flake,[60] detection of daminozide in apple,[61] detection of histamine in canned sardines, canned saury and canned mackerel,[62] detection of carbaryl in vegetables[63] and detection of trichloroacetic acid in drinking water.[64]

7.3.3 Fluorescence Sensors

Fluorescence is a promising detection method for sensor systems due to its high sensitivity and non-destructive nature. It has been widely used in environmental monitoring, food safety inspection and biomedical diagnosis.[65,66] In fluorescence sensors, fluorescence signal elements include common organic fluorescent dyes and inorganic particles,[67] while molecular recognition elements usually include antibody, aptamer, natural protein receptors and MIPs. Because of some drawbacks (*e.g.*, poor stability, short shelf-life and low availability of bio-elements (antibody, aptamer and protein)) from the conventional molecular recognition elements, using MIPs in fluorescence sensor has attracted considerable attention in the recent years. Two major ways were developed to prepare the MIPs-based fluorescence sensors: (1) directly using fluorescent monomers to synthesize fluorescent imprinted polymer sensor and (2) coating MIPs onto the surface of the fluorescence particles.

Various fluorescent monomers were designed and used to prepare various types of MIPs-based fluorescence sensors. Inoue and colleagues synthesized a novel functional monomer, *O*-acryloyl *L*-hydroxyproline conjugated with dansyl ethylenediamine (Hyp-En-Dans), bearing a rigid hydrogen-bondable site for proteins (pyrrolidine moiety) and a fluorescent probe (dansyl moiety) to detect hydrophobic regions.[68] Using this fluorescent monomer, a MIPs-based fluorescence probe was prepared for the determination of proteins. Based upon the same principle, Wu and colleagues developed a simple and rapid method for monitoring bisphenol A.[69] Firstly, fluorescent monomer dansyl methacrylate was synthesized by a chemical reaction between dansyl chloride and *N*-(2-aminoethyl) methacrylamide hydrochloride. Then, bisphenol-imprinted fluorescent polymers were prepared by the precipitation polymerization method. When the concentration of bisphenol increased from 0 to 45 $mg\,L^{-1}$, the fluorescent intensity decreased by 3-fold. The phenomenon of fluorescent quenching may be explained as the following. After binding of the phenol groups of bisphenol to the amine group of dansyl methacrylate, the electron density in the vicinity of the fluorogenic centre led to an effective quenching of the fluorescent intensity.

Although MIPs-based fluorescence sensors were successfully prepared using fluorescent monomer, a complicated synthesis and purification process is usually cited as a problem. Quantum dots (QDs) are fluorescent nanomaterials with size-dependent emission wavelengths, high fluorescence efficiency and good photostability, which have the ability to form complexes with biomelecules.[70] Therefore, the research focus on the development and application of QDs has been growing in recent years.

Semiconductor quantum dots (*e.g.*, Mn^{2+} doped ZnS QDs, CdSe QDs and CdTe QDs) are usually used in MIPs-based fluorescence sensors as a signal element. Liu and coworkers prepared a novel fluorescence sensor by anchoring a MIPs layer onto CdTe/ZnS QDs using a sol–gel with surface imprinting.[70] Firstly, water-soluble CdTe/ZnS QDs were prepared. In brief,

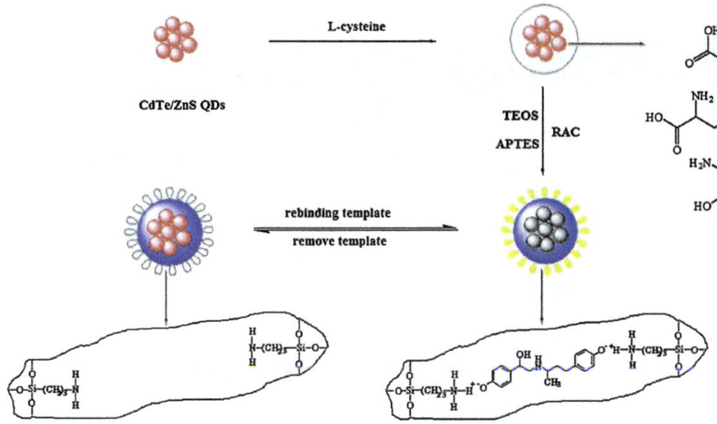

Figure 7.9 The preparation process of MIP-coated CdTe/ZnS QDs. Reproduced from *Analytica Chimica Acta*, Volume 762, H. Liu, D. Liu, G. Fang, F. Liu, C. Liu, Y. Yang and S. Wang, A novel dual-function molecularly imprinted polymer on CdTe/ZnS quantum dots for highly selective and sensitive determination of ractopamine, 76–82. Copyright 2013, with permission from Elsevier.

$CdCl_2 \cdot 2.5H_2O$ and L-cysteine were mixed in distilled water to form the cadmium precursor. After adjusting the pH value to 11 with NaOH, the freshly prepared NaHTe aqueous solution containing $NaBH_4$ and Te powder was quickly added to the mixture, followed by stirring under N_2 for 30 min. Then, the solution was heated at 100 °C for 1 h, cooled to room temperature and added dropwise to a solution of $ZnCl_2$ and Na_2S. The solution was then heated under reflux at 100 °C, and it began to generate fluorescence and the colour could be tuned by prolonging the heating time. A certain time later, L-cysteine stabilized CdTe/ZnS QDs with strong fluorescence at 570 nm were obtained. Using the prepared QDs, MIPs-based fluorescence polymers were synthesized using ractopamine (RAC) as template, 3-aminopropyltriethoxysilane (APTES) as functional monomer and tetraethoxysilane (TEOS) as cross-linker. Figure 7.9 shows the process of the preparation of coating MIPs onto the surface of CdTe/ZnS QDs. The the fluorescence quenching between ractopamine and the QDs was studied, showing that the linear range of ractopamine covered a range of 5.00×10^{-10}–3.55×10^{-7} mol L^{-1} with a correlation of 0.9934 under optimum conditions. The mechanism of the fluorescence quenching may involve charge transfer from QDs to ractopamine.

Graphene quantum dots (GQDs) are graphene sheets smaller than 100 nm with a unique sp^2 and sp^3 hybrid structure. As recently emerging carbon-based materials, GQDs have many excellent features, such as low toxicity, high fluorescent activity, robust chemical inertness and water solubility. Zhou and others synthesized a fluorescent sensor based upon GQDs and MIPs for the determination of paranitrophenol (4-NP).[71] Figure 7.10 illustrates the synthesis of a MIP-coated GQDs fluorescence sensor, which could

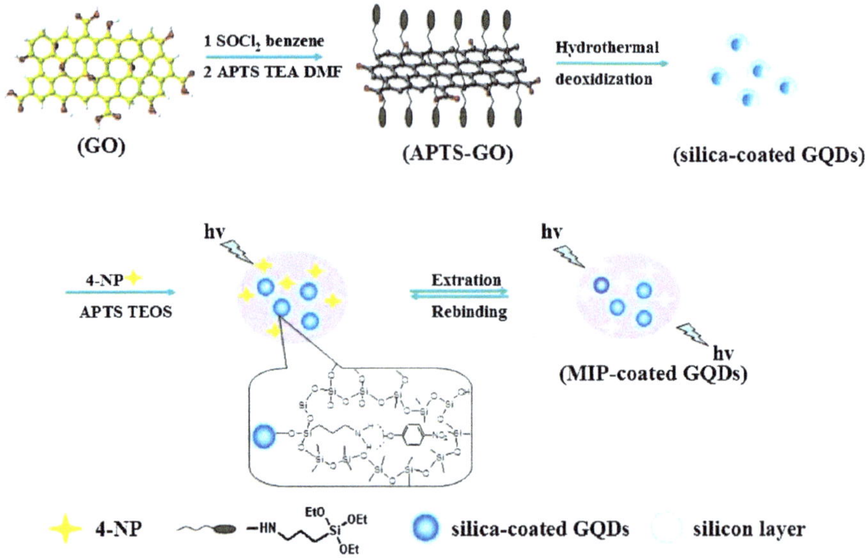

Figure 7.10 Illustration of the preparation of MIP-coated GQDs. TEA refers to triethylamine. DMF refers to *N*,*N*-dimethyl formamide.
Reproduced from *Biosensors and Bioelectronics*, Volume 52, Y. Zhou, Z. B. Qu, Y. Zeng, T. Zhou and G. Shi, A novel composite of graphene quantum dots and molecularly imprinted polymer for fluorescent detection of paranitrophenol, 317–323. Copyright 2014, with permission from Elsevier.

be divided into four steps: (1) modifying graphene oxide (GO) with APTS (3-aminopropyltriethoxysilane) to form a precursor of silica layer formed in the following reaction; (2) APTS-GO was added into silica-coated GQDs *via* a simple hydrothermal approach; (3) surface imprinting polymer was obtained by sol–gel polymerization in the presence of silica-coated FQDs using 4-NP as template, APTS as functional monomer and TEOS as cross-linker; (4) the template was removed from the silica matrix using anhydrous ethanol and distilled water as washing solution. Then, the MIP-coated GQDs fluorescence sensor was fabricated. The fluorescent quenching phenomenon between the target and the sensor might result from the resonance energy transfer from MIP-coated GQDs sensor (donor) to 4-NP molecules (acceptor). This MIP-coated GQDs sensor had a wide linear range and low detection limit, which revealed a prospective application in the analysis field.

Gold nanoclusters (AuNCs) have strong luminescence derived from the aggregation-induced emission with a quantum yield about 15%, and possess excellent merits, such as small size, easy preparation and non-poisonous feature.[72] Wu and coauthors proposed a flexible fluorescent sensing strategy for the recognition and detection of bisphenol A (BPA) based upon MIPs-coated AuNCs, combining the high selectivity of MIPs and the strong fluorescence property of AuNCs.[73] Figure 7.11 shows the process for the preparation of SiO_2@AuNCs-MIPs. Firstly, amino-functionalized SiO_2 and

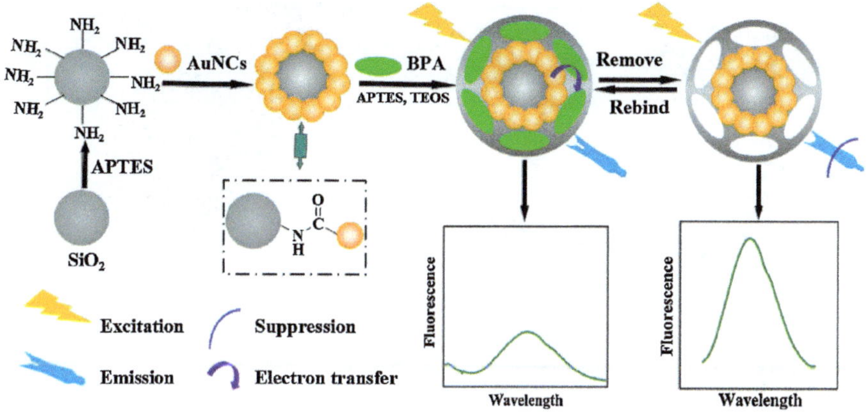

Figure 7.11 Schematic illustration of the process for the preparation and sensing of the SiO_2@AuNCs-MIPs.
Reproduced from *Sensors and Actuators B: Chemical*, Volume 211, X. Wu, Z. Zhang, J. Li, H. You, Y. Li and L. Chen, Molecularly imprinted polymers-coated gold nanoclusters for fluorescent detection of bisphenol A, 507–514. Copyright 2015, with permission from Elsevier.

carboxyl-terminated AuNCs were synthesized. They were then used to prepare SiO_2@AuNCs by making use of the powerful amido bonds. The MIPs-based SiO_2@AuNCs were fabricated by coating a MIPs layer onto the SiO_2@AuNCs surface *via* a sol–gel process using BPA as template and APTES as functional monomer. After removing the template, the fluorescence sensor was obtained with strong fluorescent emission. However, when BPA was rebound, fluorescence quenching of MIPs-based SiO_2@AuNCs would occur. Herein, the fluorescence quenching may result from electron transfer between the electron-rich aromatic ring (conjugating-OH) and electron-deficient amino groups on the surface of AuNCs. This sensor showed a linear relationship between the fluorescence intensities and the concentrations of BPA (0–13.1 $\mu mol\,L^{-1}$) with a correlation coefficient of 0.9942. Therefore, it could sense BPA in a fluorescent manner.

Upconverting particles (UCPs) doped with lanthanide have widely gained attention due to their specific characteristics, termed anti-Stokes, converting low-energy light to high-energy light through multiple photon adsorptions or energy transfers. Coating MIPs directly onto the surface of UCPs to prepare the fluorescence sensor was studied for the determination of clenbuterol and metolcarb,[6,74] showing great versatility and flexibility for future application.

To improve the sensitivity and selectivity of the sensor, a novel hybrid probe based upon aptamer-MIPs as double recognition element was grafted on upconversion nanoparticles (UCNPs) for enrofloxacin (ENR) sensing.[75] Figure 7.12 shows the preparation and recognition process of aptamer-MIP/UCNPs. Firstly, silica-modified UCNPs were functionalized with amino groups, and then immobilized with ENR aptamers onto their surface *via* the

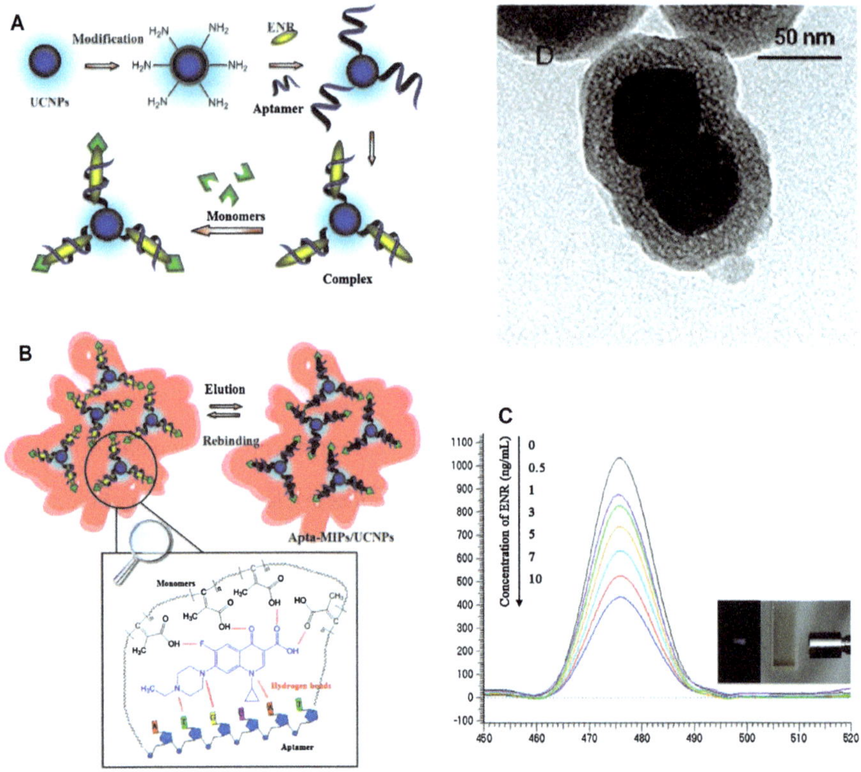

Figure 7.12 Schematic representation of the preparation (A) and recognition (B) process for aptamer-MIP/UCNPs. Luminescent spectra of aptamer-MIP/UCNPs with addition of various concentrations of ENR (C). TEM image of aptamer-MIP/UCNPs.
Reproduced from *Biosensors and Bioelectronics*, Volume 87, X. Liu, J. Ren, L. Su, X. Gao, Y. Tang, T. Ma, L. Zhu and J. Li, Novel hybrid probe based on double recognition of aptamer-molecularly imprinted polymer grafted on upconversion nanoparticles for enrofloxacin sensing, 203–208. Copyright 2017, with permission from Elsevier.

biotin-avidin affinity reaction. Adding the resulting aptamer-UCNPs into ENR solution allowed correct folding of the aptamer upon the existing targets, producing a "pre-polymerization" complex. Using this complex as a polymerization centre, the MIP was synthesized using MAA as functional monomer that interacted with the residual functional groups of ENR. The diameter of the obtained aptamer-MIP/UCNPs was about 150 nm (Figure 7.12D). This aptamer-MIP/UCNP has maximal luminescence intensity in the absence of ENR, and fluorescence quenching occurred when the increased target was bound to the sites (Figure 7.12C).

Many MIPs-based fluorescence sensors have been developed and used in sensing food safety and quality analysis, such as detection of ractopamine in pork,[6,70] detection of deltamethrin in fruit and vegetables,[76] detection of enrofloxacin in fish,[75] detection of melamine in milk,[77] detection of tetracycline in milk,[78] detection of λ-cyhalothrin in Chinese spirits,[79] detection of dimethoate in Chinese cabbage, broccoli and cucumber[80] and bifenthrin detection in honey.[81]

7.3.4 Surface Enhanced Raman Scattering Sensors

Surface enhanced Raman scattering (SERS) is a great discovery in surface science and spectroscopy due to its extremely high sensitivity and extensive applications, which can overcome the disadvantages of conventional Raman spectroscopy. After achieving single molecule detection based upon SERS in 1997,[82,83] SERS technology had been widely applied to various analytical cases. Many nanoparticles or metals (gold, silver and copper) in nanoscale with roughened surfaces can be employed as the SERS substrate to enhance the normal Raman signals of molecules. However, the mechanism of the enhancement effect is still not fully clear and is usually explained with chemical enhancement caused by charge transfer and electromagnetic enhancement when the molecule is placed in the zone of a localized surface plasmon resonance of SERS substrate.[84]

An innovative method of MIPs coupled with SERS to sensing melamine in whole milk was developed by Hu and others.[85] They firstly synthesized melamine MIPs in an aqueous environment using MAA as functional monomer and EGDMA as cross-linker. After the removal of the template melamine using the Soxhlet method, the dried MIPs were packed into a solid phase extraction (SPE) cartridge with one frit at each side to carry out the separation and purification process. The final eluent was collected and then dried using nitrogen. The residuals were redissolved in methanol, and then directly deposited onto the surface of silver dendrite (as SERS substrate) for Raman spectral collection. Figure 7.13 presents a schematic illustration of a MIPs-SERS sensor for the detection of melamine in whole milk. Regression models showed a good linear relationship between the intensity of the melamine SERS band and the concentration of melamine in the range from 0.005 to 0.05 mmol L^{-1} with the limit of detection of 0.012 mmol L^{-1}. These results confirmed the high sensitivity of this MIPs-SERS sensor.

Gao and coworkers developed a sensor combining MIPs, thin layer chromatography (TLC) and SERS for the determination of Sudan I in paprika powder.[84] Sudan I MIPs were firstly synthesized, and then used to fabricate a MIPs-TLC plate. The sample solution was dropped onto the MIPs-TLC plate. Then, TLC developing was carried out. The result showed that the template Sudan I could be successfully captured by MIPs. After putting gold colloid onto the original spot, Raman spectra were collected immediately. This MIPs-TLC-SERS sensor had fast, reliable and cost-effective features for monitoring of food chemical contamination and adulteration.

Exploring the integration of MIPs with a SERS-active film as sensing layer is very interesting and practical. Guo and coworkers presented a core-shell silver-MIPs hybrid structure to serve as a novel SERS platform.[86] The

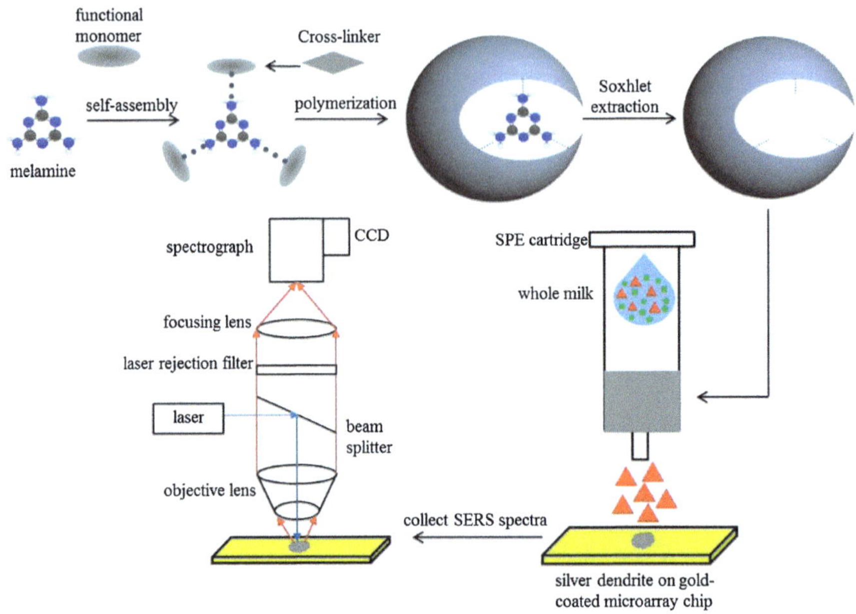

Figure 7.13 Schematic illustration of MIPs-SERS biosensor for the detection of melamine in whole milk. CCD: charge coupled device.
Reproduced from *Food Chemistry*, Volume 176, Y. Hu, S. Feng, F. Gao, E. C. Li-Chan, E. Grant and X. Lu, Detection of melamine in milk using molecularly imprinted polymers–surface enhanced Raman spectroscopy, 123–129. Copyright 2015, with permission from Elsevier.

preparation process of Ag@MIPs is shown in Figure 7.14A. Firstly, silver nanoparticles were synthesized using a liquid-phase reduction method. The diameter of the prepared Ag particle was about 800 nm. Then, the Ag particles were modified with 3-methacryloxypropyltrimethoxysilane (MPS). In order to synthesize the Ag@MIPs hybrid, rhodamine 6G (template, 25 mmol), acrylamide (functional monomer, 5 mmol), EGDMA (cross-linker, 25 mmol), AIBN (initiator, 20 mg) and Ag particles modified with MPS (200 mg) were dispersed into acetonitrile and stirred at 65 °C for 24 h under nitrogen protection. Then, Ag@MIPs was obtained after the removal of the template using methanol-acetic acid as washing solution. By contrast with TEM image of Ag spheres, a shell with a thickness of about 16 nm was clearly visualized on the Ag@MIPs surface (Figure 7.14B and C). Additionally, as-prepared core-shell nanostructure (Ag@MIPs) showed a high sensitivity and selectivity towards the template molecule. As shown in Figure 7.14D, Raman spectral fingerprints can still be clearly observed even when the concentration of rhodamine 6G was diluted to 10^{-14} M. This sensitive and selective performance would be derived from the core–shell structure and near-surface recognition sites. Because the thickness of the MIPs shell in this case was up to 16 nm, it was challenging to interpret this enhancement phenomenon by the classical theory, such as resonance energy transfer. The

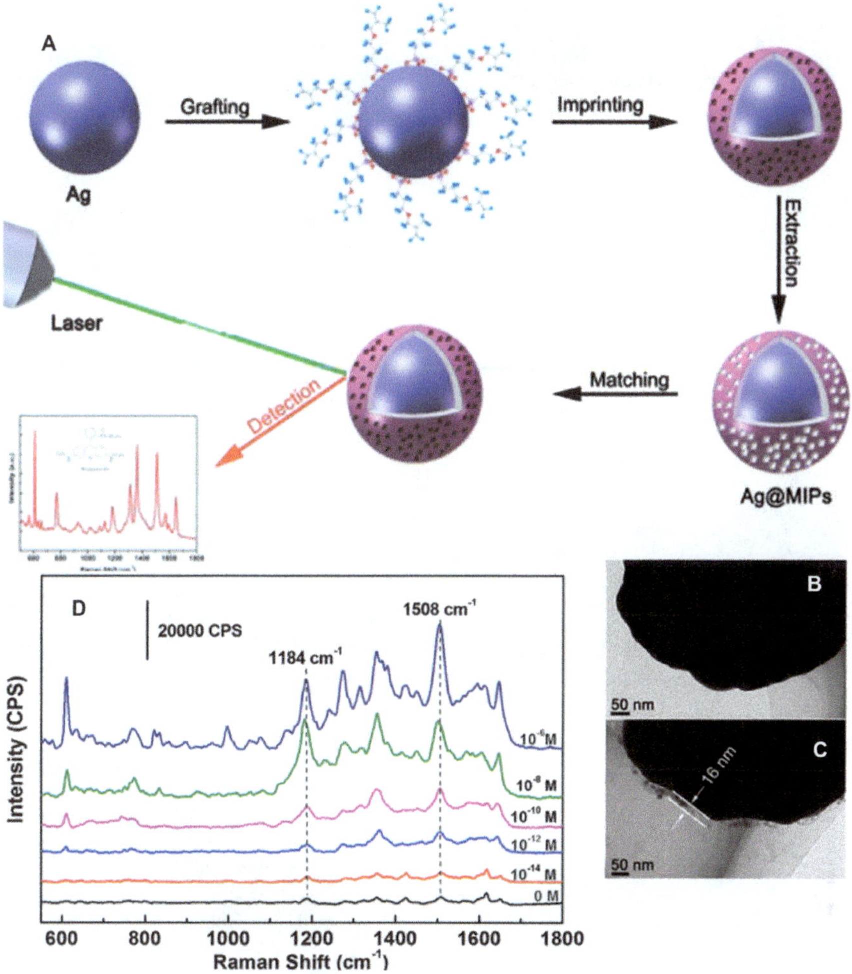

Figure 7.14 Schematic illustration of the fabrication and detection process of Ag@MIPs (A). TEM images of Ag sphere (B) and Ag@MIPs (C). Concentration-dependent SERS spectra obtained from Ag@MIPs (D). Reproduced from ref. 86 with permission from the Royal Society of Chemistry.

possible explanation for the mechanism of MIPs-based SERS is the existence of an unobserved tunnel leading chemical species to the near surface of a plasmonic metal, named the "gate effect".[86,87]

To establish a fast and simple detection method towards ciprofloxacin in a mixed sample, Guo and coworkers developed a magnetic imprinted SERS technique.[88] They firstly prepared Fe_3O_4 through a solvothermal reaction. As the core, the prepared magnetic nanoparticles were used to prepare a MIPs coated magnetic spherical nanoparticle (Figure 7.15). As shown in Figure 7.15E, the thickness of the coating of the MIPs was about 15 nm.

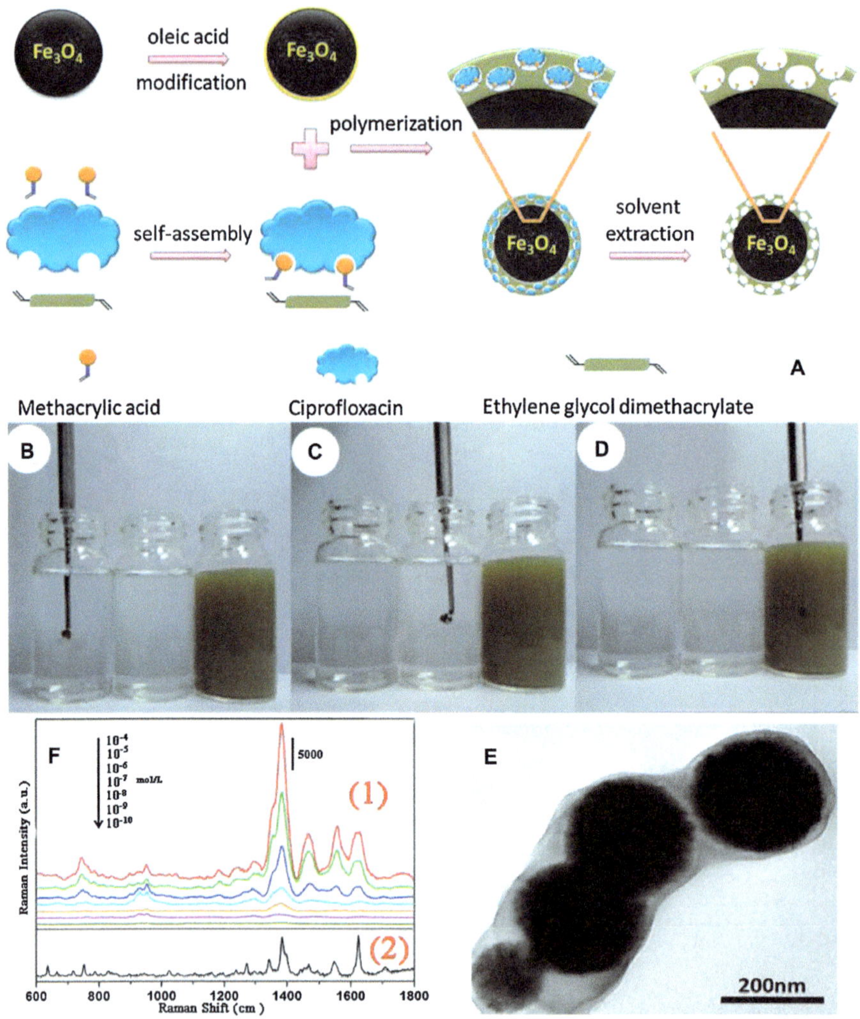

Figure 7.15 Schematic illustration of the preparation procedures of MIPs-coated magnetic spherical nanoparticles (A). Detection procedures of ciprofloxacin from the diluted bovine serum albumin by magnetic-MIPs chip. The solution in each bottle from left to right in (B), (C) and (D) is ciprofloxacin in diluted foetal bovine serum, distilled water and silver sol, respectively. TEM image of MIPs-coated magnetic nanoparticles (E). SERS spectra of ciprofloxacin adsorbed onto magnetic-MIPs (F(1)) and normal Raman spectra of bulk ciprofloxacin (F(2)).
Reproduced from ref. 88 with permission from the Royal Society of Chemistry.

Then, a magnetic-MIPs chip was invented by spreading a certain amount of magnetic-MIPs nanoparticles on the surface of one side of a cylindrical magnet (height, 1.5 mm; radius, 1.2 mm) that was integrated at the end of a bar. Two critical parameters of this design are: (1) the magnetic separation step was not required, which made the whole detection simpler and shorter, and (2) a smaller amount of magnetic-MIPs on the chip than the amount of the conventional format, making the enrichment effect of this magnetic-MIPs-SERS technique more remarkable. This magnetic-MIPs chip was portable and easy to use. To achieve sample detection by this chip, the procedures were only to sequentially put this chip in the system to be detected, followed by addition of distilled water and silver sol and blending for 2 min in each solution. After that, the SERS spectrum was directly collected on this chip. The whole detection procedure could be finished within 10 min.

Many MIPs-based SERS sensors were prepared and used in food safety and quality analysis, such as detection of ractopamine in pork and pig liver,[89] detection of melamine in milk,[85] detection of chloramphenicol in honey and milk,[90] detection of histamine in canned tuna,[91] detection of Sudan I in paprika powder[84] and detection of alpha-tocopherol in vegetable oils.[92]

7.3.5 Surface Plasmon Resonance Sensors

In recent years, the surface plasmon resonance (SPR) technique has become very popular due to its simplicity, fast response time, label-free nature, high sensitivity and low cost. Surface plasmon is the charge density oscillation of conduction electrons at the metal-dielectric interface, which are excited resonantly by *p*-polarized light incident to the metal-dielectric interface as the wave vector and frequency of the incident light matches with that of the surface plasmon.[93] Gold is usually regarded as the most practical metal because of a significant SPR dip in the near-infrared region and easy modification using gold-thiolchemistry for the immobilization of self-assembled monolayers. Functionalized gold surface results in a shift in the plasmon curve from thin to thick lines, which can be easily quantified in terms of incidence angle shift, corresponding to the shift in the positions of the reflectivity minimum before and after the derivatization.[94]

A combination of the SPR technique and molecular imprinting technology has been studied. MIPs-based SPR sensors are typically induced by using two methods, namely "grafting from" and "grafting to".[36,95] In the "grafting from" method, the initiator is first immobilized onto the surface of the substrate and polymer chains are propagated from the surface of the solid support. However, the thickness of the MIPs films may be thick and brittle due to the difficulty in immobilizing the initiator and controlling the quantity of the MIPs polymerization mixture. In the "grafting to" technique, as-prepared MIPs particles are directly attached onto a bare sensor by an intermediate step using a physical entrapment technique.

Choi and coauthors developed a MIPs-based SPR sensor for the detection of zearalenone, a representative mycotoxin identified in agri-food products.[96] Figure 7.16(a) and (b) show schematic diagrams of the setup for electropolymerization and SPR for the detection of zearalenone. Gold film (50 nm thickness) was deposited onto a glass slide (18×18 mm^2) by electron beam evaporation and used as a cathode after making contact

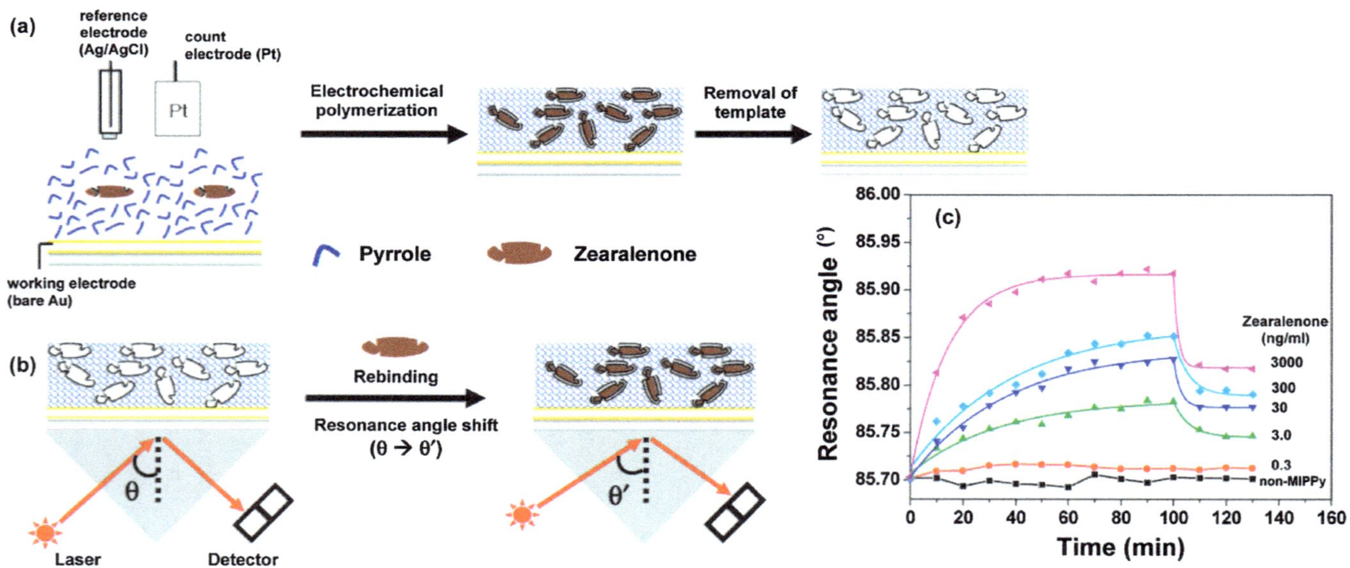

Figure 7.16 Schematic diagrams of the setup for electropolymerization and SPR to detect zearalenone. (a) Preparation of molecularly imprinted polypyrrole on bare gold using a three-electrode electrochemical system. (b) The shift in resonance angle of the SPR sensor resulting from the rebinding of zearalenone to the MIP film. Resonance angle shifts according to the concentration of zearalenone and reaction time (c), the non-MIPs curve (non-MIPPy) was obtained using a non-molecular imprinted polymer and a 3000 ng mL^{-1} concentration of zearalenone.
Reproduced with permission from S. W. Choi, H. J. Chang, N. Lee, J. H. Kim and H. S. Chun, *Journal of Agricultural and Food Chemistry*, 2009, 57, 1113–1118. Copyright 2009 American Chemical Society.

with the aluminium foil in the Teflon-made electrochemical cell for electropolymerizing MIPs with pyrrole as functional monomer and tetraethylammonium tetrafluoroborate as electrolyte. After the synthesis of MIPs, the template was extracted from the MIPs film with a constant washing procedure in acetonitrile, methanol and chloroform. Then, electrolytes with a polymeric matrix were obtained. The performance of MIPs-SPR sensors has been assessed with SPR measurement. As shown in Figure 7.15(c), an increase in the resonance angle was observed upon the interaction of the MIPs-SPR sensor with zearalenone standard solutions in comparison to a blank solution (without zearalenone). These results suggested that a combination of SPR sensing with MIPs film is a promising alternative analytical method for the detection of mycotoxin contamination in the real samples.

Verma and Gupta presented a simple and highly selective optical fibre sensor for the detection of tetracycline in foodstuffs by using the combination of SPR and MIPs.[93] This sensing probe was prepared in two steps: (1) coating metal on the optical fibre and (2) coating MIPs over the metal-coated sensing region of the optical fibre. For the first step, the cladding was removed from a 1-cm length of the 17-cm long plastic clad silica (PCS) optical fibre from the middle portion. The unclad portion was cleaned twice with acetone, and then was coated with a 40-nm thick film of silver metal by the thermal evaporation method. The metal-coated region of the fibre was further cleaned with a copious amount of deionized water. In the second step, the molecular imprinting process was carried out over the metal-coated probe and the process of preparation is shown in Figure 7.17A. Firstly, the polymerization medium was prepared by mixing appropriate amounts of the master solution (acrylamide, *N*,*N*-methylenebisacrylamide and tetracycline in Millipore water), phosphate buffer, acrylic acid, ammonium persulfate (initiator), *N*-tetramethylethylenediamine (catalyst) in a flask and diluting up to 10 mL, and then pouring into a cylindrical vessel. Then, the metal-coated cleaned fibre probe was dip-coated with the polymerization medium and kept in a vacuum oven at 60 °C for 4 h for polymerization. During the polymerization process, the polymer matrix became attached over the silver-coated fibre core. Then, the TC-MIP probe was dipped in an acetic acid solution containing 10% (w/v) sodium lauryl sulfate for 2 h at room temperature for the removal of the template. Finally, the MIP probe was dried. The experimental setup used for the characterization of TC-MIPs sensing probe is shown in Figure 7.17B. The fibre optic sensing probe was cleaved at both ends for launching maximum light, which was then fixed in a flow cell. Polychromatic light from a tungsten halogen lamp was used to launch the light into the fibre. A stabilizing circuit was used to generate a regulated current. The incident light guided through the optical fibre was collected at the other end of the optical fibre by a spectrometer interfaced with a computer. The SPR spectra of different concentrations of tetracycline samples were recorded using the TC-MIP probe in the flow cell, as shown in Figure 7.17C. The spectra were recorded within 40 seconds after pouring the sample into the flow cell. However, the sensitivity of the sensor decreased

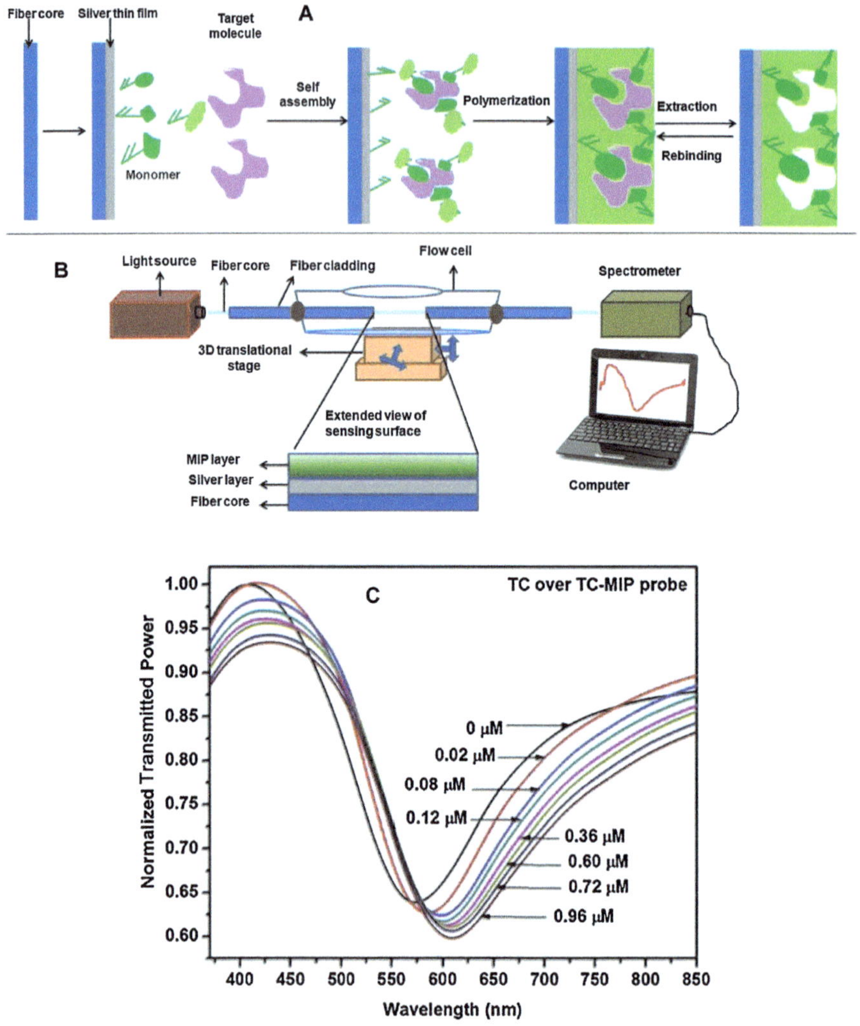

Figure 7.17 Stepwise preparation of the sensing probe (A). Schematic of the experimental setup for the characterization of the tetracycline sensors (B). Surface plasmon resonance spectra for different concentrations of tetracycline (TC) samples over the TC-MIP probe.
Reproduced from ref. 93 with permission from the Royal Society of Chemistry.

along with the increase in the concentration. Therefore, this prepared sensor is useful for the detection of low concentrations of tetracyclines in foodstuffs.

Yao and colleagues developed a method to enhance the sensitivity of detection in SPR spectroscopy integrated with a surface molecular imprinting recognition system and employing magnetic molecular imprinting polymer nanoparticles for amplifying SPR response.[97] A schematic illustration of the preparation of magnetic imprinted Fe_3O_4@PDA (polydopamine) for the

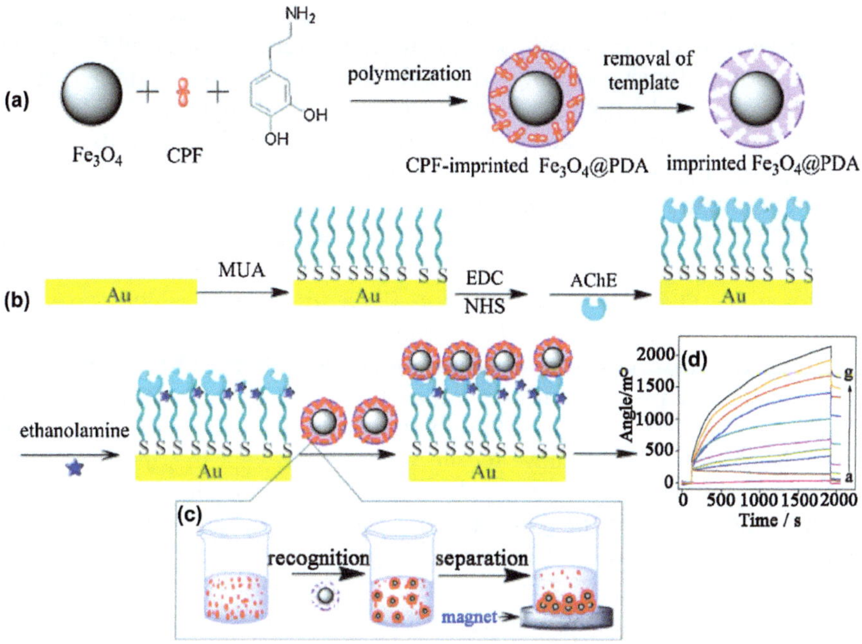

Figure 7.18 (a) Schematic illustration of the preparation of magnetic imprinted Fe_3O_4@PDA (polydopamine) for CPF (chlorpyrifos) recognition, (b) schematic illustration of the stepwise preparation process of the sensor surface for the detection of CPF and (c) illustration of recognition and separation of CPF with imprinted Fe_3O_4@PDA. SPR sensorgrams of imprinted Fe_3O_4@PDA nanoparticles binding to the AChE/MUA/gold chip after rebinding different concentrations of CPF, curves a–g represent 0, 0.001, 0.05, 0.5, 2, 5, 8, 10, 20 and 40 μM CPF.
Reproduced with permission from G. H. Yao, R. P. Liang, C. F. Huang, Y. Wang and J. D. Qiu, *Analytical Chemistry*, 2013, **85**, 11944–11951.

recognition of chlorpyrifos (CPF) is shown in Figure 7.18(a). Fe_3O_4 nanoparticles were prepared through the coprecipitation method. Then, 50 mg of Fe_3O_4 particles were dispersed in 20 mL Tris buffer (10 mM, pH = 8.0). Subsequently, 5 mL of CPF acetonitrile solution (1 mg mL^{-1}) was added and the mixture was mechanically stirred for 2 h at room temperature. After that, dopamine (25 mg) was added to the solution in an open flask to allow for continuous oxygen supply, and the reaction was continued for another 4 h at room temperature. Then, the MIP-coated Fe_3O_4 was obtained. The template was removed with a mixture of acetic acid and acetonitrile as washing solution. The sensor was prepared as follows. As shown in Figure 7.18(b), a bare gold film was initially immersed in piranha solution for 2 min to eliminate any contamination, and was then washed with absolute ethanol and ultrapure water, respectively, and dried with N_2. The cleaned surface of the gold chip was placed on the prism with mach oil and covered with the cuvette. Then, 11-mercaptoundecanoic acid (MUA) solution was injected into the cuvette to form a self-assembled monolayer with carboxyl functional group. After extensive rinsing with Tris buffer, ethyl-3-(3-dimethylamino-propylcarbodiimide) hydrochloride (EDC) and *N*-hydroxysuccinimide (NHS) were added into the cuvette to activate the terminal carboxyl group of MUA. Then, acetylcholinesterase (AChE) was added to form covalent amide linkages. The unemployed carboxyl groups were deactivated by incubating the surface with ethanolamine for 10 min. Finally, the fabricated AChE/MUA/gold film sensor interface was further used to detect CPF bound onto the imprinted Fe_3O_4@PDA. The detection procedure consisted of two steps. Firstly, the imprinted Fe_3O_4@PDA was mixed with various concentrations of CPF, and the mixture was shaken on a rocking table for 12 h at room temperature, and then was separated and enriched using a magnet, as shown in Figure 7.18(c). Secondly, the CPF-rebound imprinted Fe_3O_4@PDA solution was injected into the SPR cuvette, the solution was kept in contact with the ACeE immobilized on the AChE/MUA/gold film sensor chip for the association for 30 min, and then the solution was drained out with a peristaltic pump, and a 50 μL of Tris buffer was injected for the dissociation measurement. Under optimized experimental conditions, the CPF-rebound imprinted Fe_3O_4@PDA nanoparticles with various concentrations of CPF were injected into the SPR cuvette. The analysis of different concentrations of CPF is shown in Figure 7.18(d). The SPR angle shifts resulting from the binding of the imprinted Fe_3O_4@PDA nanoparticles gradually increased along with the increase in CPF concentration (curves a–g). These results demonstrated that the sensor made this magnetic imprinted Fe_3O_4@PDA an attractive recognition element for various types of SPR sensors to detect pesticide residuals in real samples.

Altintas and others presented a MIPs-based SPR probe to detect bacteriophage MS2.[98] A schematic illustration of the preparation of a MIPs-based SPR sensor is shown in Figure 7.19. Firstly, MIPs targeting the bacteriophage MS2 (template) were investigated using a novel solid-phase synthesis method. Briefly, the aminosilane-coated glass beads were washed with acetone under vacuum, and then incubated in glutaraldehyde solution in phosphate buffered saline (PBS, pH 7.2) for 2 h. Bacteriophage MS2 templates were then immobilized onto the surface of the bead by incubating the glass beads with a solution of the template in PBS overnight at 4 °C. The unemployed glutaraldehyde species remaining on the beads were blocked with ethanolamine. After that, the glass beads were used to prepare MIPs using *N*-isopropylacrylamide, *N*,*N*′-methylenebis(acrylamide) (BIS), *N*-tertbutylacrylamide (TBAm) and acrylic acid as monomers, *N*,*N*,*N*′,*N*′-tetramethylethylenediamine (TEMED) and ammonium persulfate (APS) as initiator. *N*-(3-Aminopropyl) methacrylamide hydrochloride (54 mg) was used as additional monomer to functionalize primary amine groups to the surface of the MIPs nanoparticles. The reaction was carried out at room temperature for 2 h, and then the temperature was adjusted to 15 °C. The MIP receptors were eluted from the affinity media by passing three fractions of 50 mL of double-distilled

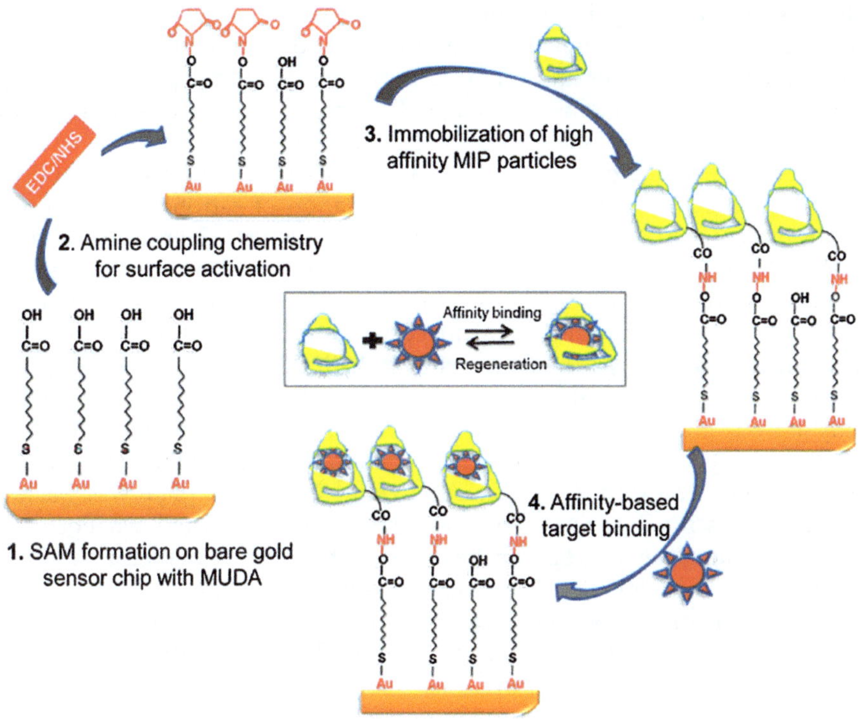

Figure 7.19 Schematic illustration of the preparation of MIPs-based SPR sensor for viral recognition.
Reproduced with permission from Z. Altintas, M. Gittens, A. Guerreiro, K. A. Thompson, J. Walker, S. Piletsky and I. E. Tothill, *Analytical Chemistry*, 2015, **87**, 6801–6807. Copyright 2015 American Chemical Society.

water at 60 °C. To immobilize the MIPs nanoparticles on the surface of the sensor, the sensor chips were initially coated with mercaptoundecanoic acid (MUDA) to obtain a self-assembled monolayer (SAM) with carboxyl groups on the surface of the sensor. After the surface of the chip was activated by an EDC/NHS mixture, the MIP receptors were deposited onto the surface *via* covalent amide linkages. Finally, a regenerative MIPs-based viral detection assay was successfully developed using a new MIPs-based SPR sensor, which provided an alternative technology for specific detection.

Many MIPs-based SPR sensors were prepared and used in food safety and quality analysis, such as detection of chloramphenicol in honey,[99] detection of zearalenone in corn,[96] detection of chlorpyrifos in apple,[97] detection of amoxicillin in egg[100] and detection of histamine in carp.[101]

7.3.6 MIPs-based Enzyme-linked Immunoassays

Enzyme-linked immunosorbent assay (ELISA) is one of the most commonly used methods for the analysis of proteins, hormones, viruses and drugs in the environmental, agri-food and forensic industries due to its rapid, sensitive and selective features. However, the stability of reagents, the high cost of producing antibodies and the need for refrigerated transport and storage are regarded as problems.[102] Using MIPs as artificial antibodies in immunoassay analysis is one of the most exciting applications.

Wang and colleagues developed a fast and new direct competitive biomimetic ELISA method for the determination of estrone using a novel molecularly imprinted film of the controlled thickness as an artificial antibody.[103] The novel imprinted film was directly polymerized onto the 96-well plate as follows. The template estrone was dissolved in a mixture of dimethyl sulfoxide and toluene, and then mixed with 1.5 mmol of 3-aminophenylboronic acid and 0.57 mL of 1-butyl-3-methylimidazolium hexafluorophosphate ($BMIM^+$ PF_6^-). After magnetic stirring for 30 min, 100 μL of the mixture was placed in the wells of a 96-well plate and mixed with 100 μL of ammonium persulfate (initiator). The reaction was carried out with shaking at room temperature for 20 min. After that, the 96-well plate was washed three times with deionized water to remove the unreacted reagents and template. Then, the template was removed with a mixture of methanol/acetic acid for 48 h by an ultrasonic cleaner, followed by methanol for 12 h. Finally, the resultant product was dried at 40 °C for 2 h. The ionic liquids ($BMIM^+$ PF_6^-) at room temperature were used to reduce the cracking and shrinking of the novel imprinted film and act as a pore template. Directly synthesizing the MIPs film onto the surface of the 96-well plate simplified the operation procedure and was particularly useful for biomimetic ELISA. The thickness of the novel film can be controlled by polymerization time. The developed biomimetic ELISA method exhibited excellent performance in real sample analysis, but the sensitivity was low.

Bi and Liu prepared a glycoprotein-imprinted 96-well microplate for ELISA by boronate affinity-based oriented surface imprinting.[104] The preparation of glycoprotein imprinting polymers is shown in Figure 7.20. The surface of the 96-well microplate was firstly modified with phenylboronic acid as follows. The wells of a 96-well microplate were filled with a H_2SO_4/HNO_3 mixture and kept for 30 min at room temperature. After being washed with water to achieve a neutral pH, the wells were filled with 5% aqueous 3-aminopropyltriethoxysilane solution, and shaken at room temperature for 2 h. After the dried wells were filled with 4-formylphenylboronic acid dissolved in anhydrous methanol, the microplate was sealed and slightly shaken at room temperature for 12 h. Subsequently, each well was supplemented with sodium cyanoborohydride methanol solution, and the microplates were sealed again and shaken for 24 h in the same conditions. Finally, the solutions in the wells were discarded, and the wells were washed with anhydrous methanol 5–10 times, followed by air drying. The obtained phenylboronic acid-modified 96-well microplate was filled with 100 μL $well^{-1}$ HRP (horseradish peroxidase, as a representative target) phosphate buffer solution, and then was sealed and shaken at room temperature for 2 h. After being washed with phosphate buffer three times,

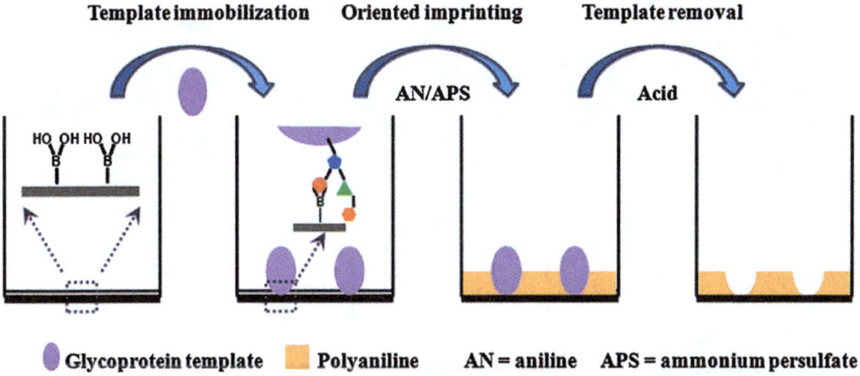

Figure 7.20 Schematic diagram of boronate affinity-based oriented surface imprinting of glycoprotein.
Reproduced with permission from X. Bi and Z. Liu, *Analytical Chemistry*, 2013, **86**, 959–966. Copyright 2013 American Chemical Society.

each well was filled with aniline dissolved phosphate buffer for 5 min. Then, ammonium persulfate was added into each well, and the microplate was sealed and shaken at room temperature for the desired time (*i.e.*, 5–80 min). After polymerization, the microplate was washed successively with water, 5% glacial acetic acid containing 0.1% Tween 20 and water to remove template, acid and reactants. The prepared HRP-imprinted microplate was dried by air and kept at 4 °C. Herein, a new version of the boronate affinity molecular imprinting method was developed to prepare high performance MIPs for ELISA towards glycoproteins, which lay the foundation for the detection of glycoprotein in foodstuffs.

Chianella and coworkers developed an ELISA for the determination of vancomycin replacing antibodies with MIPs nanoparticles.[105] The MIPs nanoparticles for vancomycin were synthesized using an automated solid-phase synthesis approach. Firstly, the template vancomycin was immobilized through its amino group by glutaraldehyde coupling to the surface of amine-derivatized glass beads. An automated reactor was utilized for the synthesis of the nano-MIPs using acrylic acid (AA), *N*-isopropylacrylamide (NIPAm) *N*,*N*′-methylene-bis-acrylamide (BIS) and *N*-*tert*-butylacrylamide (TBAm) as monomer, ammonium persulfate and tetramethylethylenediamine (TEMED) as initiator and the vancomycin-functionalized glass beads as the solid phase. After polymerization, the non-covalent interactions between the template attached to the solid-phase and the high-affinity nano-MIPs were disrupted at 60 °C, allowing the pure fraction of particles free of residual template and monomers to be collected. A schematic diagram of the preparation of nano-MIPs is shown in Figure 7.21A.

The obtained nano-MIPs were used in a competitive ELISA as follows. The wells of the microplate were coated with nano-MIPs by dispensing undiluted stock solution (40 μL well^{-1}) by evaporation overnight. After being washed with PBS (2×250 μL), each well was blocked by incubation with PBS (350 μL)

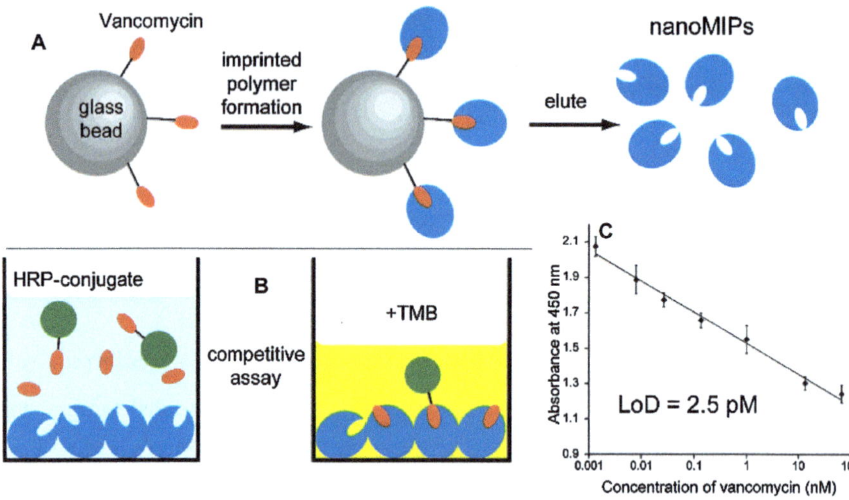

Figure 7.21 Schematic diagram of the preparation of nano-MIPs and the competitive assay.
Reproduced with permission from I. Chianella, A. Guerreiro, E. Moczko, J. S. Caygill, E. V. Piletska, I. M. P. De Vargas Sansalvador, M. J. Whitcombe, S. A. Piletsky, *Analytical Chemistry*, 2013, **85**, 8462–8468. Copyright 2013 American Chemical Society.

containing bovine serum albumin (BSA, 0.1%) and Tween 20 (1%) for 1 h. Then, the wells were washed with PBS (3×250 μL). To each well, a solution of HRP-vancomycin was added, containing free vancomycin at a final concentration of between 0.001 and 70 nM. Plates were incubated in the dark at room temperature for 1 h. Wells were then washed with PBS (3×300 μL), containing BSA (0.1%) and Tween 20 (1%), followed by addition of TMB reagent (100 μL). After incubation for 10 min, the enzymatic reaction was stopped by the addition of H_2SO_4 (0.5 M, 100 μL). Colour development was determined by measuring the absorbance of each well at 450 nm using a UV-vis microplate reader, as shown in Figure 7.21B and C. Therefore, the prepared nano-MIPs with high affinity and the lack of a requirement for cold chain logistics made them an attractive alternative to the traditional antibodies used in ELISA.

Many MIPs-based ELISA sensing techniques were developed and used in food safety and quality analysis, such as detection of metolcarb in apple juice, cucumber and cabbage,[10] detection of chloramphenicol in sea cucumber,[106] detection of ractopamine in pork,[107] detection of tribenuron-methyl in water,[108] detection of clenbuterol in water and pork[109] and detection of trichlorfon and acephate in vegetables.[102]

7.4 Conclusion

Developing and applying MIPs in sensors have been validated to be effective as a molecular recognition element prepared *via* chemical synthesis.

Fabrication methods, "grafting from" and "grafting to", have been studied and utilized in MIPs-based sensors, including electrochemical sensors, quartz crystal microbalance sensors, fluorescence sensors, surface enhanced Raman scattering sensors, surface plasmon resonance sensors and MIPs-based enzyme-linked immunoassays. These MIPs-based sensing methods have good sensitivity, selectivity, fast response time and repeatability. Compared to other biological-based recognition elements, such as antibodies, MIPs have some unique advantages that can be readily applied in the agri-food industry.

To guarantee the safety and quality of foods, applications of MIPs-based sensing methods have been widely investigated for food sensing, especially for the detection of food chemical and microbiological hazards, such as pesticide residues, trace levels of mycotoxins and pathogenic bacteria. Further research will mainly aim to continuously improve the working performance of MIPs as the recognition element in a sensing system as well as reduce the production cost of a MIPs-based sensor.

References

1. Y. Lei, *Anal. Bioanal. Chem.*, 2016, **408**, 1727–1733.
2. K. Haupt, *Anal. Chem.*, 2003, **75**, 376A–383A.
3. M. Kempe and K. Mosbach, *Anal. Lett.*, 1991, **24**, 1137–1145.
4. C. C. Hwang and W. C. Lee, *J. Chromatogr. A*, 2002, **962**, 69–78.
5. H. Zhang, L. Ye and K. Mosbach, *J. Mol. Recognit.*, 2006, **19**, 248–259.
6. Y. Tang, Z. Gao, S. Wang, X. Gao, J. Gao, Y. Ma, X. Liu and J. Li, *Biosens. Bioelectron.*, 2015, **71**, 44–50.
7. H. Taguchi, H. Sunayama, E. Takano, Y. Kitayama and T. Takeuchi, *Analyst*, 2015, **140**, 1448–1452.
8. J. He, L. Song, S. Chen, Y. Li, H. Wei, D. Zhao, K. Gu and S. Zhang, *Food Chem.*, 2015, **187**, 331–337.
9. R. Schirhagl, *Anal. Chem.*, 2013, **86**, 250–261.
10. Y. Tang, G. Fang, S. Wang, J. Sun and K. Qian, *J. AOAC Int.*, 2013, **96**, 453–458.
11. J. Zhou, N. Gan, T. Li, F. Hu, X. Li, L. Wang and L. Zheng, *Biosens. Bioelectron.*, 2014, **54**, 199–206.
12. D. Lakshmi, A. Bossi, M. J. Whitcombe, I. Chianella, S. A. Fowler, S. Subrahmanyam, E. V. Piletska and S. A. Piletsky, *Anal. Chem.*, 2009, **81**, 3576–3584.
13. C. Luo, M. Liu, Y. Mo, J. Qu and Y. Feng, *Anal. Chim. Acta*, 2001, **428**, 143–148.
14. N. T. Greene and K. D. Shimizu, *J. Am. Chem. Soc.*, 2005, **127**, 5695–5700.
15. X. A. Ton, V. Acha, P. Bonomi, B. T. S. Bui and K. Haupt, *Biosens. Bioelectron.*, 2015, **64**, 359–366.
16. O. A. Odeyemi and F. A. Bamidele, *Fut. Sci. OA*, 2016, **2**, FSO91–FSO93.
17. A. Kelly, B. Osburn and M. Salman, *Lancet Global Health*, 2014, **2**, 379–380.

18. C. Lodeiro, J. L. Capelo, E. Oliveira and C. Nuñez, *Environ. Sci. Pollut. Res.*, 2016, **23**, 24419–24421.
19. A. G. Mayes and M. J. Whitcombe, *Adv. Drug Delivery Rev.*, 2005, **57**, 1742–1778.
20. E. Turiel and A. Martín-Esteban, *Anal. Chim. Acta*, 2010, **668**, 87–99.
21. K. Haupt, *Anal. Chem.*, 2003, **75**, 376–383.
22. J. O'Mahony, A. Molinelli, K. Nolan, M. Smyth and B. Mizaikoff, *Biosens. Bioelectron.*, 2005, **20**, 1884–1893.
23. A. Nezhadali and M. Mojarrab, *Sens. Actuators, B*, 2014, **190**, 829–837.
24. Y. Tang, J. Lan, X. Gao, X. Liu, D. Zhang, L. Wei, Z. Gao and J. Li, *Food Chem.*, 2016, **190**, 952–959.
25. J. Haginaka, *J. Chromatogr. B*, 2008, **866**, 3–13.
26. R. Liang, R. Zhang and W. Qin, *Sens. Actuators, B*, 2009, **141**, 544–550.
27. R. N. Liang, D. A. Song, R. M. Zhang and W. Qin, *Angew. Chem. Int. Ed.*, 2010, **49**, 2556–2559.
28. T. Piacham, Å. Josell, H. Arwin, V. Prachayasittikul and L. Ye, *Anal. Chim. Acta*, 2005, **536**, 191–196.
29. X. Li and S. M. Husson, *Biosens. Bioelectron.*, 2006, **22**, 336–348.
30. K. Reimhult, K. Yoshimatsu, K. Risveden, S. Chen, L. Ye and A. Krozer, *Biosens. Bioelectron.*, 2008, **23**, 1908–1914.
31. J. Homola, S. S. Yee and G. Gauglitz, *Sens. Actuators, B*, 1999, **54**, 3–15.
32. M. J. Whitcombe, I. Chianella, L. Larcombe, S. A. Piletsky, J. Noble, R. Porter and A. Horgan, *Chem. Soc. Rev.*, 2011, **40**, 1547–1571.
33. V. Suryanarayanan, C. T. Wu and K. C. Ho, *Electroanalysis*, 2010, **22**, 1795–1811.
34. M. C. Blanco-López, M. J. Lobo-Castanon, A. J. Miranda-Ordieres and P. Tunon-Blanco, *TrAC, Trends Anal. Chem.*, 2004, **23**, 36–48.
35. T. Wang and C. Shannon, *Anal. Chim. Acta*, 2011, **708**, 37–43.
36. Q. Lu, X. Chen, L. Nie, J. Luo, H. Jiang, L. Chen, Q. Hu, S. Du and Z. Zhang, *Talanta*, 2010, **81**, 959–966.
37. B. Song, Y. Zhou, H. Jin, T. Jing, T. Zhou, Q. Hao, Y. Zhou, S. Mei and Y. I. Lee, *Microchem. J.*, 2014, **116**, 183–190.
38. A. Motaharian, F. Motaharian, K. Abnous, M. R. M. Hosseini and M. Hassanzadeh-Khayyat, *Anal. Bioanal. Chem.*, 2016, **408**, 6769–6779.
39. Y. Liu, Q. J. Song and L. Wang, *Microchem. J.*, 2009, **91**, 222–226.
40. M. J. Syu, T. C. Chiu, C. Y. Lai and Y. S. Chang, *Biosens. Bioelectron.*, 2006, **22**, 550–557.
41. E. Hedborg, F. Winquist, I. Lundström, L. I. Andersson and K. Mosbach, *Sens. Actuators, A*, 1993, **37**, 796–799.
42. L. J. Kong, M. F. Pan, G. Z. Fang, K. Qian and S. Wang, *Anal. Bioanal. Chem.*, 2012, **404**, 1653–1660.
43. J. Li, Y. Li, Y. Zhang and G. Wei, *Anal. Chem.*, 2012, **84**, 1888–1893.
44. M. Bougrini, A. Florea, C. Cristea, R. Sandulescu, F. Vocanson, A. Errachid, B. Bouchikhi, N. E. Bari and N. Jaffrezic-Renault, *Food Control*, 2016, **59**, 424–429.

45. Y. Liang, L. Yu, R. Yang, X. Li, L. Qu and J. Li, *Sens. Actuators, B*, 2017, **240**, 1330–1335.
46. M. H. Do, A. Florea, C. Farre, A. Bonhomme, F. Bessueille, F. Vocanson, N. T. Tran-Thi and N. Jaffrezic-Renault, *Int. J. Environ. Anal. Chem.*, 2015, **95**, 1489–1501.
47. V. L. Granado, M. Gutiérrez-Capitán, C. Fernández-Sánchez, M. T. S. Gomes, A. Rudnitskaya and C. Jimenez-Jorquera, *Anal. Chim. Acta*, 2014, **809**, 141–147.
48. A. Janshoff, H. J. Galla and C. Steinem, *Angew. Chem., Int. Ed.*, 2000, **39**, 4004–4032.
49. K. A. Marx, *Biomacromolecules*, 2003, **4**, 1099–1120.
50. S. K. Vashist and P. Vashist, *J. Sens.*, 2011, **2011**, 1–13.
51. L. Chen, P. Sun and G. Chen, *Carbohydr. Res.*, 2015, **405**, 66–69.
52. B. Ding, J. Kim, Y. Miyazaki and S. Shiratori, *Sens. Actuators, B*, 2004, **101**, 373–380.
53. M. E. Yakovleva, G. R. Safina and B. Danielsson, *Anal. Chim. Acta*, 2010, **668**, 80–85.
54. S. Buchatip, C. Ananthanawat, P. Sithigorngul, P. Sangvanich, S. Rengpipat and V. P. Hoven, *Sens. Actuators, B*, 2010, **145**, 259–264.
55. T. Eren, N. Atar, M. L. Yola and H. Karimi-Maleh, *Food Chem.*, 2015, **185**, 430–436.
56. N. Gao, J. Dong, M. Liu, B. Ning, C. Cheng, C. Guo, C. Zhou, Y. Peng, J. Bai and Z. Gao, *Analyst*, 2012, **137**, 1252–1258.
57. N. Liu, J. Han, Z. Liu, L. Qu and Z. Gao, *Anal. Methods*, 2013, **5**, 4442–4447.
58. G. Fang, G. Liu, Y. Yang and S. Wang, *Sens. Actuators, B*, 2016, **230**, 272–280.
59. W. H. Zhou, S. F. Tang, Q. H. Yao, F. R. Chen, H. H. Yang and X. R. Wang, *Biosens. Bioelectron.*, 2010, **26**, 585–589.
60. G. Fang, H. Wang, Y. Yang, G. Liu and S. Wang, *Sens. Actuators, B*, 2016, **237**, 239–246.
61. S. Yan, Y. Fang and Z. Gao, *Biosens. Bioelectron.*, 2007, **22**, 1087–1091.
62. J. Dai, Y. Zhang, M. Pan, L. Kong and S. Wang, *J. Agric. Food Chem.*, 2014, **62**, 5269–5274.
63. W. Yao, Z. Gao and Y. Cheng, *J. Sep. Sci.*, 2009, **32**, 3334–3339.
64. R. Suedee, W. Intakong and F. L. Dickert, *Talanta*, 2006, **70**, 194–201.
65. X. Ren and L. Chen, *Biosens. Bioelectron.*, 2015, **64**, 182–188.
66. X. Gao, D. Li, Y. Tong, D. Ge, Y. Tang, D. Zhang and J. Li, *Luminescence*, 2015, **30**, 1389–1394.
67. Z. Li, Y. Wang, J. Wang, Z. Tang, J. G. Pounds and Y. Lin, *Anal. Chem.*, 2010, **82**, 7008–7014.
68. Y. Inoue, A. Kuwahara, K. Ohmori, H. Sunayama, T. Ooya and T. Takeuchi, *Biosens. Bioelectron.*, 2013, **48**, 113–119.
69. Y. T. Wu, Y. J. Liu, X. Gao, K. C. Gao, H. Xia, M. F. Luo, X. J. Wang, L. Ye, Y. Shi and B. Lu, *Chemosphere*, 2015, **119**, 515–523.

70. H. Liu, D. Liu, G. Fang, F. Liu, C. Liu, Y. Yang and S. Wang, *Anal. Chim. Acta*, 2013, **762**, 76–82.
71. Y. Zhou, Z. B. Qu, Y. Zeng, T. Zhou and G. Shi, *Biosens. Bioelectron.*, 2014, **52**, 317–323.
72. J. Zheng, P. R. Nicovich and R. M. Dickson, *Annu. Rev. Phys. Chem.*, 2007, **58**, 409–431.
73. X. Wu, Z. Zhang, J. Li, H. You, Y. Li and L. Chen, *Sens. Actuators, B*, 2015, **211**, 507–514.
74. K. Qian, G. Fang and S. Wang, *RSC Adv.*, 2013, **3**, 3825–3828.
75. X. Liu, J. Ren, L. Su, X. Gao, Y. Tang, T. Ma, L. Zhu and J. Li, *Biosens. Bioelectron.*, 2017, **87**, 203–208.
76. S. Ge, J. Lu, L. Ge, M. Yan and J. Yu, *Spectrochim. Acta, Part A*, 2011, **79**, 1704–1709.
77. K. Sun, Q. Deng, T. Guo, R. Su, Y. Gu and S. Wang, *RSC Adv.*, 2015, **5**, 94084–94090.
78. J. Hou, H. Li, L. Wang, P. Zhang, T. Zhou, H. Ding and L. Ding, *Talanta*, 2016, **146**, 34–40.
79. J. Wang, L. Gao, D. Han, J. Pan, H. Qiu, H. Li, X. Wei, J. Dai, J. Yang, H. Yao and Y. Yan, *J. Agric. Food Chem.*, 2015, **63**, 2392–2399.
80. S. Li, J. Luo, G. Yin, Z. Xu, Y. Le, X. Wu, N. Wu and Q. Zhang, *Sens. Actuators, B*, 2015, **206**, 14–21.
81. X. Wei, T. Hao, Y. Xu, K. Lu, H. Li, Y. Yan and Z. Zhou, *RSC Adv.*, 2015, **5**, 79511–79518.
82. K. Kneipp, Y. Wang, H. Kneipp, L. T. Perelman, I. Itzkan, R. R. Dasari and M. S. Feld, *Phys. Rev. Lett.*, 1997, **78**, 1667–1670.
83. S. Nie and S. R. Emory, *Science*, 1997, **275**, 1102–1106.
84. F. Gao, Y. Hu, D. Chen, E. C. Li-Chan, E. Grant and X. Lu, *Talanta*, 2015, **143**, 344–352.
85. Y. Hu, S. Feng, F. Gao, E. C. Li-Chan, E. Grant and X. Lu, *Food Chem.*, 2015, **176**, 123–129.
86. Y. Guo, L. Kang, S. Chen and X. Li, *Phys. Chem. Chem. Phys.*, 2015, **17**, 21343–21347.
87. Y. Yoshimi, K. Sato, M. Ohshima and E. Piletska, *Analyst*, 2013, **138**, 5121–5128.
88. Z. Guo, L. Chen, H. Lv, Z. Yu and B. Zhao, *Anal. Methods*, 2014, **6**, 1627–1632.
89. X. Xiao, K. Yan, X. Xu and G. Li, *Talanta*, 2015, **138**, 40–45.
90. F. Gao, S. Feng, Z. Chen, E. C. Li-Chen, E. Grant and X. Lu, *J. Food Sci.*, 2014, **79**, N2542–N2549.
91. F. Gao, E. Grant and X. Lu, *Anal. Chim. Acta*, 2015, **901**, 68–75.
92. S. Feng, F. Gao, Z. Chen, E. Grant, D. D. Kitts, S. Wang and X. Lu, *J. Agric. Food Chem.*, 2013, **61**, 10467–10475.
93. R. Verma and B. D. Gupta, *Analyst*, 2013, **138**, 7254–7263.
94. S. Lépinay, K. Kham, M. C. Millot and B. Carbonnier, *Chem. Pap.*, 2012, **66**, 340–351.

95. T. Yao, X. Gu, T. Li, J. Li, J. Li, Z. Zhao, J. Wang, Y. Qin and Y. She, *Biosens. Bioelectron.*, 2016, **75**, 96–100.
96. S. W. Choi, H. J. Chang, N. Lee, J. H. Kim and H. S. Chun, *J. Agric. Food Chem.*, 2009, **57**, 1113–1118.
97. G. H. Yao, R. P. Liang, C. F. Huang, Y. Wang and J. D. Qiu, *Anal. Chem.*, 2013, **85**, 11944–11951.
98. Z. Altintas, M. Gittens, A. Guerreiro, K. A. Thompson, J. Walker, S. Piletsky and I. E. Tothill, *Anal. Chem.*, 2015, **87**, 6801–6807.
99. M. Kara, L. Uzun, S. Kolayli and A. Denizli, *J. Appl. Polym. Sci.*, 2013, **129**, 2273–2279.
100. M. L. Yola, T. Eren and N. Atar, *Sens. Actuators, B*, 2014, **195**, 28–35.
101. S. Jiang, Y. Peng, B. Ning, J. Bai, Y. Liu, N. Zhang and Z. Gao, *Sens. Actuators, B*, 2015, **221**, 15–21.
102. C. Shi, X. Liu, L. Song, X. Qiao and Z. Xu, *Food Anal. Methods*, 2015, **8**, 2496–2503.
103. S. Wang, Z. Xu, G. Fang, Y. Zhang, B. Liu and H. Zhu, *J. Agric. Food Chem.*, 2009, **57**, 4528–4534.
104. X. Bi and Z. Liu, *Anal. Chem.*, 2013, **86**, 959–966.
105. I. Chianella, A. Guerreiro, E. Moczko, J. S. Caygill, E. V. Piletska, I. M. P. De Vargas Sansalvador, M. J. Whitcombe and S. A. Piletsky, *Anal. Chem.*, 2013, **85**, 8462–8468.
106. X. Du, F. Zhang, H. Zhang, Y. Wen and T. Saren, *Food Agric. Immunol.*, 2014, **25**, 411–422.
107. G. Fang, J. Lu, M. Pan, W. Li, L. Ren and S. Wang, *Food Anal. Methods*, 2011, **4**, 590–597.
108. H. Liu, L. Ren, G. Fang, H. Li and S. Wang, *Anal. Methods*, 2013, **5**(20), 5677–5683.
109. Y. Tang, J. Gao, X. Liu, X. Gao, T. Ma, X. Lu and J. Li, *Food Chem.*, 2017, **228**, 62–69.

CHAPTER 8

Aptamer-based Sensing Techniques for Food Safety and Quality

D. N. GOUDREAU, M. SMITH, E. M. McCONNELL, A. RUSCITO, R. VELU, J. CALLAHAN AND M. C. DeROSA*

Carleton University, Department of Chemistry, 1125 Colonel By Dr., Ottawa K1S 5B6, Canada
*Email: maria.derosa@carleton.ca

8.1 Introduction

Food is intimately linked to the health and wellbeing of animals and humans, and therefore effective methods for ensuring the safety of our food supply is of global interest. Food containing natural contaminants or adulterating substances (whether added intentionally or unintentionally) can lead to illnesses that range from mild symptoms such as discomfort to more severe symptoms that may be potentially life-threatening.[1] There have been many examples of worldwide of contaminants in food that have led to sickness and/or death in animals and humans due to either lack of testing or improper testing.

As the global population grows, there will be different foods and food preparations that may increase the risk of certain contaminants, in addition to there being an increased demand for food supply.[2] There may be countless ways for natural contaminants and adulterating substances to enter the food chain, however some main pathways are described. These include

Food Chemistry, Function and Analysis No. 2
Sensing Techniques for Food Safety and Quality Control
Edited by Xiaonan Lu

Published by the Royal Society of Chemistry, www.rsc.org

improper handling and storage (mycotoxins, bacteria, bacterial toxins, viruses, food allergens), direct administration to farm animals or application to plants (pesticides, drugs, antibiotics), indirect exposure *via* animals ingesting contaminated food (pesticides, mycotoxins, bacteria, bacterial toxins, viruses), contaminated processing equipment (mycotoxins, bacteria, bacterial toxins, viruses, adulterants), intentional addition (adulterants) and natural processes such as deposition from air, soil and water (heavy metals, bacteria, bacterial toxins, mycotoxins). Technology for sensing contaminants and adulterating substances in food is critical for food safety, particularly since contamination is often not accompanied by any discernible changes to the commodity. For example, mycotoxin contamination does not always provide visible cues such as colour change or damage to crop or commodity. Similarly, early bacterial contamination may not cause illness in farm animals.[3,4]

Food safety also has an impact on international trade; while many countries may have their own protocols for growing crops, raising farm animals and preparing food that include certain pesticides, antibiotics or other, they must adhere to the standards, laws and maximum limits for contaminants in food of the country to which they wish to export. Adhering to these laws may require not using specific chemicals and/or making sure any contamination is below the maximum level permitted. For these reasons, it is imperative to test food for a wide range of possible contaminants and adulterating substances.

Although many current sensing methods are sensitive enough to ensure food safety, they may have drawbacks that can affect their widespread implementation, particularly in resource-poor scenarios. For example, the gold standard for food safety testing is typically LC-MS but the inability to allow for high-throughput screening of samples in a short amount of time and the high cost for instrumentation are serious limitations.[5] Other current testing methods include ELISAs, TLC and GC, however these approaches are either expensive, time consuming, entail tedious pre-treatments, or require skilled technicians. These problems are exacerbated in developing countries where resources are scarce and many detection methods are not available. As a result, due to forced economic circumstances, populations are ingesting unsafe food that may result in illness or death.

Alternatives to modern day food safety testing methods that are inexpensive, fast, disposable, easy to use and portable are urgently needed. Aptamer-based sensors, also known as aptasensors, may be able to meet some or all of these requirements. Aptamers are single-stranded oligonucleotides (usually $<$100 mer) that form complex three-dimensional secondary structures and are capable of binding a specific target molecule.[6] Aptamers bind to their target molecule with high affinity (typically mid nM) and selectivity by some combination of H-bonding, π-stacking interactions, electrostatic interactions and hydrophobic interactions.[7,8] Aptamers have been selected for a wide variety of food-related targets ranging from small molecule toxins to bacteria.[8] Aptamers are selected for a specific target by the iterative *in vitro* selection process termed *S*ystematic *E*volution of *L*igands by *EX*ponential enrichment (*SELEX*).[7,9,10] There have been multiple variations of this process described with specific modifications for food-related targets. Generally, the process comprises three steps: aptamer-target incubation, partitioning and amplification. In the first step, the aptamer is incubated with an oligonucleotide library containing 10^{14}–10^{16} randomized sequences. Following incubation, sequences with affinity for the target molecule are partitioned from those with little or no affinity for the target molecule. Finally, the enriched oligonucleotide pool is exponentially amplified by polymerase chain reaction to yield a second generation library.

Characterized by their high affinity for specific target molecules, aptamers have been included in a class of molecules referred to as affinity tools.[11] Also belonging to this class are antibodies, engineered binding proteins and molecularly imprinted polymers.[11–13] These affinity tools have been widely used in target detection, purification and removal applications.[14] By far, the most commonly used and commercially available affinity probes are antibodies. However, aptamers have several advantages over antibodies in food-safety and food-quality related applications.[6,11,15,16] For instance, aptamers are produced *in vitro* and do not require the use of animals or cell cultures.[11] Consequently, an aptamer can be selected under conditions that best mimic the eventual testing scenario of interest (*e.g.*, in a food matrix, *etc.*). Aptamers can be selected against a wide range of targets and are not limited to targets that are non-toxic and elicit an immune response.[6] Further, in the selection process, the cross-reactivity of aptamers to similar molecules can be controlled during counter or negative selection rounds.[17] As there can be with antibodies, there is no batch-to-batch variation with aptamers.[11] The aptamer sequence is known, and therefore synthetically reproducible.[16] The aptamer's unique three-dimensional secondary structure allows the aptamer to bind specifically to the target molecule.[8] Aptamers are intrinsically more stable and can be easily chemically modified.[15] They have a long shelf-life and can be reversibly denatured.[11] Importantly, the cost of aptamer production is significantly less than that of antibodies, reducing the overall cost of any ensuing sensing technology.[18]

8.2 Aptasensors in Food Safety

A biosensor is a quantitative and/or qualitative analytical device, which is composed of two main parts: a biological recognition element and a transducer that converts the recognition event into a readable output. The biological element is usually composed of proteins (antibodies or enzymes) or nucleic acids (DNA or RNA) and interacts strongly and selectively with the target of interest. The response from their interaction is then converted to a signal by the transducer. More and more aptamers are being used as the biological element of sensors, and thus the name aptasensor has emerged.

This chapter will focus on the recently developed (since 2011) aptasensors for food safety and quality control. It will highlight the efforts put forth by the scientific community to provide more sensitive, affordable, accurate and fast detection methods for a wide range of targets found in food. These targets include bacteria, mycotoxins, heavy metals, antibiotics, pesticides, viruses and other adulterants for which detection is imperative to protect the health of animals and humans. Tables summarize the aptamers recently selected for relevant targets and the aptasensor platforms that have been developed. The chapter sections highlight interesting aptasensor examples that show promise in one or several of the requirements for on-site food safety testing, such as ease of use, portability or ability to be multiplexed.

8.2.1 Small Molecule and Protein-based Targets

8.2.1.1 Mycotoxins

Mycotoxins are a diverse group of secondary metabolites that are produced by an array of toxigenic fungi.[19] Humans are exposed directly through food commodities such as wheat, maize, barley, coffee, nuts, *etc*., and indirectly through livestock feed.[20] Toxigenic fungi can colonize crops and predispose them to mycotoxin contaminants at all stages of production including pre-harvest and post-harvest handling and storage.[19,21] Moisture levels, temperature, pH, host-type and competition can all influence the prevalence of these metabolites. Mycotoxins and their metabolites may also be found in animal products such as milk, meats and eggs.[19,22] Mycotoxins are of great health concern as they illicit carcinogenic, teratogenic, hepatotoxic and immunosuppressive effects. Moreover, the prominence of mycotoxin contamination and its worldwide occurrence necessitates exposure prevention and the development of rapid and cost-effective detection techniques.[23] Table 8.1 lists the aptasensors that have recently been developed to detect mycotoxins in a variety of food commodities.

Mycotoxin Biosensors

Aflatoxins are a class of mycotoxins primarily produced by toxigenic filamentous fungi *Aspergillus flavus.*[23] They are responsible for producing Aflatoxins B1 and B2 primarily, named for their blue fluorescence when irradiated with UV light. Other *Aspergillus* fungi such as *Aspergillus parasiticus* are known for producing Aflatoxins B1, B2, G1 and G2.[24] First discovered in 1960, aflatoxins have become one of the most studied classes of mycotoxins. *Aspergillus flavus* can thrive in harsh conditions and has no host specificity. It can therefore contaminate many major crops. As a result, it is of great concern in many parts of the world.[24,25] Not only can it be prevalent, but aflatoxin B1 is the most toxic of the aflatoxins and indeed of all mycotoxins. It can induce hepatotoxic, teratogenic, mutagenic and immunosuppressive effects. Moreover, Aflatoxin B1 is classified as a group 1 carcinogen.[25] Chronic exposure as a result of unregulated markets can affect as many as 4.5 billion people in developing countries.[26] Table 8.1 contains a

Table 8.1 Aptasensors for mycotoxins.

Mycotoxins	Detection method	Aptasensor description	DNA/RNA	K_d	Limit of detection (LOD)	Ref.
Aflatoxin B1	Fluorescence Polarization	Protein-encoded aptamer where recycling of target *via* DNA polymerase and nicking endonuclease is achieved (Isothermal exponential amplification)	DNA	—	0.24 pM	69
	Electrochemical	Impedimetric sensor *via* aptamer functionalized SPCEs	DNA	—	0.12 ng mL^{-1} (seqA), 0.25 ng mL^{-1} (seqB)	70
	Colorimetric/ microfluidic	Target-induced disruption of NP immobilized hydrogel paired with distance-dependent read-out of a microfluidic V-chip	DNA	—	0.55 ppb	27
	Fluorometric	Fluorescent nitrogen doped carbon dots assembled onto aptamer modified AuNPs	DNA	—	5 pg mL^{-1}	71
	Electrochemical	Target recycling *via* telomerase and EXO III based two round signal amplification system	DNA	—	6×10^{-4} ppt	72
	Electrochemical	Multi-layer framework using CV and EIS *via* redox indicators	DNA	—	0.40 nM	73
	Fluorometric	FRET-based detection *via* aptamer functionalized CdTe Q-dots (donor) and GO (acceptor)	DNA	—	1.0 nM, 1.4 nM (peanut oil)	74
	Colorimetric	Structure switching split-DNAzyme probes with peroxide mimicking activity	DNA	—	0.1 ng mL^{-1}	75
	Fluorometric	RT-qPCR with aptamer recognition probe	DNA	—	25 fg mL^{-1}	76
	Fluorometric	Aptamer-based competitive dipstick assay with Cy5-modified DNA probe	DNA	<400 nM	0.1 ng mL^{-1} (buffer), 0.3 ng mL^{-1} (corn)	77
	Chemi-luminescent	Chemiluminescence competitive aptamer assay using hemin G-quadruplex HRP-mimicking DNAzyme linked AFB1 aptamer	DNA	0.2 μM[78]	0.11 ng mL^{-1}	79

Aflatoxin B2	Fluorometric	FAM-labelled aptamer immobilized onto MNPs	DNA	9.83 nM	50 ng L^{-1}	80
Aflatoxin M1	Fluorometric	RT-qPCR with aptamer recognition probe	DNA	35 nM[81]	0.03 ng L^{-1}	82
	Electrochemical	Immobilization of aptamer on SPCE for CV and EIS *via* redox indicator ferrocyanide	DNA	—	1.15 ng L^{-1}	83
	Electrochemical	Aptamer immobilized on Fe_3O_4/PANi-functionalized interdigitated electrode (IDE)	DNA	—	1.98 ng L^{-1}	84
Fumonisin B1	Electrochemical	Aptamer-modified gold nanoparticles (AuNPs) on glassy carbon electrode	DNA	—	2.0 pM	85
Ochratoxin A	Chemi-luminescent	Aptamer-linked immobilized sorbent assay (ALISA) on a modified glass substrate	DNA	1.12.2: 0.2 μM,[78] A08: 290 nM[86]	0.82 mg L^{-1} (beer)	87
	Fluorometric	Luminescence resonance energy transfer using aptamer modified UCNPs and anti-aptamer modified AuNRs	DNA	—	27 pg mL^{-1}	88
	Electrochemical	Structure-switching aptamer triggering the release of competitor-invertase and the hydrolysis of sucrose. Measured *via* personal glucometer (PGM)	DNA	0.2 μM[78]	3.31 μg L^{-1} (buffer), 3.66 μg L^{-1} (wine)	28
	Chemi-luminescent	CRET-based detection: HRP-mimicking DNAzyme-linked OTA aptamer for efficient energy transfer between luminol (donor) by dabcyl (acceptor)	DNA	0.2 μM[78]	0.22 ng mL^{-1} (coffee)	89
	Fluorometric	FAM-modified aptamer adsorbed onto single-walled carbon nanohorn (SWCNH)	DNA	0.2 μM[78]	17.2 nM	90
	Fluorometric	FRET-based detection: shielding effect-induced inhibition of FRET from Cy3-conjugated streptavidin to AuNPs	DNA	0.2 μM[78]	1.4 pg mL^{-1}	91
	Electrochemical	Impedimetric sensor *via* polypyrrole (Ppy) and polyaminoamine (PAMAM G4) functionalized gold electrode	DNA	0.2 μM[78]	2 ng L^{-1}	92
	Fluorometric	Hybridization chain reaction (HCR) amplification system using perylene composites (PAPDI/DNA)	DNA	—	0.10 pM	93

Table 8.1 (*Continued*)

Mycotoxins	Detection method	Aptasensor description	DNA/RNA	K_d	Limit of detection (LOD)	Ref.
	Colorimetric	Competitive binding assay using AuNP doped iron oxide NPs as a separation mechanism and signal indicator (HRP-mimicking)	DNA	0.2 μM[78]	30 pg mL^{-1}	94
	Colorimetric	Side-by-side assembly of gold nanorods using aptamer and linker DNA in competitive dispersion assay	DNA	—	0.22 ng mL^{-1}	95
	Colorimetric	Lateral flow strip (LFT) based competitive assay	DNA	0.2 μM[78]	1 ng mL^{-1}	96
	Electrochemical	PEG/DNA-functionalized boron-doped diamond (BDD) microcell	DNA	0.2 μM[78]	0.01 ng mL^{-1}	97
	Fluorometric	Competitive displacement assay *via* aptamer-modified carboxy-fluorescent particles	DNA	0.2 μM[78]	0.005 nM	98
	Colorimetric	Ligand binding-induced hydrogel disassembly and release of encapsulated AuNPs or Au@PtNPs	DNA	0.2 μM[78]	1.27 nM	99
	Optical	Evanescent wave all-fibre biosensing platform (EWA)	DNA	0.2 μM[78]	0.39 μg L^{-1}	100
	Electrochemical	OTA aptamer-functionalized electrode paired with electroactive GO nanoplatelets	DNA	0.2 μM[78]	310 f M–310 pM (LDR)	101
	Colorimetric	Ligand-induced aggregation of free AuNPs by polydiallyldimethylammonium chloride polymers (PDDA)	DNA	—	0.009 ng mL^{-1}	102
	Electrochemical	Immobilization of anti-OTA aptamer on SPCEs	DNA	0.2 μM[78]	0.15 ng mL^{-1}	103
	Fluorometric	FRET-based sensor utilizing dual emission ratiomeric fluorescence *via* green-emitting CdTe Q-dots and red-emitting Q-dots	DNA	0.2 μM[78]	1.67 pg mL^{-1}	104
	Electrochemical	SPCE modified with PTH and assembly of IrO_2 NPs on which the aminated aptamer selective to OTA for EIS *via* redox indicator ferri/ferrocyanide	DNA	0.2 μM[78]	5.56 ng kg^{-1}	105

Fluorometric	FRET-based sensor: 6FAM/BHQ1 modified aptamer beacon paired with RT-qPCR detection	DNA	0.2 μM[78]	19 nM (wine)	106
Fluorometric	Aptamer functionalized AuNP and streptavidin modified silica nanoparticles for fluorescence enhancement	DNA	0.2 μM[78]	0.098 nM	107
Electrochemical	Enhanced response *via* ligand binding and subsequent production of electroactive methylene blue mononucleotide	DNA	0.2 μM[78]	0.004 ng mL^{-1}	108
Fluorometric	Linkage inversion assembled nano-aptasensor (LIANA) in paper-based detection system	DNA	1.12.2: 0.2 μM,[78] A08: 290 nM[86]	130 nM	109
Optical	Evanescent wave all-fibre biosensing platform (EWA)	DNA	0.2 μM[78]	3 nM	110
Fluorometric	Fluorescence emission based on the separation of aptamer-conjugated MBs hybridized with anti-aptamer-conjugated hybrid fluorescent nanoparticles	DNA	0.2 μM[78]	5.4 pg mL^{-1}	111
Colorimetric	Ligand-induced disassembly of PEG/DNA-AuNP dimers	DNA	—	0.05 nM	112
Electrochemi-luminescent	Anti-OTA aptamer modified electrode resulting in ligand-induced hyperbranched rolling circle amplification (HRCA), and subsequent intercalation of ECL reporter	DNA	0.2 μM[78]	0.02 pg mL^{-1}	113
Fluorometric	Ligand-induced inhibition of rolling circle amplification *via* modified OTA aptamer/primer immobilized on MNPs	DNA	—	0.13 ppt	114
Surface plasmon resonance	SPR detection *via* anti-OTA aptamer immobilized sensor chip	DNA	0.2 μM[78]	0.005 nM	115
Fluorometric	Label-free (SYBR Green 1) fluorescence-based biosensing	DNA	290 nM[86]	3.6 ppb	86
Fluorometric	Aptamer-affinity column clean-up coupled with ultra-high performance liquid chromatography and fluorescence detection	DNA	0.2 μM[78]	0.5 μg kg^{-1} (ginger powder)	116

Table 8.1 (*Continued*)

Mycotoxins	Detection method	Aptasensor description	DNA/RNA	K_d	Limit of detection (LOD)	Ref.
	Electrochemical	Rolling circle amplification (RCA)-based signal-amplified aptasensor	DNA	0.2 μM[78]	0.065 pg mL^{-1}	117
	Chemi-luminescent	Ligand binding-induced strand release paired with circle amplification system and CL nanoparticle probes	DNA	0.2 μM[78]	3×10^{-13} g mL^{-1}	118
	Fluorometric	RT-qPCR amplification of complementary DNA reporter	DNA	0.2 μM[78]	1 fg mL^{-1}	119
	Chemi-luminescent	Chemically initiated electron exchange luminescence (CIEEL)	DNA	0.2 μM[78]	0.5 nM	120
	Colorimetric	HRP-mimicking DNAzyme–aptamer conjugate for H_2O_2-mediated oxidation of TMB	DNA	0.2 μM[78]	4 nM	121
	Fluorometric	Tb^{3+} signal enhancement based on structure switching aptamer functionalized MNPs	DNA	0.2 μM[78]	20 pg mL^{-1}	122
	Electrochemical	EIS: Aptamer functionalized gold electrode	DNA	1.2 nM	0.12–0.4 nM (coffee, flour, wine)	123
	Fluorometric	Target-induced structure switching signalling aptamer: 6FAM and TAMRA	DNA	0.2 μM[78]	0.8 ng mL^{-1}	124
	Fluorometric	TRF spectroscopy: OTA-terbium aptamer complex	DNA	—	0.5 μg kg^{-1}	125
	Fluorometric	FRET: 6FAM-labelled aptamer–AuNP conjugate	DNA	0.2 μM[78]	2.0 pg mL^{-1}	126
	Electrochemical	RCA-based signalling with MNP immobilized aptamers	DNA	0.2 μM[78]	0.2 pg mL^{-1}	127
	Electrochemical	Methylene blue labelled, aptamer modified gold electrode	DNA	0.2 μM[78]	0.095 pg mL^{-1}	128
	Colorimetric	HRP-mimicking DNAzyme: TMB substrate	DNA	—	2.5 nM	129
	Electrochemical	Competitive DNA aptasensor for OTA detection, using superparamagnetic nanoparticles	DNA	0.2 μM[78]	0.11 ng mL^{-1}	130

Fluorometric	Enzyme-Linked Aptamer Assays (ELAAs), based on a competition format	DNA	—	1 ng mL^{-1}	131
Electrochemical	Competitive aptasensor using functionalized paramagnetic microparticle beads (MBs) localized on SPCEs	DNA	0.2 μM[78]	0.07 ng mL^{-1}	132
HPLC-FD	DNA aptamer-based solid phase extraction column	DNA	0.2 μM[78]	23 pg g^{-1}	133
Fluorometric	Aptamer based fluorescent assay for the detection of OTA-displacement of complementary FAM-labelled DNA chain	DNA	0.2 μM[78]	1×10^{-8} g L^{-1}	134
Fluorometric	Single-walled carbon nanotubes based quenching of free FAM-aptamer	DNA	0.2 μM[78]	24.1 nM	135
Electrochemical	Impedimetric aptasensor based on polyaniline-stearic acid Langmuir–Blodgett film	DNA	0.2 μM[78]	0.1 ng mL^{-1}	136
HPLC-FD	Aptamers immobilized on CNBr-activated Sepharose oligosorbent SPE column	DNA	0.2 μM[78]	0.2 ng mL^{-1} (beer)	137
Fluorometric	PVP-coated graphene oxide for detection of OTA *via* quenching fluorescence of free FAM-aptamer	DNA	0.2 μM[78]	21.8 nM	138
Electrochemical	Amplified differential pulse voltammetry signal *via* exonuclease-catalyzed target recycling	DNA	0.2 μM[78]	1.0 pg mL^{-1}	139
Colorimetric	Competitive aptamer-based chromatographic strip assay	DNA	0.2 μM[78]	1 ng mL^{-1}	140
Fluorometric	An aptamer-quantum-dot-based chromatographic strip assay	DNA	0.2 μM[78]	1.9 ng mL^{-1}	141
Fluorometric	Aptamer-functionalized magnetic nanoparticle-based bioassay for the detection of ochratoxin A using upconversion nanoparticles as labels	DNA	0.2 μM[78]	1×10^{-13} g mL^{-1}	142
HPLC-FD	Aptamer-targeted magnetic nanospheres as a solid-phase extraction sorbent	DNA	0.2 μM[78]	0.3 μg kg^{-1} (wheat flour), 0.5 μg kg^{-1} (coffee), 0.3 μg kg^{-1} (cereal)	143

Table 8.1 (*Continued*)

Mycotoxins	Detection method	Aptasensor description	DNA/RNA	K_d	Limit of detection (LOD)	Ref.
	Colorimetric	Aptamer-based colorimetric biosensing using unmodified gold nanoparticle reporter	DNA	0.2 μM[78]	20 nM	144
Patulin	Colorimetric	Enzyme-chromogenic substrate system using aptamer and glucose oxidase immobilized AuNPs	DNA	21.83 nM	48 pg mL^{-1}	145
T-2	Fluorometric	Aptamer-based fluorescent bioassay with FAM-labelled aptamer and GO acceptor	DNA	20.8 nM	0.4 μM (beer)	146
Versicolorin A	Electrochemical	Differential pulse voltammetry (DPV) *via* aptamer-modified Au electrode and ferrocyanide indicator	DNA	9.26 μM	10 pg mL^{-1}	147
Zearalenone	Fluorometric	Magnetic separation *via* aptamer-functionalized SA-coated magnetic beads	DNA	41 nM	0.785 nM	148
OTA/AFB1	Fluorometric	DNA-scaffolded silver nanoclusters and dual aptamer labelled magnetic beads for the Zn(II) signal-enhanced detection of OTA and AFB1	DNA	200 nM (OTA)[78]	0.2 pg mL^{-1}	149
	Surface-enhanced Raman scattering	SERS label embedded Ag@Au core-shell nanoparticle satellite assemblies	DNA	200 nM (OTA), 75 nM (AFB1)[78]	0.006 ng mL^{-1} (OTA), 0.03 ng mL^{-1} (AFB1)	150
	Fluorometric	FRET: Multiplexed UCNPs with graphene oxide as acceptor	DNA	0.2 μM[78]	0.02 ng mL^{-1} (OTA), 0.1 ng mL^{-1} (FB1)	151

summary of the recently developed aptasensors for a variety of mycotoxin targets.

Recently, Ma and colleagues developed an AFB1-responsive hydrogel aptasensor.[27] The sensor utilizes polyacrylamide-DNA conjugates P-SA and P-SB by first synthesizing DNA strands with an acrydite modification and inducing polymerization. Hydrogel preparation is then performed using an aptamer linker and either gold or platinum nanoparticles (AuNPs or PtNPs, respectively). The resulting product is a three-strand complex held together by complementary interactions between the DNA on polyacrylamide and the aptamer linkers (see Figure 8.1). When aflatoxin B1 is added to the hydrogel, aptamer-ligand binding causes a disruption in the hydrogel and a release of the NPs trapped within. This is easily observed *via* a dose-dependent dispersion of NPs in solution. The group also designed a V-chip for a quantitative read-out from the PtNP containing target-responsive hydrogel. Similarly, when the target is present the hydrogel is disrupted. The release of PtNPs then catalyzes the decomposition of H_2O_2, thus increasing the pressure in the ink bar of a microfluidic chip, and providing simple distance-based read-out. The target-responsive hydrogel system using AuNPs provided a visual colour change over the 0.25–40 μM range, while the target-responsive hydrogel system paired with the distance read-out V-chip provided a limit of detection of 0.55 ppb (1.77 nM). The method was found to be very effective when applied to a beer sample, and showed little to no cross-reactivity with other common mycotoxins including other aflatoxins (G1, B2, M1).

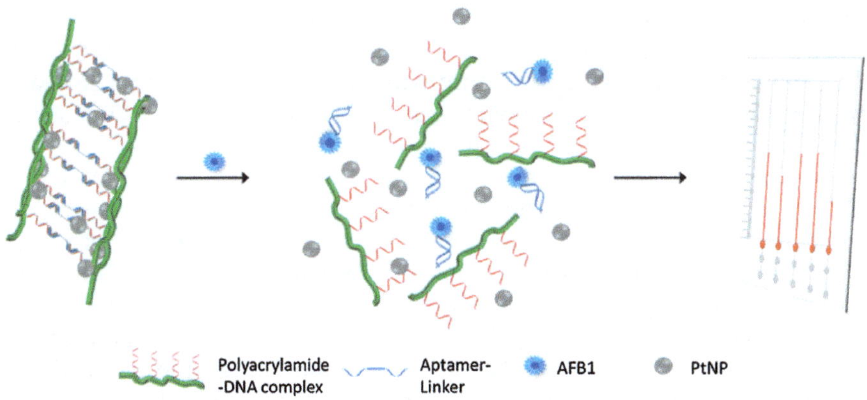

Figure 8.1 Polyacrylamide hydrogel was assembled using aptamer/polyacrylamide complexes and partially complementary sequences as linkers. The hydrogel is loaded with platinum nanoparticles. Upon AFB1 binding, the disruption in the hydrogel causes ligand-induced disassembly. The released platinum nanoparticles then catalyze the decomposition of peroxide into oxygen. A microfluidic chip allows for a straightforward dose-dependent distance read-out.
Adapted from ref. 27 with permission from the Royal Society of Chemistry.

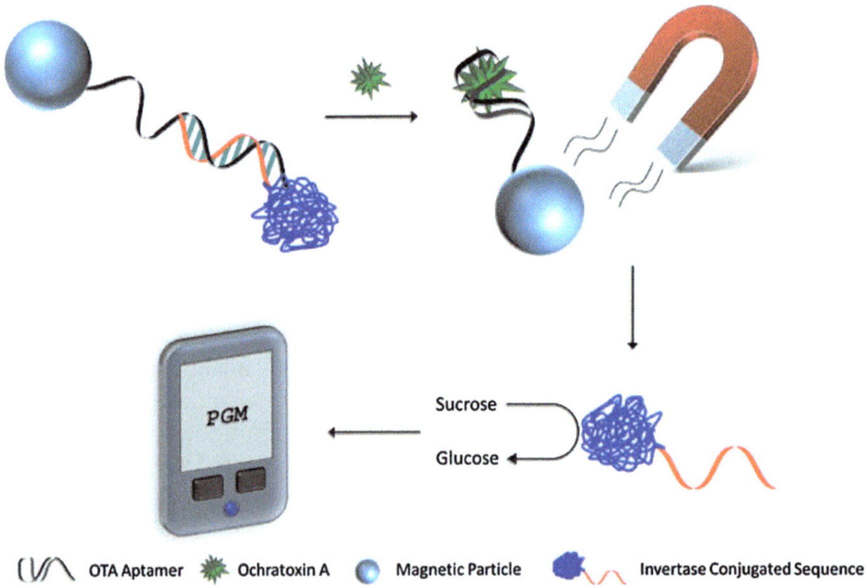

Figure 8.2 OTA aptasensor based on ligand-induced disassembly of invertase probe and relative personal glucose meter (PGM) read-out as a result of sucrose hydrolysis.
Adapted from ref. 28 with permission from the Royal Society of Chemistry.

Low cost and portability were two characteristics that were emphasized in the development of the structure-switching competitive assay by Gu and others.[28] The sensor was recently developed for ochratoxin A (OTA), another mycotoxin target of great importance for food safety (See Figure 8.2). In short, an OTA-binding aptamer is immobilized directly onto magnetic beads *via* streptavidin-biotin interactions. A second complementary sequence is synthesized and conjugated to the enzyme invertase. The two strands are then hybridized to form the aptamer-competitor complex. The introduction of OTA induces a structural change in the aptamer and, thus, the invertase probe is released. Following magnetic separation, the supernatant containing the released invertase DNA probe is used to catalyze the hydrolysis of added sucrose to glucose. This can then be read using a commercially available personal glucometer (PGM). The limit of detection for ochratoxin A was determined to be 3.31 and 3.66 $\mu g\ L^{-1}$ in buffer and wine, respectively. Gu and others developed a novel sensor that may be the first to combine the affinity and specificity of an aptamer with the ease of use and portability of a PGM for the detection of a mycotoxin food contaminant.

8.2.2 Bacterial Toxins

Diverse bacterial strains are responsible for an ever-growing library of toxic compounds. Toxigenesis can yield two primary types of harmful bacterial toxins. These include lipopolysaccharides, known as endotoxins, and protein-based toxins, known as exotoxins. Endotoxins are typically associated with Gram-negative bacteria, while exotoxins are produced predominantly by Gram-positive bacteria. Many bacterial exotoxins are among some of the most potent toxins known. It is estimated that a single milligram of the potent exotoxin botulinum could fulfil the lethal dose of 1000 people. In the context of food contamination and human ingestion, many types of exotoxins tend to be rendered inactive by the time they reach their location of toxicity. This inactivation is often due to the contact with acidic digestive enzymes or thermal denaturation *via* standard food preparation procedures such as boiling. It should be noted that some exotoxins can become more potent through partial proteolysis by digestive enzymes. Exotoxins can travel quite far in the body before generating their toxicological effects. This has led to the formation of three classifications based on the site of toxicity, which include enterotoxins, neurotoxins and cytotoxins. Enterotoxins generate toxicity in the small intestine and include examples such as choleragen, secreted by *Vibrio cholerae* to produce well-known symptoms of cholera infection. Another example is Shiga toxin produced primarily by *Shigella dysenteriae* and some serotypes of Shiga toxin-producing *Escherichia coli*. Bacterial neurotoxins target the nervous system and include examples such as the potent alpha-toxin produced by *Staphylococcus aureus*, which can open large Ca^{2+} pores in lipid bilayers and generate rapid cell death. Bacterial cytotoxins are toxic to a broad spectrum of eukaryotic cells types and typically act by targeting regulators involved in the control of the cytoskeleton or the modification of actin directly.[29]

Due to their potent toxicity, bacterial toxins are of great interest to food production industries. Many handling and food treatment procedures revolve around the prevention of bacterial growth and reduction of bacterial toxin levels. However, while many common food treatment procedures serve to sterilize pathogenic bacteria, bacterial toxins can be more resilient and pose a health hazard if allowed to accumulate.[30] *S. aureus*, *E. coli* and *Bacillus cereus* all produce toxins that are heat stable.[31] This gives rise to the necessity for the detection of bacterial toxins in complex biological media. This task is well suited for aptasensing, as aptamers have proven themselves extremely capable in performing as recognition elements in a wide variety of sensitive sensor designs. Table 8.2 lists the recent examples of bacterial toxin detecting aptasensors.

Biosensors for Bacterial Toxins

Microcystin-LR (MCLR) is a cyanobacterial toxin of special consideration in the food industries due to its hazardous presence in contaminated waters, potential to accumulate in fish and toxicity to terrestrial mammals.[2] The MCLR toxin has been shown to inhibit serine/threonine protein phosphatases 1 and A2.[32] The inhibition of this enzyme has widespread effects on

Table 8.2 Aptasensors for bacterial toxins.

Target	Assay	Sensor description	DNA/RNA	K_d	Limit of detection (LOD)	Ref.
Alicyclobacillus spores	Fluorometric	Fluorescent Alexa488 labelled aptamer. Fluorescent measure of aptamer strands bound	DNA	11.3 nM	—	152
Anatoxin-A	Cyclic voltammetry	Disulfide modified aptamer mounted to a gold electrode showed reduced electrical resistance in the event of target binding	DNA	81.3 nM	0.5 nM	153
Bacillus Cereus spores	Fluorometric	Fluorescence microscopy used to visualize the target aptamer interaction. Real time PCR and SPR assays were also employed	DNA	5.7 nM	—	154
Botulinum toxin type A	PCR	PCR based assay to detect proteolytic activity of botulinum toxin using protein-aptamer assembly	DNA	—	100 fg	155
	Surface plasmon resonance	Shifts in SPR are observed when target interacts with immobilized aptamer	RNA	87 nM	5.8 $ng\,mL^{-1}$	156
Botulinum toxins	Fluorometric	5′-TYE 665 fluorophore and 3′-Iowa Black RQ quencher labelled DNA beacons	DNA	—	1 $ng\,mL^{-1}$	157
	Electrochemical	Methylene blue modified aptamer is bound to electrode. Binding of target moves methylene blue away from electrode and generates a current change	DNA	0.4 nM	—	158
Brevetoxin-2	Electrochemical	Target is mounted to a gold electrode surface with aptamer free in solution. Aptamer interaction with surface bound target generates an electrical signal. When analyte target is added to the ambient solution, the aptamer shows less interest in the surface bound target and the reduced number of interactions can be detected	DNA	42 nM	106 $pg\,ml^{-1}$	159
Clostridium difficile toxin B	ELISA	Amino-modified aptamer bound maleic anhydride on surface. In the presence of target-aptamer binding, a primary antibody can interact with the bound target. Secondary antibody conjugated with HRP can then mount and produce signal	DNA	47.3 nM	50 nM	160

Lipo-polysaccharide (endotoxin)	Electrochemical	Cyclic voltammetry: aptamer functionalized Au electrodes	DNA	12 nM	1 ng mL^{-1}	161
Microcystin-LR	UV-Vis absorbance	Target binding interactions would generate a split in GNP dimers resulting in an absorbance shift	DNA	50 nM	0.05 nM	35
	Surface-enhanced Raman scattering	An aptamer connects a gold nanoparticle to a nanoflower through complimentary binding to two DNA strands. Aptamer and target binding disrupts this connection and the change is observable *via* SERS	DNA	—	8.6 pM	162
	Photo-electrochemical	Vertically aligned titanium dioxide nanotubes photoanode. Target interactions were detected through aptamer π–π stacking interaction with nanotubes	DNA	—	0.03 pM	163
	Photo-electrochemical	BiOBr nanoflakes/N-doped graphene p–n heterojunction electrode. Target interactions were detected through aptamer π–π stacking interaction with electrode	DNA	50 nM	0.5 fM	164
	Electrochemical	Aptamer thiol mounted to gold electrode. Target interactions reduced impedance	DNA	—	1.8×10^{-11} mol L^{-1}	165
	Colorimetric	Salt-induced aggregation of AuNP is prevented by aptamer coating. Target binding removes coating and allows aggregation	DNA	—	0.37 nM	166
Microcystin-RR	Fluorometric	Immobilized aptamer is hybridized to a fluorescent strand of complimentary DNA. On target binding, the complimentary strand is release and washed away generating a reduced fluorescence	DNA	45.7 nM	80 pg mL^{-1}	167
Microcystin-LR and okadaic acid	Fluorometric	Lanthanide-doped $KGdF_4$ fluorescent nanoparticles and gold nanoparticle quenchers tethered together *via* aptamers and hybridized DNA	DNA	—	0.025 and 0.05 ng mL^{-1}	168
Okadaic acid	Electrochemical	Aptamer thiol mounted to gold electrode. Target interactions reduced impedance	DNA	77 nM	70 pg mL^{-1}	169
	ELLA	Direct competitive enzyme-linked aptamer assay. Lack of target binding allowed enzyme mounted complimentary strand to bind to aptamer	DNA	42 nM	0.01 ng mL^{-1}	170
Staphylococcus enterotoxin B	Piezoresistive microcantilever	Gold coated piezoresistive microcantilevers were thiol aptamer mounted. Target interactions were detected by a piezoresistor	DNA	—	6 ng mL^{-1}	171

Table 8.2 (*Continued*)

Target	Assay	Sensor description	DNA/ RNA	K_d	Limit of detection (LOD)	Ref.
	Fluorometric	DNA probes with fluorescent and quenching components experienced FRET when a complimentary aptamer hybridized. Target binding removes aptamer and prevents quenching. Removed aptamer is vulnerable to exonuclease 1 mediated target recycling	DNA	—	0.3 pg mL^{-1}	172
	ALISSA	Hybrid sandwich-ALISA employing chicken IgY and aptamer	DNA	—	50 ng mL^{-1}	173
	Electrochemical	Capture DNA is thiol bound to modified Au electrode. Aptamer is hybridized to capture DNA and removed by target. Unbound capture DNA is then bound to horseradish peroxidase. Enzyme catalyzed electrochemical signal is detected	DNA	—	0.24 ng mL^{-1}	174
	SERS	Functionalized magnetic Au nanorod SERS probes in sandwich assay	DNA	0.8 μM	224 aM	175
	Colorimetric	Salt-induced aggregation of AuNP is prevented by aptamer coating. Target binding removes coating and allows aggregation	DNA	—	10 ng mL^{-1}	176
Staphylococcus aureus enterotoxin A	Fluorometric	Graphene oxide (GO) quenches fluorescent aptamer. On target binding GO is displaced and fluorescence returns	DNA	48.57–6.52 nM	8.7×10^{-3} μg mL^{-1}	177
Staphylococcus aureus enterotoxin C1	Fluorometric	Fluorescent aptamer is exposed to graphene oxide (GO) and quenched on contact. Target removes aptamer from GO and fluorescence is restored	DNA	65.14 nmol L^{-1}	6 ng mL^{-1}	178
Staphylococcus aureus alpha toxin	ELISA	Amino-modified aptamer bound maleic anhydride on surface. In the presence of target-aptamer binding, a primary antibody can interact with the bound target. Secondary antibody conjugated with HRP can then mount and produce signal	DNA	93.7 nM	200 nM	179

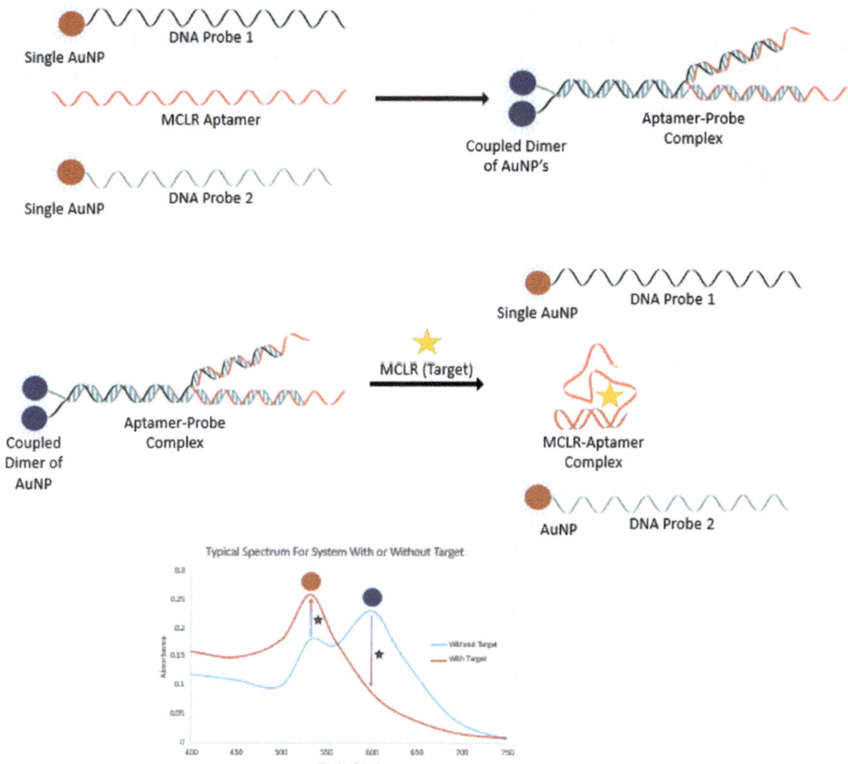

Figure 8.3 Colourimetric sensor for the detection of microcystin-LR. The method utilizes a Y-shaped DNA/AuNP assembly which, when hybridized, involves a controlled aggregation of the gold nanoparticles. The addition of target promotes the disassembly of the AuNP duplexes, resulting in a colour change from blue to red.
Reprinted from *Biosensors and Bioelectronics*, Volume 68, F. Wang, S. Liu, M. Lin, X. Chen, S. Lin, X. Du, H. Li, H. Ye, B. Qiu, Z. Lin, L. Guo and G. Chen, Colorimetric detection of microcystin-LR based on disassembly of orient-aggregated gold nanoparticle dimers, 475-780, Copyright 2015, with permission from Elsevier.

many biological pathways. The net effect is the generation of a build-up of phosphorylated proteins in the liver. This can eventually cause severe liver damage.[33] This effect has also been associated with potential tumour-promoting activities. Other adverse effects of MCLR include the formation of reactive oxygen and nitrogen species, the mutation of DNA (typically large deletions) and inhibition of DNA repair mechanisms.[34] All microcystins are cyclic heptapeptides containing two variable domains, of which MCLR had been found to be the most toxic.[35] Microcystins are produced by cyanobacteria, which are generally found in aquatic environments such as lakes or slow streams.

The World Health Organization (WHO) has evaluated MCLR and determined its estimated lethal dose at 50 $\mu g\,kg^{-1}$.[34] The WHO also provided a tolerable daily intake at 0.04 $\mu g\,kg^{-1}\,day^{-1}$. For drinking water, the WHO provided a safe drinking limit in water of 1.5 $\mu g\,L^{-1}$.[35] Due to its potency, rapid on-site detection of MCLR is of interest to professionals within environmental protection agencies and the fisheries and agriculture industries. A 2015 sensor designed by Wang and others may suit this need (See Figure 8.3).[35] The sensor uses a MCLR aptamer to disassemble pre-formed orient-aggregated gold nanoparticle (AuNP) dimers on target binding. This sensor yielded results rapidly with a processing time of only ~5 minutes, while demonstrating a limit of detection of 0.05 nM (~0.05 $\mu g\,L^{-1}$) in the binding buffer solution.

8.2.3 Antibiotics, Drugs and Other Residues

In an attempt to prevent and treat diseases in farmed animals, antibiotics are most often used. Although exceptionally useful, overuse or misuse of antibiotics may lead to residues left behind in the farmed animals. Ultimately, consumption of antibiotic-tainted meat by humans and animals can result in widespread antibiotic resistance, thereby reducing the effects of drugs that can cure illnesses. To avoid this, the FDA has only allowed antibiotic use for treatment of infections.[36] Pesticides on the other hand, are used to kill and/or deter pests that can damage crops. Yet, these pesticides sometimes linger in the crops and may be consumed by humans and animals. If uncontrolled or misused, many of these pesticides can become health hazards and can have a negative impact on the environment used to grow additional food. Thus, the pesticide type and the amount are typically highly regulated.[37] Due to the possible health implications from consuming food with antibiotics and pesticides, it is necessary to monitor and detect levels in food to ensure food safety and quality. To do so, aptasensors that are sensitive and fast are ideal devices. Table 8.3 lists the recent aptasensors that have been designed for this purpose.

Biosensors for Antibiotics, Drugs and Other Residues

A novel method for detecting the antibiotic streptomycin was developed by Taghdisi and colleagues. Streptomycin (STR) is an aminoglycoside antibiotic used to treat humans and animals for diseases associated with Gram-negative bacteria. STR residues in food could cause nephrotoxicity and ototoxicity side effects in humans.[38] A label-free fluorescence aptasensor has been developed that uses exonuclease III, which selectively digests the 3′-end of dsDNA, SYBR Gold for fluorescence and an aptamer complementary strand (See Figure 8.4). The methodology is as follows: the aptamer is bound to the complementary strand and the aptamer dissociates from the complement strand and forms a complex with STR in the presence of STR. The complex formation results in protection against Exo III digestion. Upon

Table 8.3 Aptasensors for antibiotics.

Antibiotics	Assay	Aptasensor description	DNA/RNA	K_d	Limit of detection (LOD)	Ref.
Amino-glycosidic antibiotics	Electrochemical	Folding-based sensors utilizing structure-switching aptamers	RNA	0.22–42 μM	—	180
	—	Selex	DNA	—	3.9 μM (KANA)	181
	Fluorometric	Microsomal triglyceride transfer protein (MTP) assay	DNA	5 μmol L^{-1} (KANA in real waste water)	2.8 μM (KANA)	
	—	Selex	RNA	10–100 nM	—	182
	—	Selex	DNA	3.9 μM	—	183
Ampicillin	Electrochemical	Based on dual recycling amplification strategy	DNA	—	4.0 pM	184
Ampicillin/Kanamycin	Colorimetric	Aptamer–gold nanoparticle based using SERS	DNA	13.4 nM (ampicillin), 78.8 nM (KAN)	Not specified (ampicillin), 300 nM (KAN)	185
Ampicillin/Kanamycin	Electrochemical	Conductive polymer	DNA	13.4 nM (ampicillin), 78.8 nM (KAN)	10^{-10} M (ampicillin), 10^{-6} M (KAN)	186
Ampicillin	Fluorometric-colorimetric	Gold nanoparticle based	DNA	9.4 nM	0.5 ng mL^{-1} (fluorescence), 5 ng mL^{-1} (colorimetry) by sensor; 2 ng mL^{-1} (fluorescence), 10 ng mL^{-1} (colorimetry) in milk and distilled water	187
Chlor-amphenicol	Fluorometric	DIL-encapsulated liposome as nanotracer	DNA	0.766 μM	1 pM	188
	Fluorometric	FRET between quantum dots as donor and the magnetic SiO_2@Au NPs probe as acceptor with exonuclease-assisted target recycling	DNA	0.766 μM	0.0002 ng mL^{-1}	189
	Fluorometric	Switch-on sensor based on a magnetic composite probe with aptamer and hemin/G-quadruplex coimmobilized nano-Pt–luminol as signal tracer	DNA	0.766 μM	0.0005 ng mL^{-1}	190

Table 8.3 (*Continued*)

Antibiotics	Assay	Aptasensor description	DNA/RNA	K_d	Limit of detection (LOD)	Ref.
Chlor-amphenicol and PCB72	Electrochemical	Magnetic and aptamer-modified quantum dot-encoded dendritic nanotracers for signal amplification	DNA	0.766 μM (CAP), 66 nM (PCB72)	0.33 pg mL^{-1} (CAP), 0.35 pg mL^{-1} (PCB72)	191
Chlor-amphenicol	Colorimetric	Based on an indirect competitive enzyme-free method	DNA	0.766 μM	451 pM, 697 pM (milk)	192
	Fluorometric	"Off–on" fluorescence assay using vesicle quantum dot–gold colloid composite probes	DNA	0.766 μM	0.3 pM	193
	Colorimetric	Gold nanoparticle based assay	DNA	0.766 μM	0.55 pg mL^{-1}	194
	—	Selex	DNA	0.10162 and 0.03224 μM	—	195
	Electrochemi-luminescent	Based on the ratio of working signal to internal standard signal	DNA	0.766 μM	0.03 nM	196
Chlor-amphenicol/ Streptomycin/ Tetracycline	Electrochemical	Based on aptamers and quantum dots	DNA	0.766 μM (CAP), 199.1 nM (STR), 483.5 nM (TET)	5 nM (CAP), 10 nM (STR), 20 nM (TET)	197
Chlor-amphenicol Oxytetracycline	Electrochemical	Based on high-capacity magnetic hollow porous nanotracers coupling exonuclease-assisted cascade target recycling	DNA	—	0.15 ng mL^{-1} (CAP), 0.10 ng mL^{-1} (OTC)	198
Chlor-amphenicol/ Kanamycin	Electrochemical	Nanoscale metal organic framework	DNA	0.766 μM (CAP), 78.8 nM (KANA)	0.19 pM (CAP), 0.16 pM (KANA)	199
Chlor-amphenicol	Chemi-luminescent	Based on *N*-(4-aminobutyl)-*N*-ethylisoluminol-functionalized flower-like gold nanostructures and magnetic nanoparticles	DNA	—	0.01 ng mL^{-1} (buffer), 1 ng mL^{-1} (milk)	200
	Photo-electrochemical	Based on nitrogen-doped graphene quantum dots	DNA	0.766 μM	3.1 nM	201
	Fluorometric	Based on upconversion nanoparticles	DNA	0.766 μM	0.01 ng mL^{-1}	202
	Electrochemical	Based on aptamer incorporated gelatine	DNA	—	1.83×10^{-10} M	203

	Colorimetric	Based on double-stranded DNA antibody labelled enzyme-linked polymer nanotracers for signal amplification	DNA	0.766 μM	0.015 ± 0.003 ng mL^{-1}	204
	Colorimetric	Polymeric HRP-antibody conjugate for signal amplification	DNA	0.766 μM	3 pg mL^{-1}	205
	Electrochemical	Based on gold nanocubes-modified screen-printed gold electrode	DNA	0.766 μM	4.0 nM	206
	Electrochemical impedimetric	Changes of the interfacial properties of the electrode after the interaction of aptamer and target	DNA	1.16 μM	1.76 nM	207
	Electrochemical	Target-induced strand release	DNA	—	0.29 nM	208
	Electrochemical	Functionalized gold electrode	DNA	—	1.6 nM	209
	—	Selex	DNA	0.766 μM	—	210
Danofloxacin	—	Selex	RNA	3.36 nM	—	211
Enrofloxacin	Luminescent	Aptamer functionalized magnetic nanoparticle conjugated with upconversion nanoprobes	DNA	188 nM	0.06 ng mL^{-1}	212
	Chemi-luminescent	Chemiluminescent enzyme immunoassay associating with biotin-streptavidin	DNA	188 nM	2.26 ng mL^{-1}	213
Kanamycin	Electrochemical	Based on peroxidase-like activity of gold nanoparticles	DNA	78.8 nM	0.06 nM	214
	Electrochemical	Based on exonuclease III-assisted autocatalytic DNA biosensing platform	DNA	78.8 nM	0.74 pM	215
	Electrochemical	Based on thionine functionalized graphene and hierarchical nanoporous PtCu	DNA	78.8 nM	0.42 pg mL^{-1}	216
	Electrochemical	Based on exonuclease recycling cleavage	DNA	78.8 nM	1 pM	217
	Fluorometric	Aptamer-mediated $CuInS_2$ quantum dots@graphene oxide nanocomposites-based "turn off–on" nanosensor	DNA	78.8 nM	0.070 μg L^{-1}	218

Table 8.3 (*Continued*)

Antibiotics	Assay	Aptasensor description	DNA/RNA	K_d	Limit of detection (LOD)	Ref.
	Electrochemical	Signal amplification	DNA	78.8 nM	0.005 μg L^{-1}	219
	Fluorometric	Based on catalytic recycling activity of exonuclease III and gold nanoparticles	DNA	78.8 nM	321 pM	220
	Electrochemical	Signalling-probe displacement	DNA	78.8 nM	1 nM	221
	Fluorometric	Based on silica nanoparticles (SNPs) coated with streptavidin	DNA	78.8 nM	612 pM	222
	Fluorometric	Aptamer-based signal-on bio-assay using gold nanoparticles	DNA	78.8 nM	0.3 nM	223
	Electrochemical	Based on MWCNTs–$HMIMPF_6$ and nanoporous PtTi alloy	DNA	78.8 nM	3.7 pg mL^{-1}	224
	Electrochemical	Gold electrode and aptamer conformational shift	DNA	78.8 nM	0.014 nM	225
	Electrochemical	Based on MWCNTs–$BMIMPF_6$ and amino functionalized graphene nanocomposite films	DNA	78.8 nM	0.87 nM	226
Kanamycin/Oxytetracycline	Fluorometric	Aptamer-functionalized magnetic nanoparticles	DNA	78.8 nM (KAN)	0.92 ng mL^{-1} (KAN), 0.85 ng mL^{-1} (OTC)	227
Kanamycin	Colorimetric	Gold nanoparticles (AuNPs) based sensor	DNA	78.8 nM	1 nM	228
	Colorimetric	Aptamer-mediated "turn-off/turn-on" nanozyme activity of gold nanoparticles	DNA	78.8 nM	1.49 nM	229
	Electrochemical	Cantilever array sensor	DNA	78.8 nM	50 μM	230
	Electrochemical	Based on thionine, graphene–polyaniline composite film and gold nanoparticles	DNA	78.8 nM	8.6 nM	231
	Electrochemical	Using graphene–polyaniline and polyamidoamine dendrimer–gold nanoparticle nanocomposites on glass carbon electrode	DNA	78.8 nM	4.6×10^{-6} μg mL^{-1}	232

	Fluorometric	Based on upconversion nanoparticles	DNA	78.8 nM	9 pM	233
Kanamycin/Sulfa-dimethoxine	Colorimetric	Unmodified gold nanoparticle based assay	DNA	78.8 nM (KAN), 84 nM (SDM)	100 ng mL^{-1} (KAN), 500 ng mL^{-1} (SDM) visually	234
Kanamycin	Luminescence	Utilizes a square-planar luminescent platinum(II) complex	DNA	78.8 nM	143 nM	235
	Electrochemical	Immobilized aptamer on DPB(AuNP) conducting nanocomposite	DNA	78.8 nM	9.4 nM	236
	Colorimetric	Gold nanoparticle-based through salt aggregation	DNA	78.8 nM (KANA), 84.5 nM (KANB), 103 nM (TOB) (all same aptamer)	25 nM	237
Lysozyme	Electrochemical	Aptamer-antibody sandwich assay	DNA	2.8 nM	4.3 fM	238
	Electrochemical	Based on the specific recognition by the Apt immobilized on screen-printed carbon electrodes *via* diazonium coupling reaction	DNA	Aptamer 1: 2.8 nM, aptamer 2: 31–65 nM (unbiotinylated + biotinylated target, respectively)	aptamer 1: 25 nM, aptamer 2: 100 nM	239
	Electrochemical	Based on graphene oxide coated on the surface of a glassy carbon electrode	DNA	31–65 nM (unbiotinylated + biotinylated target, respectively)	6 fM	240
Neomycin b	Fluorometric	Self-assembly of the RNA aptamer/Neomycin B complex and fluorescence quenching by AuNPs	RNA	—	0.01 μM	241
Ofloxacin	—	Selex	DNA	130.1 nM	—	242
Oxytetracycline	Colorimetric	Aptamer binding-mediated anti-aggregation of positively charged gold nanoparticles	DNA	—	0.227 μg mL^{-1}	243
	Fluorometric	Based on fluorescence quenching of DNA aptamer-templated silver nanoclusters	DNA	56.84 nM	0.1 nM	244
	Fluorometric	Based on target-responsive graphene oxide hydrogel	DNA	—	25 μg L^{-1} (LOQ)	245

Table 8.3 (*Continued*)

Antibiotics	Assay	Aptasensor description	DNA/RNA	K_d	Limit of detection (LOD)	Ref.
	Luminescence	Based on upconversion and magnetic nanoparticles	DNA	—	0.036 ng mL^{-1}	246
	Luminescence	Based on luminescence resonance energy transfer (LRET) from $NaYF_4$:Yb,Tm upconversion nanoparticles (UCNPs) to SYBR Green I	DNA	—	0.054 ng mL^{-1}	247
	Colorimetric	Using gold nanoparticles	DNA	—	1 nM (buffer and tap water)	248
	Colorimetric	Based on combining magnetic separation and direct competitive ELAA	DNA	—	0.88 ng mL^{-1}	249
	Colorimetric	Indirect competitive enzyme-linked aptamer assay	DNA	4.7 nM	12.3 μg L^{-1} (in milk)	250
	Fluorometric	Using a fluorescein-labelled long-chain aptamer assembled onto reduced graphene oxide	DNA	56.84 nM	10 nM	251
	Electrochemical	Based on a microfabricated cantilever array. The sensing cantilevers in the array are functionalized with self-assembled monolayers (SAMs) of OTC-specific aptamer, which acts as a recognition molecule for OTC	DNA	—	0.2 nM	252
PCB72/PCB106	—	Selex	DNA	2.5 nM (PCB72), 99 nM (PCB106)	—	253
Penicillin	Electrochemical	Using magnetic graphene nanocomposite (GR–Fe_3O_4NPs) and a poly(3,4-ethylenedioxythiophene)–gold nanoparticles composite (PEDOT–AuNPs) as the platform	DNA	—	0.057 ng mL^{-1}	254

Streptomycin	Electrochemical	Based on gold nanoparticle-functionalized magnetic multi-walled carbon nanotubes and nanoporous PtTi alloy	DNA	199.1 nM	7.8 pg mL^{-1}	255
	Colorimetric and fluorescence quenching	Based on double-stranded DNA and gold nanoparticles	DNA	199.1 nM	73.1 nM (colorimetric) and 47.6 nM (fluorescence quenching)	256
	Electrochemical	Based on arch-shape structure of aptamer-complimentary strand conjugate and exonuclease I	DNA	199.1 nM	11.4 nM (in milk)	257
	Fluorometric	Based on exonuclease III (Exo III), SYBR Gold and aptamer complimentary strand	DNA	199.1 nM	54.5 nM	38
	Colorimetric	Using unmodified gold nanoparticles	DNA	132.3 nM	47.2 nM	258
	Colorimetric	Gold nanoparticles (AuNPs)-based	DNA	199.1 nM, 221.3 nM	0.2–1.2 μM	259
Sulfa-dimethoxine	Colorimetric	Based on the inhibition and reactivation of the peroxidase-like activity of gold nanoparticles	DNA	84 nM	10 ng mL^{-1}	260
	Colorimetric	Using unmodified gold nanoparticles	DNA	84 nM	50 ng mL^{-1}	261
	Fluorometric	Coordination polymer nanobelt-based aptasensor	DNA	84 nM (in dH_2O and milk)	10 ng mL^{-1}	262
Tetracycline	Electrochemical	Sensor based on modified carbon paste electrode (CPE), oleic acid (OA) and aptamer (CPE/OA/anti-TET)	DNA	63.6 nM	3.0×10^{-13} M (EIS), 2.9×10^{-11} M (DPV)	263
		Sensor based on carbon paste electrode, oleic acid, a magnetic bar carbon paste electrode (MBCPE) with Fe_3O_4 magnetic nanoparticles and oleic acid, and aptamer (MBCPE/Fe_3O_4NPs/OA/anti-TET)			3.8×10^{-15} M (EIS), 3.1×10^{-13} M (DPV)	

Table 8.3 (*Continued*)

Antibiotics	Assay	Aptasensor description	DNA/RNA	K_d	Limit of detection (LOD)	Ref.
	Electrochemical	Amino-modified aptamer attached to an activated gold electrode. The ferricyanide solution was used as an electrochemical indicator	DNA	52.5 nM	10 ng ml^{-1}	264
	Electrochemical	Using a screen-printed carbon electrode modified with an alginate film containing reduced graphene oxide and magnetite (Fe_3O_4) nanoparticles	DNA	63.6 nM	0.6 nM	265
	Electrochemical	Sensor modified by ionic liquid–Fe_3O_4 based on microarray electrodes	DNA	—	1×10^{-9} M	266
	Electrochemical	Using a glassy carbon modified with nanosheets of graphene oxide	DNA	63.6 nM	29 fM	267
	Electrochemical	Signal amplification of multi-walled carbon nanotubes-chitosans and Fe_3O_4-chitosans	DNA	—	10^{-9} M	268
	Electrochemical	Based on M-shape structure of aptamer-complementary strands of aptamer complex, exonuclease I (Exo I) and gold electrode	DNA	63.6 nM	450 pM, 740 pM (milk)	269
	Electrochemical	Sensor modified with electrodeposited gold nanoparticles and methylene blue	DNA	63.6 nM	0.42×10^{-11} M	270
	Colorimetric	Competitive enzyme-linked aptamer assay	DNA	Aptamer 1: 63.6 nM, aptamer 2: 2.94 μM	Not reported	271

	Colorimetric	Triple-helix molecular switch and gold nanoparticles	DNA	1.067 nM	266 pM	272
	Colorimetric	Using cysteamine-stabilized gold nanoparticles as probe	DNA	63.6 nM	0.039 $\mu g\,mL^{-1}$	273
	Fluorometric	Triple-helix molecular switch	DNA	100.6 nM	2.09 nM	274
	Electrochemical	Multi-walled carbon nanotubes and graphene for tetracycline detection	DNA	63.6 nM	0.56×10^{-11} M	275
	Electrochemical	Gold electrode and aptamer conformational shift	DNA	5.18×10^{-5} mol L^{-1}	1.0 $ng\,mL^{-1}$	276
Tetracycline/ oxytetracycline	—	Selex	DNA	1.067 nM (TET), 1.104 nM (OTC)	—	277
Tetracycline	Colorimetric	Uses aptamer, unmodified gold nanoparticles and hexadecyltri-methylammonium bromide			122 nM	278
	Colorimetric	Assembly of cationic conjugated polymer-aggregated gold nanoparticles	DNA	63.6 nM	1 μM (naked eye), 45.8 nM (colorimetric detector)	279
	Colorimetric	Competitive enzyme linked aptamer assay	DNA	DNA aptamer: 63.6 nM, RNA aptamer: 770 pM	2.10×10^{-8} M (milk)	280
	Electrochemical	Glassy carbon electrode with aptamer modified multi-walled carbon nanotubes	DNA	63.6 nM	5 nM (milk)	281
Tobramycin	Electrochemical and ratiometric	Aptamer structure-switching	RNA	23 μM (parent aptamer), 45 ± 3 nM (mutant aptamer)	5–2381 $\mu A \mu M\,(cm^2)^{-1}$ (Sensitivity)	282

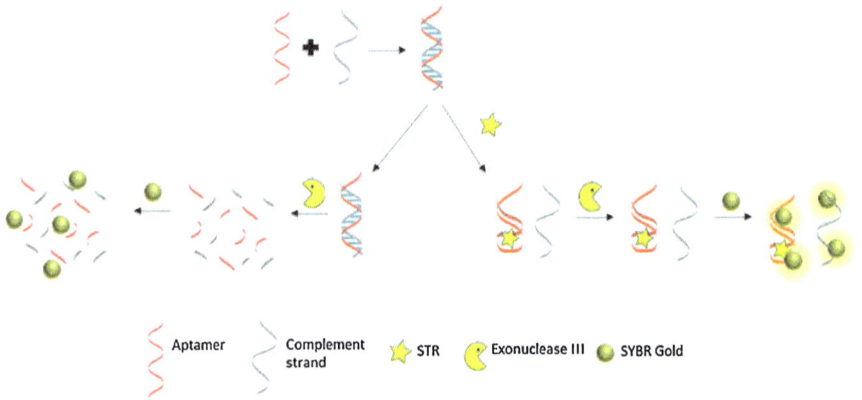

Figure 8.4 Fluorescence assay for detection of streptomycin (STR). In the absence of target, the aptamer-complement complex is degraded by exonuclease III, and little fluorescence is observed upon the addition of SYBR Gold. In the presence of target, the aptamer-complement complex dissociates and the aptamer binds the target. Aptamer and complement are not degraded by exonuclease III, and thus the addition of SYBR Gold results in a strong fluorescence.
Reprinted from *Food Chemistry*, Volume 203, S. M. Taghdisi, N. M. Danesh, M. A. Nameghi, M. Ramezani and K. Abnous, A label-free fluorescent aptasensor for selective and sensitive detection of streptomycin in milk and blood serum, 145-149. Copyright 2016, with permission from Elsevier.

addition of SYBR Gold, which binds the aptamer, a strong fluorescence intensity is observed. In the absence of STR, the double-stranded aptamer-complement complex does not separate and is digested by Exo III. As a result, SYBR Gold does not have a platform to which to bind and the consequence is little to no fluorescence observed. In buffer, this aptasensor has a low limit of detection (LOD) of 54.5 nM and this sensor was also shown to work with real samples, namely blood serum and milk. Even with a complicated sample matrix, there was no interference in detecting STR, and the sensor had low LODs of 71 nM and 76.06 nM in serum and milk, respectively. These LODs are much lower than the amount of STR that would be toxic in blood and the allowable limit of STR in milk. Finally, the selectivity of the sensor was tested using the antibiotics amoxicillin, kanamycin, ciprofloxacin and gentamycin. Little to no fluorescence was observed for those antibiotics, especially compared to the high fluorescence intensity observed for STR, which shows that this assay has great selectivity. This sensor is sensitive, fast, selective for STR detection and compatible for detection in complex matrices.

Another interesting sensor was developed by Liu and coauthors for acetamiprid; acetamiprid is part of a new neonicotinoid class of systemic broad-spectrum insecticides and is toxic because it inhibits the normal conduction of the central nervous system.[39] This novel fluorescence sensor by Liu and coauthors involves a highly stable triple-helix DNA. The triple helix is a non-traditional DNA structure, which *via* Hoogsteen hydrogen bonds, has a third strand in the major groove of the DNA duplex. The fluorescent sensing platform uses a triple-helix molecular switch (THMS) that is composed of two DNA probes; a recognition probe, which is a label-free aptamer flanked by two segments, and a signal transduction probe (STP), which is a hairpin-shaped oligonucleotide. The STP labelled with a fluorophore on the 3′-end and a quencher on the 5′-end. In the absence of acetamiprid, the segments flanking the aptamer bind to the complement sequence on the loop of the STP, which opens the STP. Because it is opened, the fluorescence of the THMS is on. However, in the presence of acetamiprid, the aptamer preferentially binds the acetamiprid over the STP, which disassembles and releases the STP. The free STP refolds into a hairpin structure, and the fluorescence is quenched. In this study, it was found that the unquenched fluorescence intensity was proportional to the concentration of acetamiprid in the range from 100 to 1200 nM with a limit of detection (LOD) of 9.12 nM (Table 8.4).

To determine the sensor's effectiveness and applicability with real samples, it was tested with spiked Chinese cabbage. It was found that the recovery of acemiprid was satisfactory with the Chinese cabbage, and the results were comparable to those obtained by LC-MS. In addition, there were no obvious interferences from other substances in the real sample. Finally, the selectivity of the sensor was evaluated by testing the sensor with molecules structurally similar to acetamiprid such as imidacloprid, chlorpyrifos, 2,4-dichlorophenoxyacetic acid, glufosinate-ammonium, methamidophos, carbaryl, atrazine and nicotine. It was found that the sensor was highly selective as there were little to no fluorescent intensity changes observed for these other molecules. This sensor is convenient, simple to use, sensitive, fast and highly generalizable since the aptamer sequence can be altered for use with another target without changing the triple-helix structure (Table 8.5).

8.2.4 Heavy Metals

Heavy metals are inorganic metallic elements that are characterized by a relatively high density and potent toxicity.[40] Examples include copper (Cu), lead (Pb), mercury (Hg) and silver (Ag). The chemical properties of arsenic (As) also render it a heavy metal ion contaminant.[41] Heavy metal ions can contaminate a variety of foodstuffs including fish and fish products, meat and meat products, milk and milk products, fats, oils and animal feed. As heavy metal contaminants are both difficult to eliminate and readily accumulate throughout the food chain, heavy metal contamination has become a major environmental concern in several countries worldwide.[42,43] Adverse health effects such as cancer and neurodegenerative diseases have been associated with chronic exposure to heavy metal food contaminants.[44] Therefore, the detection of heavy metal ions in foodstuffs is an important

Table 8.4 Aptasensors for pesticides.

Pesticides	Assay	Aptasensor description	DNA/RNA	K_d	Limit of detection (LOD)	Ref.
Acetamiprid	Computational/Colorimetric	Using silver nanoparticles	DNA	4.98 μM	89.8 nM (computationally)	283
	Fluorometric	Turn on sensor with quantum dots functionalized with aptamer	DNA	4.98 μM	0.7 nM	284
	Chemi-luminescent	Aptamer recognition and gold nanoparticles morphology effect on catalysis	DNA	4.98 μM	62 pM	285
	Colorimetric	Aptamer-wrapped gold nanoparticles assay	DNA	4.98 μM (full length aptamer)	0.4 μM	286
	Fluorometric	Based on the inner filter effect of gold nanoparticles on the fluorescence of CdTe quantum dots	DNA	4.98 μM	7.29 nM	287
	Fluorometric	Fluorescence resonance energy transfer between NH_2-$NaYF_4$: Yb, Ho@SiO_2 and gold nanoparticles	DNA	4.98 μM	3.2 nM	288
	Fluorometric	Triple-helix molecular switch	DNA	4.98 μM	9.12 nM	289
	Resonance light scattering	Using gold nanoparticles	DNA	4.98 μM	1.2 nM	290
	Electrochemical	Based on a dual signal amplified strategy by employing a gold-polyaniline modified screen-printed electrode and an enzyme-linked label	DNA	4.98 μM	Not reported	291
	Colorimetric	Based on the peroxidase-like activity of hemin-functionalized reduced graphene oxide	DNA	4.98 μM	40 nM	292
	Electrochemical	Silver nanoparticles anchored on nitrogen-doped graphene	DNA	4.98 μM	3.3×10^{-14} M	293

	Colorimetric	Using inherent peroxidase-like nanozyme activity of gold nanoparticles	DNA	4.98 μM	0.1 ppm	294
	Colorimetric	Inter-particle plasmon coupling during the aggregation of gold nanoparticles	DNA	4.98 μM	5 nM	295
	Electrochemical	Aptamer immobilization on gold nanoparticles electrodeposited on bare gold electrode surface	DNA	23.41 nM	1 nM	296
	—	Selex	DNA	4.98 μM	—	297
Atrazine	—	Selex	DNA	0.62 nM	—	298
Bromacil	—	Selex	DNA	9.6 nM	—	299
Carbofuran	Chemi-luminescent	Using fullerene-loaded gold nanoparticles and carbon dots	DNA	—	8.8×10^{-13} mol L^{-1}	300
Diazinon	Colorimetric/Computational	Silver nanoparticle based assay	DNA	—	17.903 nM	301
Fluoro-acetamide	—	Selex	DNA	1 and 1.2 μM	—	302
Iprobenfos/Edifenphos	Colorimetric	Using gold nanoparticles	DNA	1.67 μM (iprobenfos), 38 nM (edinfenphos)	10 nM (iprobenfos), 5 nM (edinfenphos)	303
	Fluorometric	Fluorescent resonance energy transfer	DNA	—	11.4 μM (isocarbophos); 14.0 μM (profenofos)	304
	Electrochemical	Chitosan-iron oxide nanocomposite	DNA	—	0.001 ng mL^{-1}	305
	Colorimetric	Using unmodified gold nanoparticles and polydiallyldimethylammonium chloride	DNA	—	0.06 pM	306
	Spectroscopy	Based on polymer-gold nanoparticle composite microspheres	DNA	—	3.3 μg mL^{-1}	307

Table 8.4 (*Continued*)

Pesticides	Assay	Aptasensor description	DNA/RNA	K_d	Limit of detection (LOD)	Ref.
Omethoate	Colorimetric	Gold nanoparticle based assay	DNA	—	0.1 µmol L^{-1}	308
Organo-phosphorous pesticides	Fluorometric	Using quantum dots with capillary electrophoresis	DNA	—	0.10–0.23 µM	309
	Fluorometric	Using a gold-based nanobeacon probe	DNA	—	0.035–2.35 µM	310
	Colorimetric	Gold nanoparticle-based assay	DNA	0.83 µM (isocarbophos)	Not reported	311
	Spectroscopy	Single aptamer-based capture procedure	DNA	0.83–2.5 µM	0.4–24 µM	312
	Fluorometric	Fluorescence polarization	DNA	Aptamer 1: 0.83–2.5 µM, Aptamer 2: 0.9–2 µM	13.4–23.4 nM (LOQ)	313
	—	Selex	DNA	Aptamer 1: 0.83–2.5 µM, Aptamer 2: 0.9–2 µM	—	314
Phorate	Colorimetric	Aggregation of the gold nanoparticles	DNA	1.11 µM	0.01 nM	315

Table 8.5 Aptasensors for drug and other residues.

Drugs	Assay	Aptasensor description	DNA/RNA	K_d	Limit of detection (LOD)	Ref.
17 beta-estradiol	Colorimetric	Gold nanoparticle based assay	DNA	Parent: 25 nM, Aptamer 1: 14 nM, Aptamer 2: 11 nM	200 pM	316
	Electrochemi-luminescent	Competition assay with cDNA	DNA	0.13 μM	1.1×10^{-12} M	317
	Electrochemical	Based on vanadium disulfide nanoflowers and gold nanoparticles	DNA	0.13 μM	1.0×10^{-12} M	318
	Colorimetric	Using split DNA aptamers immobilized on unmodified gold nanoparticles	DNA	0.13 μM	0.1 ng mL^{-1}	319
	Fluorometric	Using a quantum dot-bioconjugate as a nano-bioprobe		0.13 μM	0.22 nM	320
	Electrochemical	Aptamer functionalized gold electrode	DNA	0.13 μM	2.0 pM	321
	Fluorometric	Immobilized 17 beta-estradiol-BSA on optical fibre sensor	DNA	0.13 μM	2.1 nM	322
	Separation	Aptamer-anchored isothiocyanate-modified beads	DNA	35.19 μM	Not reported	323
Beta agonists	Electrochemical	Aptamer-agonist-based gold electrode	DNA	3.34×10^{-5} mol L^{-1} (phenylethanol-amine)	0.35–40 ng mL^{-1}	324
Cocaine	Electrochemical	Based on glassy carbon electrode modified with platinum nanoparticles and using rutin as a redox probe	DNA	—	100 pM	325
	Electrochemical	Based on H-shape structure of aptamer-complimentary strands conjugate	DNA	—	0.228 nM	326
	Electrochemical	Target-responsive aptamer release from manganese dioxide nanosheets	DNA	—	32 pM	327
	Chemi-luminescent	Based on stimulus-responsive mesoporous silica nanoparticle		—	1.43 μM	328

Table 8.5 (*Continued*)

Drugs	Assay	Aptasensor description	DNA/RNA	K_d	Limit of detection (LOD)	Ref.
Cocaine/ benzoylecgonine- (metabolite)	Fluorometric	Using quantum dots and gold nanoparticles	DNA	—	0.138 nM (cocaine), 1.66 μM (benzoylecgonine)	329
Cocaine	Photo-electrochemical	Based on an electropolymerized ruthenium-intercalator complex		3.8 mM	10 nM	330
	Raman scattering	Using gated mesoporous silica nanoparticles loaded with crystal violet and SERS	DNA	—	10 nM	331
	Fluorometric	Based on a G-quadruplex and ruthenium polypyridyl complex molecular light switch	DNA	—	5 nM	332
	Fluorometric	Based on graphene oxide and exonuclease III for signal amplification	DNA	—	0.1 nM	333
	Fluorometric	Using unmodified AuNPs	DNA	—	30 nM	334
	Fluorometric	Rolling circle amplification of short DNA strand separated by magnetic beads	DNA	—	0.48 nM	335
	Electrochemical	Electrode surface decorated with an aptamer probe-pendant tetrahedral DNA nanostructure that facilitates cocaine-induced fusion of the aptamer	DNA	—	33 nM	336
	Interferometric	Aptamer functionalized microcantilever	DNA	—	5 μM	337

and growing concern. Interest in reliable, fast and inexpensive on-site detection methods has prompted the development of aptasensors for heavy metal ion detection. Table 8.6 summarizes the recent advances in aptasensors for this class of food contaminant.

Heavy Metal Biosensors

Arsenic (As) is a carcinogenic heavy metal that is toxic even at very low concentrations. Although there are four oxidation states of As (−3, 0, +3 and +5), arsenite(III) and arsenate(V) are the two most common states of this heavy metal contaminant. The World Health Organization (WHO) provides a guideline of 10 ppb, however exposure beyond these limits occurs worldwide.[45] Ingestion can occur directly through contaminated drinking water or indirectly through consumption of crop grown in contaminated soil. Arsenate (As(III)) has been found to be 60 times more toxic than As(V), leading to adverse health effects such as cancer and skin and circulatory damage.[45,46] Similar to those for other heavy metal targets, current detection methods are complex, expensive and time-consuming.[45] In 2012, Wu and others developed a biosensor to detect arsenic(III) in buffer solution (see Figure 8.5). Gold nanoparticles (AuNPs) are used to provide the colorimetric aspect of the developed biosensor. The aggregation behaviour of AuNPs can be used as a signal for target detection through a visible colour change in solution. Water-soluble cationic polymers have been used to control the aggregation of AuNPs.[47] Furthermore, there is an electrostatic interaction between these cationic polymers and the negatively charged phosphate backbone in DNA.[48] These interactions have been employed to serve as the platform of the developed colorimetric biosensor. Kim and others selected an arsenic(III) binding aptamer with a reported dissociation constant (K_d) of 7.05 nM.[46] This aptamer, Ars-3, was used as the bio-recognition element for the sensitive and specific detection of arsenic(III). Incorporated into the biosensor, along with Ars-3, were AuNPs for colorimetric signalling and poly(diallyldimethylammonium) (PDDA) as the cationic polymer for controlled AuNP aggregation.[45] In this sensor, PDDA both promotes AuNP aggregation and interacts electrostatically with free aptamer in solution. Without As(III) present in solution, the aptamer is free in solution and available to form a complex with the PDDA. Since PDDA is not available to promote AuNP aggregation when in a complex, the AuNPs remain dispersed in solution yielding a wine-red colour.[45] However, when As(III) is present in solution, the Ars-3 aptamer preferentially binds this target molecule leaving PDDA free in solution. PDDA promotes AuNP aggregation, which causes a colour change from wine-red to blue. Therefore, the detection of As(III) is signalled by a visible change in colour. Aggregation of AuNPs upon presence of As(III) was characterized through UV-Vis absorption spectra and transmission electron microscopy (TEM). Over a linear concentration range of 5 to 3000 ppb As(III), a detection limit of 5.3 ppb was determined.[45] The use of the cationic polymer PDDA for the control of AuNP aggregation allowed

Table 8.6 Aptasensors for heavy metals.

Heavy metals	Detection method	Aptasensor description	DNA/RNA	K_d	Limit of detection (LOD)	Ref.
Arsenic(III)	Fluorometric	Fluorometric aptasensor based on cystamine stabilized CdTe/ZnS quantum dots aggregation	DNA	7.05 nM	1.3 pmol L^{-1}	338
	Colorimetric	Aptamer-conjugated silver nanoparticles using response surface methodology	DNA	7.05 nM[46]	6 μg L^{-1}	339
	Surface enhanced Raman scattering (SERS)	Aptamer modified Au@Ag nanoparticles for the detection of arsenic through SERS	DNA	7.05 nM[46]	0.1 ppb	340
	Surface enhanced Raman scattering (SERS)	Target induced aggregation of modified Au@Ag nanoparticles for SERS	—	—	59 ppt	341
	Rayleigh scattering	Rayleigh scattering using aptamer modified nanogold probe	DNA	7.05 nM[46]	1.9 ng mL^{-1}	342
	Surface enhanced Raman scattering (SERS)	Aptamer modified gold nanoparticles with Rodamine 6G for the detection of arsenic through SERS	DNA	—	0.1 ng mL^{-1}	343
	Colorimetric	Aptamer-based gold nanoparticle colorimetric sensing assay	DNA	7.05 nM[46]	1.26 ppb	344
	Colorimetric	Aptamer and PDDA mediated AuNP aggregation	DNA	7.05 nM[46]	5.3 ppb	45
	Rayleigh scattering	Nanoparticle assembly by aptamers and crystal violet for arsenic(III) detection in aqueous solution based on a resonance Rayleigh scattering spectral assay	DNA	7.05 nM[46]	0.2 ppb	345

Copper(II)	Electrochemical	Electrochemical aptasensor based on a reagentless signal-on architecture and amplification by gold nanoparticles	DNA	—	0.1 pM	346
Lead(II)	Graphene field effect transistor (FET)	A label-free and portable graphene FET aptasensor	DNA	—	37.5 ng L^{-1}	347
	Electrochemiluminescence	Electrochemiluminescence utilizing co-reaction signal enhancement	DNA	—	4.0×10^{-14} M	348
	Fluorometric	Combination of a microfluidic sample pre-treatment device with a DNA aptamer linked photoluminescent graphene oxide quantum dot (GOQD) sensor	DNA	—	0.64 nM	349
	Fluorescence	A fluorescent nanosensor based on graphene quantum dots–aptamer probe and graphene oxide platform	DNA	—	0.6 nM	350
	Colorimetric	Colorimetric photonic hydrogel aptasensor for the screening of heavy metal ions	DNA	—	—	351
Mercury(II)	Fluorometric/Colorimetric	Label free G-quadruplex-based turn-off fluorescent biosensor	DNA	—	—	352
	Magneto-elastic	Streptavidin and gold nanoparticles-based dual signal amplification for sensitive magnetoelastic sensing of mercury using a specific aptamer probe	DNA	—	0.885 nM	353
	Colorimetric	Aptamer-functionalized one-dimensional photonic crystals consisting of TiO_2 and poly(*N*-isopropylacrylamide-acrylic acid), P(NIPAM-AA)	DNA	—	—	354

Table 8.6 (*Continued*)

Heavy metals	Detection method	Aptasensor description	DNA/RNA	K_d	Limit of detection (LOD)	Ref.
	Colorimetric	Smart-phone to digitally quantify mercury concentration using a plasmonic gold nanoparticle (Au NP) and aptamer based colorimetric transmission assay	DNA	—	3.5 ppb	355
	SERS	Nanoporous gold/aptamer Surface enhanced Raman scattering (SERS) sensor	DNA	—	1 pM	356
	Colorimetric	Aptamer-functionalized hydrogel microparticles	DNA	—	10 nM	357
	Chemi-luminescence	Chemiluminescent and (CRET) detection using hemin/G-quadruplexes and CdSe/ZnS quantum dots	DNA	—	10 nM	358
	Optical	Reusable evanescent wave DNA biosensor	DNA	—	2.1 nM	359
	Colorimetric	Colorimetric assay based upon DNA-linked AuNPs	DNA	—	5.4 nM	360
Multiplex: As(II), Hg(II)	Colorimetric	Colorimetric photonic hydrogel aptasensor for the screening of heavy metal ions	DNA	—	—	351
Multiplex: Hg(II), Ag(I)	Fluorometric	Graphene oxide-based homogeneous fluorescent sensor for multiplex detection	DNA	—	200 nM (Ag) 400 nM (Hg)	361

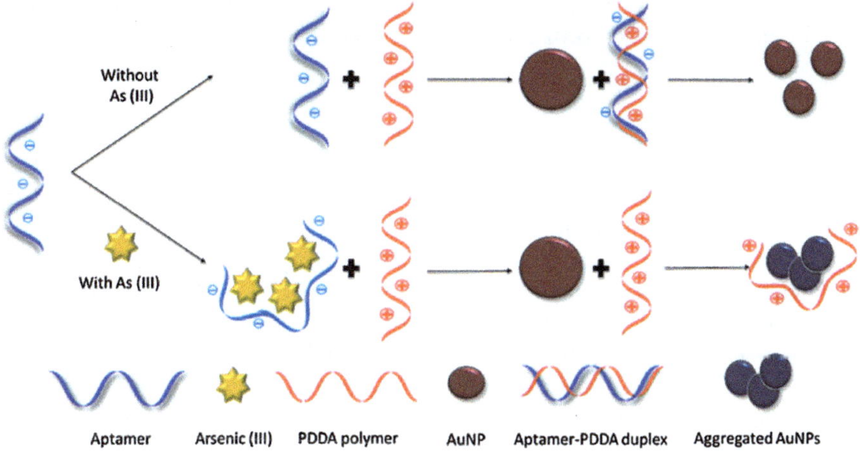

Figure 8.5 Detection of As(III) with As-3 aptamer (blue ribbon) as the molecular recognition element of a AuNP-based sensor utilizing cationic PDDA polymer (red ribbon) mediated aggregation. In the absence of the target molecule, As(III), the aptamer and cationic PDDA polymer form a complex in solution. However, when As(III) is present in solution, the aptamer preferentially binds to the target molecule. This leaves the cationic PDDA polymer free in solution to promote AuNP aggregation resulting in a visible colour change from wine-red to blue when target is present. Adapted from ref. 45 with permission from the Royal Society of Chemistry.

for a detection limit below the regulatory limit to be achieved. The colorimetric sensor was shown to be highly selective for As(III) compared to other metal ions (Calcium (Ca(II)), Iron (Fe(III) and Fe(II)), Copper (Cu(II)), Nickel (Ni(II)), Manganese (Mn(II)), Zinc (Zn(II)), Magnesium (Mg(II)), Silver (Ag(I)), Mercury (Hg(II)), Cadmium (Cd(II)), Lead (Pb(II))), with the exception of Arsenite (As(V)). Although the developed sensor was not as selective against As(V) compared to other metal ions, there was still a marginal difference between As(III) and As(V).

8.3 Cellular Targets

8.3.1 Bacteria

As one of the foremost causes of food safety related incidents worldwide, the timely and cost-effective detection of bacterial pathogens is especially important.[49] The detection of pathogenic bacteria in food is complicated by differences in the standards of food-processing and handling procedures as well as the onset of newly emerging and evolving strains.[50] Although developing countries are more affected by foodborne diseases generally, up to 30% of the population of developed countries are also affected.[51] As such, detection methods for common bacterial pathogens such as *Campylobacter*, *E. coli*, *Listeria*, *Salmonella*, *Shigella*, *Staphylococcus*, *Vibrio* and *Yersinia* must be versatile and high throughput.[51,52] Table 8.7 lists the recent aptasensors developed for the detection of bacterial targets in food.

Bacterial Aptasensors

Kurt and coauthors (2016) developed a dual-excitation sensing method for the multiplex detection of *Salmonella enterica* serovar Typhimurium and *S. aureus* based on aptamer-functionalized upconverting nanoparticles and quantum dots, respectively (see Figure 8.6).[53] *S. Typhimurium* and *S. aureus* are foodborne pathogens, which have been implicated in food poisoning. In the first step of the assay described in this work, aptamer-nanoparticle conjugates were immobilized on magnetic beads by hybridization of a partially complementary strand tethered to the surface of the magnetic bead. In the presence of the respective target bacteria, aptamer-target binding led to a decrease in luminescent signal intensity that was relative to the bacteria concentration. In PBS buffer, the limit of detection for *S. Typhimurium* and *S. aureus* were 28 CFU mL^{-1} and 16 CFU mL^{-1}, respectively. Further, the specificity of the sensor for the specific target bacteria compared to *E. coli* 0157:H7, *L. monocytogenes* and *P. aeruginosa* was shown. An advantage of this sensor design was the judicious choice of fluorescent nanoparticles with non-overlapping emissions. This sensor design could be expanded for the simultaneous detection of multiple target analytes.

8.3.2 Viruses

Viruses are another type of pathogenic microorganism. The viruses are endemic in many areas of the world.[54] Viral infections are difficult to distinguish at the onset, highlighting a need for rapid and accurate testing to ensure the safety of food and water supplies. In recent years, screening foodborne virus-specific aptamers has included aptamers for the norovirus,[55,56] avian influenza virus[57] and hepatitis C virus.[58] Quick, simple and inexpensive detection methods for viruses can play a vital role in controlling infectious spread. More recently, the developed aptamers for viruses have been incorporated into fluorometric, SPR and electrochemical biosensors among others. The details can be found in Table 8.8.

Viral Aptasensors

Norovirus is classified as a world leading cause for foodborne disease outbreaks. As one of the leading causes of acute gastroenteritis, norovirus can pose a serious health concern. Individuals who face the highest risk are young children, the elderly, as well as those that have suppressed immune function.[59] Limitations regarding cell cultivation in the past have undoubtedly influenced the epidemical nature of norovirus infections.[60] Additionally, noroviruses have low infectious doses, high environmental stability and multiple routes of transmission including the contamination of

Table 8.7 Aptasensors for bacterial targets.

Bacteria	Assay	Aptasensor description	DNA/RNA	K_d	Limit of detection (LOD)	Ref.
Campylobacter jejuni	Fluorometric	Aptamer magnetic capture assay in conjunction with real-time PCR	DNA	292.8 nM	1.1 log 10 (300 mL)$^{-1}$ cells	362
Escherichia coli	Electrochemical	A paper-based graphene-nanocauliflower hybrid composite	RNA	110 nM	~4 CFU mL^{-1}	363
Escherichia coli O157:H7	Fluorometric	A sandwich-type assay based on quantum dot/aptamer bioconjugates	DNA	—	10^2 CFU mL^{-1}	364
Escherichia coli	Electrochemical	Based on rolling circle amplifications coupled peroxidase-mimicking DNAzyme amplification	DNA	12.4 nM	8 CFU mL^{-1}	365
Enterotoxigenic Escherichia coli (ETEC) K88	Fluorometric	Based on graphene oxide-dependent nanoquencher and Klenow fragment-triggered target cyclic amplification	DNA	28 nM	1×10^2 CFU mL^{-1}	366
Escherichia coli O157:H7	Colorimetric	Lateral flow assay based on aptamer mediated strand displacement amplification	DNA	—	10 CFU mL^{-1}	367
Escherichia coli	Fluorometric	Microfluidic device with FITC-stained bacteria cells	DNA	24.8 nM	Target cells captured at 232 ± 15 cell (mm^2)$^{-1}$	368
Escherichia coli	—	Selex	DNA	12.4 nM	—	369
Francisella tularensis	Fluorometric	Pulldown assay	DNA	—	Detection range of 1–10^6 CFU mL^{-1}	370
Lactobacillus acidophilus	Spectroscopic	Aptamer-decorated porous silicon biosensors	DNA	13 nM	10^6 cells mL^{-1}	371

Table 8.7 (*Continued*)

Bacteria	Assay	Aptasensor description	DNA/RNA	K_d	Limit of detection (LOD)	Ref.
Lactobacillus acidophilus/ Streptococcus pyogenes	—	Selex	DNA	4 nM	—	372
Listeria monocytogenes	Colorimetric	Using Fe_3O_4 nanoparticle cluster modified aptamer	DNA	—	5.4 10^3 CFU mL^{-1}	373
Listeria monocytogenes	Fluorometric	Aptamer-based sandwich assay	DNA	Aptamer 1: 2.01×10^{-12} M, Aptamer 2: 1.56×10^{-10} M	20 CFU mL^{-1}	374
Listeria monocytogenes	—	Selex	DNA	Aptamer 1:106.4 nM, Aptamer 2:74.4 nM	—	375
Listeria monocytogenes	—	Selex	DNA	60.01 nM	—	376
Mycobacterium tuberculosis secreted antigen CFP10-ESAT6	Electrochemical	Gold-IDE/CFP10-ESAT6 aptamer/DNA-gold nanoparticles MSPQC	DNA	—	10^3 CFU mL^{-1}	377
Salmonella	Electrochemical	Aptamer immobilized onto reduced GO-MWCNT composite on glassy carbon electrode	DNA	—	25 CFU mL^{-1}	378
Salmonella O8	—	Selex	DNA	32.04 nM	—	379
Salmonella	QCM	QCM sensor coupled to aptamer functionalized core–shell type magnetic separation	DNA	1.73 µM	10 CFU mL^{-1}, 100 CFU mL^{-1} (in milk)	380
Salmonella	Electrochemical	Based on a glassy carbon electrode modified with graphene oxide and gold nanoparticles	DNA	—	3 CFU mL^{-1}	381

Salmonella enteritidis	Fluorometric	Using enzyme-induced cascade two-stage toehold strand-displacement-driven assembly of G-quadruplex DNA	DNA	—	60 CFU mL^{-1}	382
Salmonella enteritidis	Colorimetric	Gold nanoparticle-based lateral flow assay	DNA	—	10^1 CFU	383
Salmonella enteritidis	Fluorometric	Based on fluorescence-switch signalling graphene oxide	DNA	6.3 nM	40 CFU mL^{-1}	384
Salmonella choleraesuis (flagellin)	Fluorometric	Using FITC-labelled aptamer	DNA	41 nM	10^3 CFU mL^{-1}	385
Salmonella paratyphi A	Fluorometric	DNase I-mediated cyclic signal amplification	DNA	27 nM	1×10^2 cells mL^{-1}	386
Salmonella paratyphi A (flagellin)	Fluorometric	DNAzyme-aptamer sensor with single-walled nanotubes	DNA	27 nM	5 ng mL^{-1} by fluorescence and 20 ng mL^{-1} by spectrophotometry; in milk: 105 CFU mL^{-1} by fluorescence and 106 CFU mL^{-1} by spectrophotometry	387
Salmonella typhimurium	Impedimetric	Diazonium-based sensor	DNA	—	6 CFU mL^{-1}	388
Salmonella typhimurium	Raman spectroscopy	Au@Ag core/shell nanoparticles functionalized with aptamer	DNA	6 nM	15 CFU mL^{-1}	389
Salmonella typhimurium/ *Staphylococcus aureus*	Fluorometric	Dual-excitation upconverting nanoparticle and quantum dot	DNA	35 nM (*S.a.*), N/A (*S.t.*)	16 CFU mL^{-1} (*S.a.*), 28 CFU mL^{-1} (*S.t.*)	390
Salmonella typhimurium	Colorimetric	Based on aptamer recognition and gold nanoparticles	DNA	—	10 CFU mL^{-1}	391

Table 8.7 (*Continued*)

Bacteria	Assay	Aptasensor description	DNA/RNA	K_d	Limit of detection (LOD)	Ref.
Salmonella typhimurium/*Staphylococcus aureus*	Fluorometric	Fe_3O_4 magnetic nanoparticles and aptamers labelled with multi-colour lanthanide-doped time-resolved fluorescence nanoparticles	DNA	N/A (*S.t.*), 210.703 nM (EC_{50}, *S.a.*)	20 CFU mL^{-1} (*S.a.*), 15 CFU mL^{-1} (*S.t.*)	392
Salmonella typhimurium	Colorimetric	Based on target-triggered isothermally exponential amplification	DNA	6 nM	80 CFU mL^{-1}	393
Salmonella typhimurium	Fluorometric	MoS_2 based digital response platform	DNA	6.33 nM	10 CFU mL^{-1}	394
Salmonella typhimurium	Fluorometric	RuBPY-doped silica nanoparticles based assay	DNA	6 nM	Not reported	395
Salmonella typhimurium	Fluorometric	Using aptamer-conjugated carbon dots	DNA	—	50 CFU mL^{-1}	396
Salmonella typhimurium	Impedimetric	Based on poly(pyrrole-*co*-3-carboxyl-pyrrole) copolymer supported aptamer	DNA	—	3 CFU mL^{-1}	397
Salmonella typhimurium	Colorimetric	Based on peroxidase-like activity of magnetic nanoparticles	DNA	—	7.5×10^5 CFU mL^{-1}	398
Salmonella typhimurium	Colorimetric	Nanogold-based	DNA	0.36 nM	10^6 CFU mL^{-1}	399
Salmonella typhimurium/*Staphylococcus aureus*/*Vibrio parahaemolyticus*	Fluorometric	Based on fluorescence resonance energy transfer	DNA	16.88 nM (*V.p.*), 210.70 nM (EC_{50}, *S.a.*), 6.33 nM (*S.t.*)	25 CFU mL^{-1} (*V.p.*), 50 CFU mL^{-1} (*S.a.*) and 50 CFU mL^{-1} (*S.t.*)	400

Salmonella typhimurium/*Vibrio parahaemolyticus*	Fluorometric	Using quantum dots and carbon nanoparticles	DNA	16.88 nM (*V.p.*), 6.33 nM (*S.t.*)	25 CFU mL^{-1} (*V.p.*), 35 CFU mL^{-1} (*S.t.*)	401
Salmonella typhimurium/*Lactobacillus acidophilus*/*Pseudomonas aeruginosa*	SPR	Aptamer-immobilized LSPR-based sensor with a multi-spot gold-capped nanoparticle array	DNA	13 nM (*L.a.*), N/A (*S.t.*), 17.27 nM (*P.a.*)	30 CFU $assay^{-1}$	402
Salmonella typhimurium/*Staphylococcus aureus*	Raman spectroscopy	Using Fe_3O_4 magnetic gold nanoparticles and gold nanoparticles modified with Raman molecules	DNA	N/A (*S.t.*), 210.70 nM (EC_{50}, *S.a.*)	35 CFU mL^{-1} (*S.a.*), 15 CFU mL^{-1} (*S.t.*)	403
Salmonella typhimurium or *Escherichia coli*	Fluorometric	Aptamer-based magnetic separation system	DNA	—	100 CFU mL^{-1} for both bacteria in spiked-milk samples	404
Salmonella typhimurium/*Salmonella enteritidis*	Fluorometric	Polyvalent directed aptamer polymer	DNA	4.66 μM and 3.83 μM (*S.t.*), 0.530 μM (*S.e.*)	~100-fold increase in affinity (57.86 ± 14.5 nM) after creation of PDAP	405
Salmonella typhimurium	Fluorometric	Magnetic nanoparticles-based assay	DNA	6.33 nM	25 CFU mL^{-1}	406
Salmonella typhimurium	—	Selex	DNA	1.73 μM (lowest)	—	407
Salmonella typhimurium	Colorimetric	Sandwich-type assay with aptamer/bacteria/aptamer-gold nanoparticles complexes	DNA	—	7 CFU mL^{-1}	408
Salmonella enterica serovar Typhimurium	Fluorometric	Molecular-beacon based real-time PCR assay targeting invA gene	DNA	—	100 CFU mL^{-1}	409
Shigella flexneri	Fluorometric	Based on graphene oxide and fluorescent aptamer probe	DNA	29 nM	100 CFU mL^{-1}	410

Table 8.7 (*Continued*)

Bacteria	Assay	Aptasensor description	DNA/RNA	K_d	Limit of detection (LOD)	Ref.
Shigella sonnei	Fluorometric	Dual aptamer sandwich assay	DNA	Aptamer 1: 5.980 nM, Aptamer 2: 14.32 nM	30 CFU mL^{-1}	411
Staphylococcus aureus	Commercial glucose meter	Aptamer coupled with a signal amplification method based on invertase-promoted hydrolysis of sucrose	DNA	61.50 nM (EC_{50})	1.0×10^5 CFU mL^{-1}	412
Staphylococcus aureus	Raman spectroscopy	Silver-coated magnetic nanoparticles and SERS-based assay	DNA	Aptamer 1:70.86 nM (EC_{50}), Aptamer 2: 210.70 nM (EC_{50})	10 cells mL^{-1}	413
Staphylococcus aureus	Magnetoelastic	Aptamer modified magnetoelastic wireless detection platform	DNA	210.70 nM (EC_{50})	5 CFU mL^{-1} (lowest)	414
Staphylococcus aureus	—	Selex	DNA	3.49 nM	—	415
Staphylococcus aureus	Electrochemical	Aptamer/graphene interdigitated gold electrode piezoelectric sensor	DNA	210.70 nM (EC_{50})	41 CFU mL^{-1}	416
Staphylococcus aureus Protein A (SpA)	—	Selex	RNA	38.9 nM	—	417
Staphylococcus aureus	Electrochemical	Aptamer-conjugated silver nanoparticles in dual-aptamer-based sandwich assay	DNA	35 nM	1.0 CFU mL^{-1}	418

Staphylococcus aureus	Fluorometric	Using aptamer-coated magnetic beads and antibiotic-capped gold nanoclusters	DNA	210.70 nM (EC_{50})	16 CFU mL^{-1}	419
Staphylococcus aureus	Fluorometric	SpA slow off-rate aptamers (SOMAmers) in flow cytometry	DNA	0.03 nM	10^5–10^6 cells mL^{-1}	420
Staphylococcus aureus	Fluorometric/ Resonance light scattering	Aptamer conjugated gold nanoparticle assay	DNA	35 nM (SA17), 129 nM (SA61); 3.03 nM (SA17-GNPs), 9.9 nM (SA61-GNPs)	Instrument: 63 ± 21 GNPs μL^{-1} (100 nm particles), 508 ± 176 GNPs μL^{-1} (60 nm particles). 312 cells (SA17-GNPs), 1250 cells (SA61- GNPs)	421
Streptococcus pneumoniae	Colorimetric	DNA transducer-triggered signal switch	DNA	—	156 CFU mL^{-1}	422
Streptococcus pyogenes M-type 11	—	Selex	DNA	7 nM	—	423
Vibrio parahaemolyticus	Surface enhanced Raman spec.	Using SiO_2@Au core/shell nanoparticles paired with Cy3 modified aptamer	DNA	16.88 nM	10 CFU mL^{-1}	424
Vibrio parahaemolyticus	—	Selex	DNA	16.88 nM	—	425
Xanthomonas axonopodis protein FtsZ	—	Selex	DNA	44.5 nM	2.84 μM (IC_{50}), 121.84 μM (MIC_{50})	426

foodstuffs.[59,60] For the detection of the norovirus outbreak strain GII.3, Giamberardino and coauthors reported the selection of aptamers against murine norovirus (MNV), which show high affinity for both MNV and lab-synthesized capsids for human norovirus. The selected aptamer, AG3, was utilized in an electrochemical biosensor (See Figure 8.7). The electrochemical sensor consisted of thiolated AG3 aptamer immobilized onto gold nanoparticle-modified screen-printed carbon electrodes (GNP-SPCEs). The electrochemical studies included cyclic voltammetry (CV) and

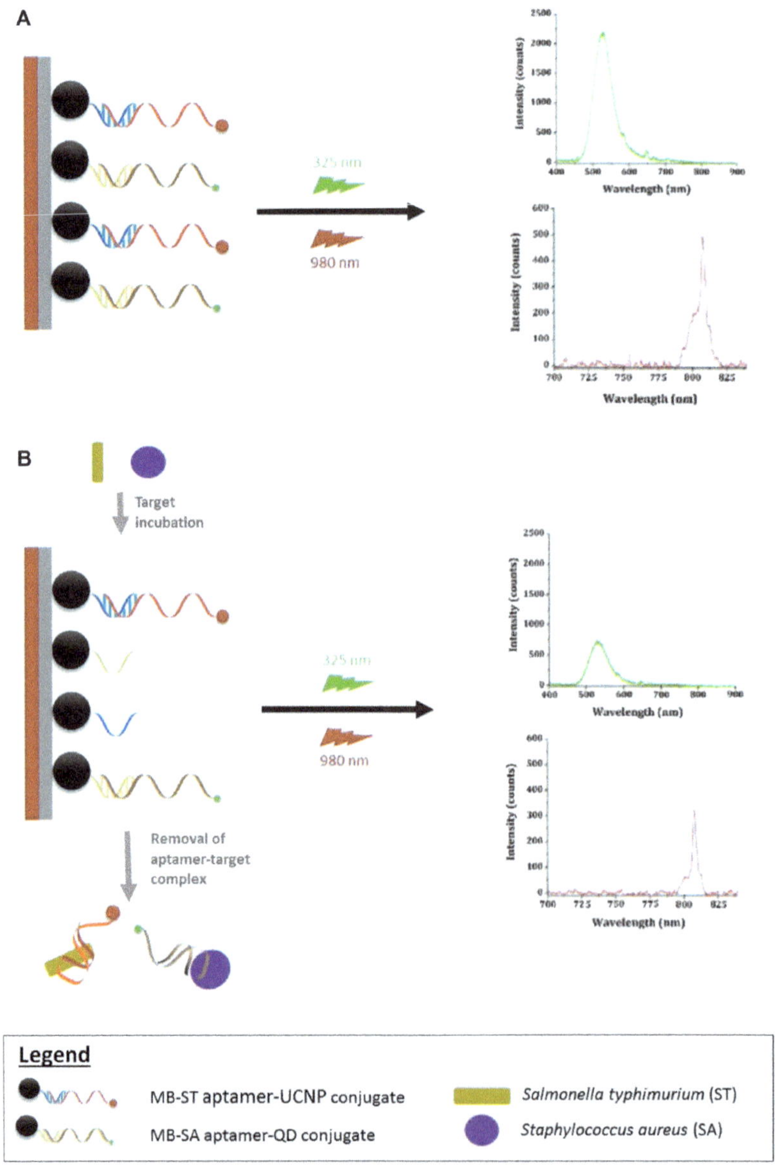

electrochemical impedance spectroscopy (EIS). The limit of detection was reported at 180 virus particles.[56] Selectivity experiments were performed using viruses of similar size and structure to norovirus. Those tested included vesicular stomatitis virus, HSA and vaccinia virus. The sensing system offers rapidity, ease of preparation and high sensitivity, making it a valuable tool for on-site application.

8.4 Aptasensors for Food Quality: Adulterants, Additives and Allergens

The contamination of food by deleterious compounds has become a great concern for regulatory bodies and consumers around the world.[61,62] As previously touched upon in this text, food contamination can originate from natural sources (pathogenic bacteria, bacterial toxins, mycotoxins, *etc.*) as well as man-made substances (pesticides, antibiotics, veterinary drugs, packaging materials, *etc.*). Unintentional contaminations by natural products can occur prior to food processing, while other chemical contaminants will often result from food processing and storing practices. Incidents involving sickness and death have attracted considerable attention to food contaminants that exhibit acutely toxic effects. Accordingly, detection methods for a vast array of contaminants have been extensively researched in the goal of continually producing more sensitive, rapid and cost effective sensing platforms.

In recent years, there has been an increase in public awareness regarding dietary choices. Consequently, consumers are more mindful of health effects and food quality, and therefore scrutinize the validity of food products and

Figure 8.6 The multiplex detection of *Salmonella typhimurium* (ST) and *Staphylococcus aureus* (SA) by dual-excitation of an upconverting nanoparticle (UCNP) and quantum dot (QD) based aptasensor is represented schematically. (A) The aptamer-nanoparticle conjugates were immobilized to magnetic beads (MB: black spheres) *via* hybridization of the aptamer to its respective complement. The UCNP (red sphere) conjugated ST aptamer (red ribbon) was immobilized to the MB through the ST-aptamer complement (blue ribbon). Similarly, the QD (green sphere) modified SA aptamer (brown ribbon) was immobilized to the MB by hybridization with the SA aptamer complement (yellow ribbon). Excitation of the QD (325 nm) and UCNP (980 nm) by UV and NIR light respectively provided the initial intensity count. (B) Following the incubation of the MB-ST aptamer-UCNP conjugate and MB-SA aptamer QD conjugate with the ST and SA, aptamer-target complex was removed and discarded. The intensities of the UCNP and QD from the remaining MB-ST aptamer-UCNP conjugate and MB-SA aptamer QD conjugate were measured. When aptamer-target binding occurred, a decrease in intensity was observed. Portions of the figure were reproduced with permission from *Biosensors and Bioelectronics*, Volume 81, H. Kurt, M. Yüce, B. Hussain and H. Budak, Dual-excitation upconverting nanoparticle and quantum dot aptasensor for multiplexed food pathogen detection, 280–286. Copyright 2016, with permission from Elsevier.

Table 8.8 Aptasensors for viruses.

Foodborne pathogen	Detection method	Aptasensor description	DNA/RNA	K_d	Limit of detection (LOD)	Ref.
Hepatitis B	Chemi-luminescence	Aptamer label Fe_3O_4 MNPs-AMPPD	DNA	—	0.1 ng mL^{-1}	427
Hepatitis C	Colorimetric	Streptavidin-biotin aptamer conjugated-AuNP in a lateral flow device	DNA	—	10 pg–100 pg mL^{-1}	57
	Octet platform	Streptavidin-biotin RNA aptamer conjugated	RNA	—	700 pg mL^{-1}	428
H5N1	Electrochemical	Gold microelectrode functionalized *via* biotin-labelled aptamer	DNA	—	0.25 HAU (pure virus), 1 HAU (spiked chicken)	429
	Fluorometric	Aptamer label QDs-quencher-hydrogel	DNA	—	0.4 HAU	430
	Fluorometric	Aptamer label Ag-SiO_2-TO dye	DNA	—	2 ng mL^{-1} in buffer and 3.5 ng mL^{-1} in human serum	431
	Electrochemical	Gold microelectrode-biotin-label aptamer	DNA	4.65 nM	0.0128 HAU	432
	Fluorometric	Label free method SG-aptamer-graphene oxide	DNA	—	2.5 µg mL^{-1}	58
H1N1-H5N1	Colorimetric	Aptamer on-AuNS (gold nanoshell)	RNA	—	3×10^8 viral particles	433
H1N1	Surface plasmon resonance	Biotin-aptamer on Au surface	DNA	—	0.128 HAU	434
	Electrochemical	Polymer (PEDOT-OH:TsO) microelectrode array was developed	DNA	—	103 pfu mL^{-1}	435
Norovirus	Fluorometric	Aptamer-label SiO_2-QDs-BHQ	RNA	—	9.3 copies mL^{-1}	55
	Fluorometric	RT-qPCR-FAM-labelled aptamer	DNA	—	10 RNA copies (lettuce)	436
	Electrochemical	Electrochemical sensor based on AuNP modified SPCE	DNA	—	180 virus particles	56

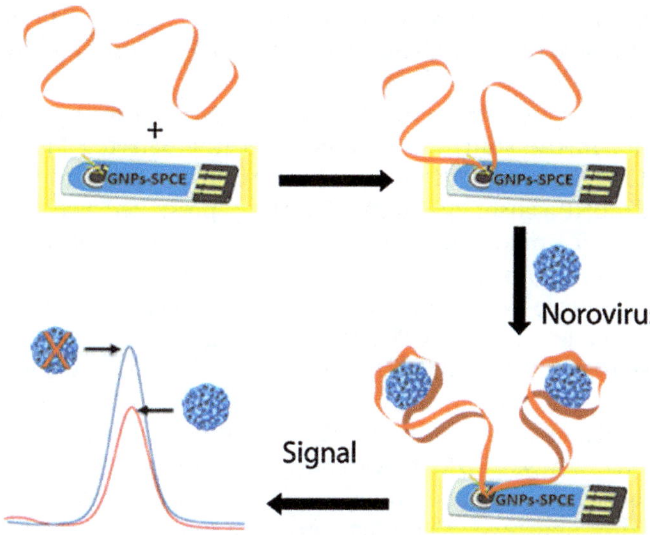

Figure 8.7 Schematic representation of the electrochemical detection of murine norovirus: A thiolated norovirus aptamer was self-assembled onto a gold nanoparticles-modified screen-printed carbon electrode (GNPs-SPCE). Detection of the virus causes a decrease in the redox current, measured *via* square wave voltammetry.
Reproduced from A. Giamberardino, M. Labib, E. M. Hassan, J. A. Tetro, S. Springthorpe, S. A. Sattar, M. V. Berezovski and M. C. DeRosa, *PLoS ONE*, 2013, **8**, e79087. © 2013 Giamberardino *et al.*

their contents.[62] While not as extensively examined, economically motivated adulterants, undeclared additives and allergen contamination in foods can pose a serious risk when dealing with globally sourced food commodities.[61,63] Intentional adulteration is frequently seen in food products that have relatively high market values.[64] These commodities attract the use of cheaper materials as fillers. In general, these substances do not exhibit acute toxicity and are colourless, odourless, tasteless and easily obtainable. Though not all immediately toxic, many adulterants can have lasting consequences in consumer health.

Milk and milk products are heavily consumed in many parts of the world. Undeclared substances in milk products can have a potentially harmful effect on a large portion of the world's population. The adulteration of milk products can be carried out for several reasons. Water is often used to dilute the product. Starches, sugar, maltodextrin, salt and detergents are used to alter the taste, thickness and frothiness of the milk. They can also be used as inexpensive fillers for powdered products. Substances such as bicarbonates and sodium hydroxide are used to neutralize the acidity resulting from soured milk. Boric acid, hydrogen peroxide and salicylic acid can all be used as preservatives. In addition, nitrogenous substances such as urea are the most commonly used form of protein adulteration.[63,65] Recent incidents involving milk adulteration in India have attracted more attention on the matter. Milk and milk product adulteration in developing countries can most often be attributed to large populations and insufficient supply.[63,66]

In 2015, Kumar and others developed an aptasensor for the detection of urea in milk. In the study, an aptamer was first selected for urea using FluMag-SELEX. The high-affinity aptamer for urea was then utilized in an unmodified AuNP sensor platform. The study presented two methods for detecting urea in milk. The first was a ligand-mediated, salt-induced aggregation of AuNPs resulting in a colorimetric read-out. The second was a FRET-based sensor, which utilized AuNPs as efficient quenchers for a fluorescein labelled aptamer. The reported limit of detection for the colorimetric sensor was 20 mM, and the linear dynamic range was 20 to 150 nM. The urea aptasensor was also compared to two separate commercially available urease test kits. The results of the comparison were similar in many ways (LDR, repeatability and timescale). However, since most urea biosensors depend on urease enzymes, this sensor is believed to be much more robust when used as an on-site tool.[63]

Unintentional adulteration in food also emerges occasionally. For example, common allergens such as peanut, milk, egg, shellfish, soy, wheat and sulfites can contaminate foods and pose a risk for individuals that are hypersensitive. Extra care must therefore be taken when guaranteeing the authenticity of food and the accuracy of labelling. Lysozyme is a highly studied protein, which was discovered in the early 1900s. It is a low-molecular-weight enzyme that has been used to model enzyme catalysis, protein interactions and amyloid aggregate formation. It is found abundantly in the egg whites of birds, and is known to be an egg allergen.[67,68] Lysozyme is commonly used in the production of cheeses. It is used in beer production for its antibacterial properties, since it has the ability to hydrolyze the cell walls of certain Gram-positive bacteria. During the second stage of fermentation lysozyme is used to control microbial activity and the conversion of L-malic acid to L-lactic acid. It is used as an alternative for more hazardous sulfites. Its antibacterial properties have also been exploited in extending the shelf-lives of many raw food products.[67,68] Sensing of lysozyme in consumer dedicated food products is important as it can reduce the risk of accidental exposure for individuals that are sensitive to these allergens. One example of this is the detection of lysozyme in wine samples. Recently, Mihai and others developed an aptamer-based SPR sensor for the detection of immunogenic protein lysozyme. The gold SPR interfaces were first functionalized with PEG-COOH, followed by NeutrAvidin. Biotinylated aptamer was then immobilized by high affinity capture. SPR measurements were recorded using the portable system, which utilized a Spreeta chip and multi-flow cell. The sensor provided a limit of detection of 2.4 nM in spiked red and white wine samples.[68] This sensor could allow the detection of lysozyme throughout different stages of wine making. Current methods of detection for allergenic proteins are generally immunological methods.[62] Food samples can vary tremendously in terms of composition. The intrinsically lower stability of antibodies provides some limitation when dealing with high salinity and varying pHs. These can be overcome through the use of aptasensors.

8.5 Conclusions and Future Directions

This review describes the recent advances in aptasensors for the detection of food-safety related target molecules. These target molecules are contaminants such as chemical and bacterial toxins, antibiotics, pesticides, drugs, bacteria, viruses, mycotoxins and heavy metals. Aptamers can be chemically modified, which is advantageous for the development of optical and electrical aptasensors. Furthermore, the *in vitro* condition for aptamer selection can provide molecular recognition elements for diverse targets, which are found to contaminate foodstuffs and pose health risks. In recent years, there has been a notable increase in the selection of aptamers for targets of growing interest in food safety. For example, the selection and incorporation of aptamers for mycotoxin detection has shown to be particularly more prominent in recent literature. Likewise, aptamers for viral food pathogens, such as norovirus, have progressed and shown promise as molecular recognition elements in practical biosensor applications. Although many of the reviewed aptasensors have been tested in complex food matrices displaying selectivity and reliability, there remains a challenge for the detection of food-safety related targets. In terms of food safety, the practicality and real-world applications of recently developed aptamer platforms must be considered, particularly when discussing the need for inexpensive, on-site detection methods. Additionally, focus should be placed on simultaneous multiple analyte detection. Presently, there is an expanding interest in target molecules associated with food quality. With the economically driven use of cheaper materials to act as fillers, the validity of certain food products is in question. Further exploration of aptamers for food quality and authenticity assessment is warranted. A continued understanding and improvement of aptasensors for food safety will encourage the development of similar detection methods for food quality. Aptamers continue to be used as sensitive, reliable and practical molecular recognition elements for a variety of food safety related target molecules. In recent years, there have been considerable advancements for detection of targets, which were once underrepresented. Progress in both selection of aptamers and biosensor development has allowed for selective detection of food-safety target molecules in complex, real-world matrices at low concentrations. These technologies will hopefully serve to improve access to safe, high-quality, healthy food for consumers around the world.

References

1. Centers for Disease Control and Prevention, *Foodborne Outbreaks*, 2015.
2. S. Scarano, S. Mariani and M. Minunni, *ACTA IMEKO*, 2016, **5**, 36–44.

3. Ł. Stępień, A. Waśkiewicz and M. Urbaniak, *Microb. Ecol.*, 2016, **71**, 927–937.
4. S. Eckard, F. E. Wettstein, H.-R. Forrer and S. Vogelgsang, *Toxins*, 2011, **3**, 949–967.
5. W. Liao and X. Lu, *Trends Food Sci. Technol.*, 2016, **54**, 103–113.
6. M. McKeague, E. M. McConnell, J. Cruz-Toledo, E. D. Bernard, A. Pach, E. Mastronardi, X. Zhang, M. Beking, T. Francis, A. Giamberardino, A. Cabecinha, A. Ruscito, R. Aranda-Rodriguez, M. Dumontier and M. C. DeRosa, *J. Mol. Evol.*, 2015, **81**, 150–161.
7. C. Tuerk and L. Gold, *Science*, 1990, **249**, 505–510.
8. S. K. Silverman, in *Functional NucleicAcids for Analytical Applications*, ed. L. Yingfu and L. Yi, Springer, New York, NY, 2009, pp. 47–108.
9. A. D. Ellington and J. W. Szostak, *Nature*, 1990, **346**, 818–822.
10. D. L. Robertson and G. F. Joyce, *Nature*, 1990, **344**, 467–468.
11. V. J. B. Ruigrok, M. Levisson, M. H. M. Eppink, H. Smidt and J. van der Oost, *Biochem. J.*, 2011, **436**, 1–13.
12. P. Szatkowska, M. Koba, P. Koslinski and M. Szablewski, *Mini-Rev. Org. Chem.*, 2013, **10**, 400–408.
13. H. K. Binz and A. Plückthun, *Curr. Opin. Biotechnol.*, 2005, **16**, 459–469.
14. Y. Dong, Y. Xu, W. Yong, X. Chu and D. Wang, *Crit. Rev. Food Sci. Nutr.*, 2014, **54**, 1548–1561.
15. F. Tolle and G. Mayer, *Chem. Sci.*, 2013, **4**, 60–67.
16. G. Mayer, *Angew. Chem., Int. Ed.*, 2009, **48**, 2672–2689.
17. J. Wu, Y. Zhu, F. Xue, Z. Mei, L. Yao, X. Wang, L. Zheng, J. Liu, G. Liu, C. Peng and W. Chen, *Microchim. Acta*, 2014, **181**, 479–491.
18. H. Sun and Y. Zu, *Molecules*, 2015, **20**, 11959–11980.
19. C. Ji, Y. Fan and L. Zhao, *Anim. Nutr.*, 2016, **2**, 127–133.
20. H. S. Hussein and J. M. Brasel, *Toxicology*, 2001, **167**, 101–134.
21. J. D. Miller, *Food Addit. Contam., Part A,*, 2008, **25**, 219–230.
22. F. Cheli, D. Battaglia, R. Gallo and V. Dell'Orto, *Food Control*, 2014, **37**, 315–325.
23. J. Robens and K. Cardwell, *J. Toxicol. Toxin Rev.*, 2003, **22**, 139–152.
24. J. Yu, *Toxins*, 2012, **4**, 1024–1057.
25. M. Abrar, F. M. Anjum, M. S. Butt, I. Pasha, M. A. Randhawa, F. Saeed and K. Waqas, *Crit. Rev. Food Sci. Nutr.*, 2013, **53**, 862–874.
26. P. Villers, *Front. Microbiol.*, 2014, **5**, 1–6.
27. Y. Ma, Y. Mao, D. Huang, Z. He, J. Yan, T. Tian, Y. Shi, Y. Song, X. Li, Z. Zhu, L. Zhou and C. J. Yang, *Lab Chip*, 2016, **16**, 3097–3104.
28. C. Gu, F. Long, X. Zhou and H. Shi, *RSC Adv.*, 2016, **6**, 29563–29569.
29. K. Aktories and J. T. Barbieri, *Nat. Rev. Microbiol.*, 2005, **3**, 397–410.
30. U.S. Department of Health and Human Services. Center for Food Safety and Applied Nutrition, 2011.
31. J. S. Henkel, M. R. Baldwin and J. T. Barbieri, *EXS*, 2010, **100**, 1–29.
32. B. Žegura, *Mini-Rev. Med. Chem.*, 2016, **16**, 1042–1062.
33. N. Butler, J. C. Carlisle, R. Linville and B. Washburn, Office of Environmental Health Hazard Assessment, *Microcystins: A brief*

overview of their toxicity and effects, with special reference to fish, wildlife, and livestock, 2009.
34. *Toxic Cyanobacteria in Water: A Guide to Their Public Health Consequences, Monitoring, and Management*, ed. I. Chorus and J. Bartram, E & FN Spon, London, New York, 1999.
35. F. Wang, S. Liu, M. Lin, X. Chen, S. Lin, X. Du, H. Li, H. Ye, B. Qiu, Z. Lin, L. Guo and G. Chen, *Biosens. Bioelectron.*, 2015, **68**, 475–480.
36. U.S. Food and Drug Administration, 2011.
37. Health Canada, *Consum. Prod. Saf.*, 2013.
38. S. M. Taghdisi, N. M. Danesh, M. A. Nameghi, M. Ramezani and K. Abnous, *Food Chem.*, 2016, **203**, 145–149.
39. X. Liu, Y. Li, J. Liang, W. Zhu, J. Xu, R. Su, L. Yuan and C. Sun, *Talanta*, 2016, **160**, 99–105.
40. J. O. Duruibe, M. O. C. Ogwuegbu, J. N. Egwurugwu and others, *Int. J. Phys. Sci.*, 2007, **2**, 112–118.
41. A. K. Wanekaya, *Analyst*, 2011, **136**, 4383.
42. P. Wu, T. Zhao, S. Wang and X. Hou, *Nanoscale*, 2014, **6**, 43–64.
43. H. Kang, L. Lin, M. Rong and X. Chen, *Talanta*, 2014, **129**, 296–302.
44. G. Tsekenis, M. K. Filippidou, M. Chatzipetrou, V. Tsouti, I. Zergioti and S. Chatzandroulis, *Sens. Actuators, B*, 2015, **208**, 628–635.
45. Y. Wu, S. Zhan, F. Wang, L. He, W. Zhi and P. Zhou, *Chem. Commun.*, 2012, **48**, 4459.
46. M. Kim, H.-J. Um, S. Bang, S.-H. Lee, S.-J. Oh, J.-H. Han, K.-W. Kim, J. Min and Y.-H. Kim, *Environ. Sci. Technol.*, 2009, **43**, 9335–9340.
47. Y. Ofir, B. Samanta and V. M. Rotello, *Chem. Soc. Rev.*, 2008, **37**, 1814.
48. H. Peng, C. Soeller and J. Travas-Sejdic, *Chem. Commun.*, 2006, 3735.
49. J. Chen and B. Park, *J. Food Prot.*, 2016, **79**, 1055–1069.
50. K. S. Gracias and J. L. McKillip, *Can. J. Microbiol.*, 2004, **50**, 883–890.
51. Y. Dong, Y. Xu, W. Yong, X. Chu and D. Wang, *Crit. Rev. Food Sci. Nutr.*, 2014, **54**, 1548–1561.
52. A. Rohde, J. A. Hammerl, B. Appel, R. Dieckmann and S. Al Dahouk, *Food Microbiol.*, 2015, **46**, 395–407.
53. H. Kurt, M. Yüce, B. Hussain and H. Budak, *Biosens. Bioelectron.*, 2016, **81**, 280–286.
54. M. M. Patel, A. J. Hall, J. Vinjé and U. D. Parashar, *J. Clin. Virol.*, 2009, **44**, 1–8.
55. O. Adegoke, M.-W. Seo, T. Kato, S. Kawahito and E. Y. Park, *Biosens. Bioelectron.*, 2016, **86**, 135–142.
56. A. Giamberardino, M. Labib, E. M. Hassan, J. A. Tetro, S. Springthorpe, S. A. Sattar, M. V. Berezovski and M. C. DeRosa, *PLoS One*, 2013, **8**, e79087.
57. C. Wang, L. Zhang and X. Shen, *Nucleosides Nucleotides Nucleic Acids*, 2013, **32**, 59–68.
58. X. F. Keyi Liu, *J. Nanomed. Nanotechnol.*, 2015, **6**, 1–6.
59. H. L. Koo, N. Ajami, R. L. Atmar and H. L. DuPont, *Discovery Med.*, 2010, **10**, 61–70.

60. L. Barclay, G. W. Park, E. Vega, A. Hall, U. Parashar, J. Vinjé and B. Lopman, *Clin. Microbiol. Infect.*, 2014, **20**, 731–740.
61. J. Draher, V. Pound and T. M. Reddy, *J. Chromatogr. A*, 2014, **1373**, 106–113.
62. R. Krska, A. Becalski, E. Braekevelt, T. Koerner, X.-L. Cao, R. Dabeka, S. Godefroy, B. Lau, J. Moisey, D. F. K. Rawn, P. M. Scott, Z. Wang and D. Forsyth, *Anal. Bioanal. Chem.*, 2012, **402**, 139–162.
63. P. Kumar, P. Ramulu Lambadi and N. Kumar Navani, *Biosens. Bioelectron.*, 2015, **72**, 340–347.
64. T. Ferreira, A. Farah, T. C. Oliveira, I. S. Lima, F. Vitório and E. M. M. Oliveira, *Food Chem.*, 2016, **199**, 433–438.
65. A. S. Luna, J. S. A. Pinho and L. C. Machado, *Anal Methods*, 2016, **8**, 7204–7208.
66. R. Early, *The Technology of Dairy Products*, Blackie Academic & Professional, Newport, Shropshire, 2nd edn, 1992.
67. A. Vasilescu, Q. Wang, M. Li, R. Boukherroub and S. Szunerits, *Chemosensors*, 2016, **4**, 10.
68. I. Mihai, A. Vezeanu, C. Polonschii, C. Albu, G.-L. Radu and A. Vasilescu, *Sens. Actuators, B*, 2015, **206**, 198–204.
69. H. Huang, J. Qin, K. Hu, X. Liu, S. Zhao and Y. Huang, *RSC Adv.*, 2016, **6**, 86043–86050.
70. K. Yugender Goud, G. Catanante, A. Hayat, S. M. K. Vengatajalabathy Gobi and J. L. Marty, *Sens. Actuators, B*, 2016, **235**, 466–473.
71. B. Wang, Y. Chen, Y. Wu, B. Weng, Y. Liu, Z. Lu, C. M. Li and C. Yu, *Biosens. Bioelectron.*, 2016, **78**, 23–30.
72. W. Zheng, J. Teng, L. Cheng, Y. Ye, D. Pan, J. Wu, F. Xue, G. Liu and W. Chen, *Biosens. Bioelectron.*, 2016, **80**, 574–581.
73. G. Castillo, K. Spinella, A. Poturnayová, M. Šnejdárková, L. Mosiello and T. Hianik, *Food Control*, 2015, **52**, 9–18.
74. Z. Lu, X. Chen, Y. Wang, X. Zheng and C. M. Li, *Microchim. Acta*, 2015, **182**, 571–578.
75. Y. Seok, J.-Y. Byun, W.-B. Shim and M.-G. Kim, *Anal. Chim. Acta*, 2015, **886**, 182–187.
76. X. Guo, F. Wen, N. Zheng, Q. Luo, H. Wang, H. Wang, S. Li and J. Wang, *Biosens. Bioelectron.*, 2014, **56**, 340–344.
77. W.-B. Shim, M. J. Kim, H. Mun and M.-G. Kim, *Biosens. Bioelectron.*, 2014, **62**, 288–294.
78. J. A. Cruz-Aguado and G. Penner, *J. Agric. Food Chem.*, 2008, **56**, 10456–10461.
79. W.-B. Shim, H. Mun, H.-A. Joung, J. A. Ofori, D.-H. Chung and M.-G. Kim, *Food Control*, 2014, **36**, 30–35.
80. X. Ma, W. Wang, X. Chen, Y. Xia, N. Duan, S. Wu and Z. Wang, *Food Control*, 2015, **47**, 545–551.
81. S. Malhotra, A. K. Pandey, Y. S. Rajput and R. Sharma, *J. Mol. Recognit.*, 2014, **27**, 493–500.
82. X. Guo, F. Wen, N. Zheng, S. Li, M.-L. Fauconnier and J. Wang, *Anal. Bioanal. Chem.*, 2016, **408**, 5577–5584.

83. G. Istamboulié, N. Paniel, L. Zara, L. R. Granados, L. Barthelmebs and T. Noguer, *Talanta*, 2016, **146**, 464–469.
84. B. H. Nguyen, L. D. Tran, Q. P. Do, H. L. Nguyen, N. H. Tran and P. X. Nguyen, *Mater. Sci. Eng. C*, 2013, **33**, 2229–2234.
85. X. Chen, Y. Huang, X. Ma, F. Jia, X. Guo and Z. Wang, *Microchim. Acta*, 2015, **182**, 1709–1714.
86. M. McKeague, R. Velu, K. Hill, V. Bardóczy, T. Mészáros and M. DeRosa, *Toxins*, 2014, **6**, 2435–2452.
87. F. Costantini, C. Sberna, G. Petrucci, M. Reverberi, F. Domenici, C. Fanelli, C. Manetti, G. de Cesare, M. DeRosa, A. Nascetti and D. Caputo, *Sens. Actuators, B*, 2016, **230**, 31–39.
88. S. Dai, S. Wu, N. Duan and Z. Wang, *Microchim. Acta*, 2016, **183**, 1909–1916.
89. E.-J. Jo, H. Mun, S.-J. Kim, W.-B. Shim and M.-G. Kim, *Food Chem.*, 2016, **194**, 1102–1107.
90. L. Lv, C. Cui, C. Liang, W. Quan, S. Wang and Z. Guo, *Food Control*, 2016, **60**, 296–301.
91. Y. Liu, J. Yu, Y. Wang, Z. Liu and Z. Lu, *Sens. Actuators*, 2016, **222**, 797–803.
92. N. Mejri-Omrani, A. Miodek, B. Zribi, M. Marrakchi, M. Hamdi, J.-L. Marty and H. Korri-Youssoufi, *Anal. Chim. Acta*, 2016, **920**, 37–46.
93. B. Wang, Y. Wu, Y. Chen, B. Weng, L. Xu and C. Li, *Biosens. Bioelectron.*, 2016, **81**, 125–130.
94. C. Wang, J. Qian, K. Wang, X. Yang, Q. Liu, N. Hao, C. Wang, X. Dong and X. Huang, *Biosens. Bioelectron.*, 2016, **77**, 1183–1191.
95. X. Xu, C. Xu and Y. Ying, *RSC Adv.*, 2016, **6**, 50437–50443.
96. W. Zhou, W. Kong, X. Dou, M. Zhao, Z. Ouyang and M. Yang, *J. Chromatogr. B*, 2016, **1022**, 102–108.
97. A. Chrouda, A. Sbartai, A. Baraket, L. Renaud, A. Maaref and N. Jaffrezic-Renault, *Anal. Biochem.*, 2015, **488**, 36–44.
98. A. Hayat, R. K. Mishra, G. Catanante and J. L. Marty, *Anal. Bioanal. Chem.*, 2015, **407**, 7815–7822.
99. R. Liu, Y. Huang, Y. Ma, S. Jia, M. Gao, J. Li, H. Zhang, D. Xu, M. Wu, Y. Chen, Z. Zhu and C. Yang, *ACS Appl. Mater. Interfaces*, 2015, **7**, 6982–6990.
100. L. Liu, X. Zhou and H. Shi, *Biosens. Bioelectron.*, 2015, **72**, 300–305.
101. A. H. Loo, A. Bonanni and M. Pumera, *ChemElectroChem*, 2015, **2**, 743–747.
102. Y. Luan, J. Chen, C. Li, G. Xie, H. Fu, Z. Ma and A. Lu, *Toxins*, 2015, **7**, 5377–5385.
103. R. K. Mishra, A. Hayat, G. Catanante, C. Ocaña and J.-L. Marty, *Anal. Chim. Acta*, 2015, **889**, 106–112.
104. J. Qian, K. Wang, C. Wang, M. Hua, Z. Yang, Q. Liu, H. Mao and K. Wang, *Analyst*, 2015, **140**, 7434–7442.
105. L. Rivas, C. C. Mayorga-Martinez, D. Quesada-González, A. Zamora-Gálvez, A. de la Escosura-Muñiz and A. Merkoçi, *Anal. Chem.*, 2015, **87**, 5167–5172.

106. S. Sanzani, M. Reverberi, C. Fanelli and A. Ippolito, *Toxins*, 2015, **7**, 812–820.
107. S. M. Taghdisi, N. M. Danesh, H. R. Beheshti, M. Ramezani and K. Abnous, *Nanoscale*, 2016, **8**, 3439–3446.
108. Y. Tan, X. Wei, Y. Zhang, P. Wang, B. Qiu, L. Guo, Z. Lin and H.-H. Yang, *Anal. Chem.*, 2015, **87**, 11826–11831.
109. R. Velu, N. Frost and M. C. DeRosa, *Chem. Commun.*, 2015, **51**, 14346–14349.
110. R. Wang, Y. Xiang, X. Zhou, L. Liu and H. Shi, *Biosens. Bioelectron.*, 2015, **66**, 11–18.
111. Y.-K. Wang, Q. Zou, J.-H. Sun, H. Wang, X. Sun, Z.-F. Chen and Y.-X. Yan, *J. Agric. Food Chem.*, 2015, **63**, 136–141.
112. R. Xiao, D. Wang, Z. Lin, B. Qiu, M. Liu, L. Guo and G. Chen, *Anal Methods*, 2015, **7**, 842–845.
113. L. Huang, J. Wu, L. Zheng, H. Qian, F. Xue, Y. Wu, D. Pan, S. B. Adeloju and W. Chen, *Anal. Chem.*, 2013, **85**, 10842–10849.
114. L. Yao, Y. Chen, J. Teng, W. Zheng, J. Wu, S. B. Adeloju, D. Pan and W. Chen, *Biosens. Bioelectron.*, 2015, **74**, 534–538.
115. Z. Zhu, M. Feng, L. Zuo, Z. Zhu, F. Wang, L. Chen, J. Li, G. Shan and S.-Z. Luo, *Biosens. Bioelectron.*, 2015, **65**, 320–326.
116. X. Yang, W. Kong, Y. Hu, M. Yang, L. Huang, M. Zhao and Z. Ouyang, *J. Sep. Sci.*, 2014, **37**, 853–860.
117. L. Huang, J. Wu, L. Zheng, H. Qian, F. Xue, Y. Wu, D. Pan, S. B. Adeloju and W. Chen, *Anal. Chem.*, 2013, **85**, 10842–10849.
118. X. Hun, F. Liu, Z. Mei, L. Ma, Z. Wang and X. Luo, *Biosens. Bioelectron.*, 2013, **39**, 145–151.
119. W. Ma, H. Yin, L. Xu, Z. Xu, H. Kuang, L. Wang and C. Xu, *Biosens. Bioelectron.*, 2013, **42**, 545–549.
120. L. Park, J. Kim and J. H. Lee, *Talanta*, 2013, **116**, 736–742.
121. C. Yang, V. Lates, B. Prieto-Simón, J.-L. Marty and X. Yang, *Talanta*, 2013, **116**, 520–526.
122. J. Zhang, X. Zhang, G. Yang, J. Chen and S. Wang, *Biosens. Bioelectron.*, 2013, **41**, 704–709.
123. G. Castillo, I. Lamberti, L. Mosiello and T. Hianik, *Electroanalysis*, 2012, **24**, 512–520.
124. J. Chen, Z. Fang, J. Liu and L. Zeng, *Food Control*, 2012, **25**, 555–560.
125. A. De Girolamo, L. Le, G. Penner, R. Schena and A. Visconti, *Anal. Bioanal. Chem.*, 2012, **403**, 2627–2634.
126. N. Duan, S. Wu, X. Ma, X. Chen, Y. Huang and Z. Wang, *Anal. Lett.*, 2012, **45**, 714–723.
127. P. Tong, W.-W. Zhao, L. Zhang, J.-J. Xu and H.-Y. Chen, *Biosens. Bioelectron.*, 2012, **33**, 146–151.
128. J. Wu, H. Chu, Z. Mei, Y. Deng, F. Xue, L. Zheng and W. Chen, *Anal. Chim. Acta*, 2012, **753**, 27–31.
129. C. Yang, V. Lates, B. Prieto-Simón, J.-L. Marty and X. Yang, *Biosens. Bioelectron.*, 2012, **32**, 208–212.

130. L. Barthelmebs, J. Jonca, A. Hayat, B. Prieto-Simon and J.-L. Marty, *Food Control*, 2011, **22**, 737–743.
131. L. Barthelmebs, A. Hayat, A. W. Limiadi, J.-L. Marty and T. Noguer, *Sens. Actuators B Chem.*, 2011, **156**, 932–937.
132. L. Bonel, J. C. Vidal, P. Duato and J. R. Castillo, *Biosens. Bioelectron.*, 2011, **26**, 3254–3259.
133. A. De Girolamo, M. McKeague, J. D. Miller, M. C. DeRosa and A. Visconti, *Food Chem.*, 2011, **127**, 1378–1384.
134. N. Duan, S.-J. Wu and Z.-P. Wang, *Chin. J. Anal. Chem.*, 2011, **39**, 300–304.
135. Z. Guo, J. Ren, J. Wang and E. Wang, *Talanta*, 2011, **85**, 2517–2521.
136. N. Prabhakar, Z. Matharu and B. D. Malhotra, *Biosens. Bioelectron.*, 2011, **26**, 4006–4011.
137. A. Rhouati, N. Paniel, Z. Meraihi and J.-L. Marty, *Food Control*, 2011, **22**, 1790–1796.
138. L. Sheng, J. Ren, Y. Miao, J. Wang and E. Wang, *Biosens. Bioelectron.*, 2011, **26**, 3494–3499.
139. P. Tong, L. Zhang, J.-J. Xu and H.-Y. Chen, *Biosens. Bioelectron.*, 2011, **29**, 97–101.
140. L. Wang, W. Ma, W. Chen, L. Liu, W. Ma, Y. Zhu, L. Xu, H. Kuang and C. Xu, *Biosens. Bioelectron.*, 2011, **26**, 3059–3062.
141. L. Wang, W. Chen, W. Ma, L. Liu, W. Ma, Y. Zhao, Y. Zhu, L. Xu, H. Kuang and C. Xu, *Chem. Commun.*, 2011, **47**, 1574–1576.
142. S. Wu, N. Duan, Z. Wang and H. Wang, *Analyst*, 2011, **136**, 2306.
143. X. Wu, J. Hu, B. Zhu, L. Lu, X. Huang and D. Pang, *J. Chromatogr. A*, 2011, **1218**, 7341–7346.
144. C. Yang, Y. Wang, J.-L. Marty and X. Yang, *Biosens. Bioelectron.*, 2011, **26**, 2724–2727.
145. S. Wu, N. Duan, W. Zhang, S. Zhao and Z. Wang, *Anal. Biochem.*, 2016, **508**, 58–64.
146. X. Chen, Y. Huang, N. Duan, S. Wu, Y. Xia, X. Ma, C. Zhu, Y. Jiang and Z. Wang, *J. Agric. Food Chem.*, 2014, **62**, 10368–10374.
147. H.-L. Jiang, X.-Y. Liu, Y. Qiu, D.-S. Yao, C.-F. Xie and D.-L. Liu, *Food Control*, 2015, **56**, 202–210.
148. X. Chen, Y. Huang, N. Duan, S. Wu, X. Ma, Y. Xia, C. Zhu, Y. Jiang and Z. Wang, *Anal. Bioanal. Chem.*, 2013, **405**, 6573–6581.
149. J. Zhang, Y.-K. Xia, M. Chen, D.-Z. Wu, S.-X. Cai, M.-M. Liu, W.-H. He and J.-H. Chen, *Sens. Actuators, B*, 2016, **235**, 79–85.
150. Y. Zhao, Y. Yang, Y. Luo, X. Yang, M. Li and Q. Song, *ACS Appl. Mater. Interfaces*, 2015, **7**, 21780–21786.
151. S. Wu, N. Duan, X. Ma, Y. Xia, H. Wang, Z. Wang and Q. Zhang, *Anal. Chem.*, 2012, **84**, 6263–6270.
152. T. Hünniger, C. Fischer, H. Wessels, A. Hoffmann, A. Paschke-Kratzin, I. Haase and M. Fischer, *J. Agric. Food Chem.*, 2015, **63**, 2189–2197.
153. R. Elshafey, M. Siaj and M. Zourob, *Biosens. Bioelectron.*, 2015, **68**, 295–302.

154. C. Fischer, T. Hünniger, J.-H. Jarck, E. Frohnmeyer, C. Kallinich, I. Haase, U. Hahn and M. Fischer, *J. Agric. Food Chem.*, 2015, **63**, 8050–8057.
155. A. V. Kolesnikov, A. V. Kozyr, A. K. Ryabko and I. G. Shemyakin, *Pathog. Dis.*, 2016, **74**, ftv112.
156. P. Janardhanan, C. M. Mello, B. R. Singh, J. Lou, J. D. Marks and S. Cai, *Talanta*, 2013, **117**, 273–280.
157. J. G. Bruno, A. M. Richarte, M. P. Carrillo and A. Edge, *Biosens. Bioelectron.*, 2012, **31**, 240–243.
158. L. Fetter, J. Richards, J. Daniel, L. Roon, T. J. Rowland and A. J. Bonham, *Chem. Commun.*, 2015, **51**, 15137–15140.
159. S. Eissa, M. Siaj and M. Zourob, *Biosens. Bioelectron.*, 2015, **69**, 148–154.
160. K. L. Hong, E. Maher, R. M. Williams and L. J. Sooter, *J. Nucleic Acids*, 2015, **2015**, 1–12.
161. S.-E. Kim, W. Su, M. Cho, Y. Lee and W.-S. Choe, *Anal. Biochem.*, 2012, **424**, 12–20.
162. Y. Zhao, X. Yang, H. Li, Y. Luo, R. Yu, L. Zhang, Y. Yang and Q. Song, *Chem. Commun.*, 2015, **51**, 16908–16911.
163. X. Du, D. Jiang, L. Dai, L. Zhou, N. Hao, J. Qian, B. Qiu and K. Wang, *Biosens. Bioelectron.*, 2016, **81**, 242–248.
164. M. Liu, J. Yu, X. Ding and G. Zhao, *Electroanalysis*, 2016, **28**, 161–168.
165. Z. Lin, H. Huang, Y. Xu, X. Gao, B. Qiu, X. Chen and G. Chen, *Talanta*, 2013, **103**, 371–374.
166. X. Li, R. Cheng, H. Shi, B. Tang, H. Xiao and G. Zhao, *J. Hazard. Mater.*, 2016, **304**, 474–480.
167. S. Wu, Q. Li, N. Duan, H. Ma and Z. Wang, *Microchim. Acta*, 2016, **183**, 2555–2562.
168. S. Wu, N. Duan, H. Zhang and Z. Wang, *Anal. Bioanal. Chem.*, 2015, **407**, 1303–1312.
169. S. Eissa, A. Ng, M. Siaj, A. C. Tavares and M. Zourob, *Anal. Chem.*, 2013, **85**, 11794–11801.
170. H. Gu, N. Duan, S. Wu, L. Hao, Y. Xia, X. Ma and Z. Wang, *Sci. Rep.*, 2016, **6**, 21665.
171. R. Zhao, Y. Wen, J. Yang, J. Zhang and X. Yu, *J. Microelectromech. Syst.*, 2014, **23**, 1054–1062.
172. S. Wu, N. Duan, X. Ma, Y. Xia, H. Wang and Z. Wang, *Anal. Chim. Acta*, 2013, **782**, 59–66.
173. V. Mudili, S. S. Makam, N. Sundararaj, C. Siddaiah, V. K. Gupta and P. V. L. Rao, *Sci. Rep.*, 2015, **5**, 15151.
174. R. Deng, L. Wang, G. Yi, E. Hua and G. Xie, *Colloids Surf. B Biointerfaces*, 2014, **120**, 1–7.
175. E. Temur, A. Zengin, İ. H. Boyacı, F. C. Dudak, H. Torul and U. Tamer, *Anal. Chem.*, 2012, **84**, 10600–10606.
176. A. Liu, Y. Zhang, W. Chen, X. Wang and F. Chen, *Eur. Food Res. Technol.*, 2013, **237**, 323–329.

177. Y. Huang, X. Chen, Y. Xia, S. Wu, N. Duan, X. Ma and Z. Wang, *Anal Methods*, 2014, **6**, 690–697.
178. Y. Huang, X. Chen, N. Duan, S. Wu, Z. Wang, X. Wei and Y. Wang, *Food Chem.*, 2015, **166**, 623–629.
179. K. Hong, L. Battistella, A. Salva, R. Williams and L. Sooter, *Int. J. Mol. Sci.*, 2015, **16**, 2794–2809.
180. L. R. Schoukroun-Barnes, S. Wagan and R. J. White, *Anal. Chem.*, 2014, **86**, 1131–1137.
181. N. Nikolaus and B. Strehlitz, *Sensors*, 2014, **14**, 3737–3755.
182. N. Derbyshire, S. J. White, D. H. J. Bunka, L. Song, S. Stead, J. Tarbin, M. Sharman, D. Zhou and P. G. Stockley, *Anal. Chem.*, 2012, **84**, 6595–6602.
183. R. Stoltenburg, N. Nikolaus and B. Strehlitz, *J. Anal. Methods Chem.*, 2012, **2012**, 1–14.
184. X. Wang, S. Dong, P. Gai, R. Duan and F. Li, *Biosens. Bioelectron.*, 2016, **82**, 49–54.
185. S. Pang and L. He, *Anal. Methods*, 2016, **8**, 1602–1608.
186. J. Daprà, L. H. Lauridsen, A. T. Nielsen and N. Rozlosnik, *Biosens. Bioelectron.*, 2013, **43**, 315–320.
187. K.-M. Song, E. Jeong, W. Jeon, M. Cho and C. Ban, *Anal. Bioanal. Chem.*, 2012, **402**, 2153–2161.
188. Y.-B. Miao, H.-X. Ren, N. Gan, Y. Cao, T. Li and Y. Chen, *Biosens. Bioelectron.*, 2016, **81**, 454–459.
189. Y. Miao, N. Gan, T. Li, Y. Cao, F. Hu and Y. Chen, *Sens. Actuators, B*, 2016, **222**, 1066–1072.
190. Y.-B. Miao, N. Gan, H.-X. Ren, T. Li, Y. Cao, F. Hu and Y. Chen, *Talanta*, 2016, **147**, 296–301.
191. M. Chen, N. Gan, H. Zhang, Z. Yan, T. Li, Y. Chen, Q. Xu and Q. Jiang, *Microchim. Acta*, 2016, **183**, 1099–1106.
192. K. Abnous, N. M. Danesh, M. Ramezani, A. S. Emrani and S. M. Taghdisi, *Biosens. Bioelectron.*, 2016, **78**, 80–86.
193. Y.-B. Miao, H.-X. Ren, N. Gan, Y. Zhou, Y. Cao, T. Li and Y. Chen, *Anal. Chim. Acta*, 2016, **929**, 49–55.
194. R. Sharma, K. V. Ragavan, K. S. M. S. Raghavarao and M. S. Thakur, in *Biotechnology and Biochemical Engineering*, ed. P. B. D. S. N. Gummadi and P. V. Vadlani, Springer, Singapore, 2016, pp. 187–195.
195. Y. Duan, Z. Gao, L. Wang, H. Wang, H. Zhang and H. Li, *Appl. Biochem. Biotechnol.*, 2016, **80**, 1644–1656.
196. X. Feng, N. Gan, S. Lin, T. Li, Y. Cao, F. Hu, Q. Jiang and Y. Chen, *Sens. Actuators, B*, 2016, **226**, 305–311.
197. J. Xue, J. Liu, C. Wang, Y. Tian and N. Zhou, *Anal Methods*, 2016, **8**, 1981–1988.
198. Z. Yan, N. Gan, T. Li, Y. Cao and Y. Chen, *Biosens. Bioelectron.*, 2016, **78**, 51–57.
199. M. Chen, N. Gan, Y. Zhou, T. Li, Q. Xu, Y. Cao and Y. Chen, *Sens. Actuators B Chem.*, 2016, **242**, 1201–1209.

200. L. Hao, N. Duan, S. Wu, B. Xu and Z. Wang, *Anal. Bioanal. Chem.*, 2015, **407**, 7907–7915.
201. Y. Liu, K. Yan, O. K. Okoth and J. Zhang, *Biosens. Bioelectron.*, 2015, **74**, 1016–1021.
202. S. Wu, H. Zhang, Z. Shi, N. Duan, C. Fang, S. Dai and Z. Wang, *Food Control*, 2015, **50**, 597–604.
203. E. Hamidi-Asl, F. Dardenne, R. Blust and K. De Wael, *Sensors*, 2015, **15**, 7605–7618.
204. Y. Miao, N. Gan, T. Li, H. Zhang, Y. Cao and Q. Jiang, *Sens. Actuators, B*, 2015, **220**, 679–687.
205. H. Gao, D. Pan, N. Gan, J. Cao, Y. Sun, Z. Wu and X. Zeng, *Microchim. Acta*, 2015, **182**, 2551–2559.
206. A. Bagheri Hashkavayi, J. Bakhsh Raoof, R. Ojani and E. Hamidi Asl, *Electroanalysis*, 2015, **27**, 1449–1456.
207. S. Pilehvar, T. Dierckx, R. Blust, T. Breugelmans and K. De Wael, *Sensors*, 2014, **14**, 12059–12069.
208. L. Yan, C. Luo, W. Cheng, W. Mao, D. Zhang and S. Ding, *J. Electroanal. Chem.*, 2012, **687**, 89–94.
209. S. Pilehvar, J. Mehta, F. Dardenne, J. Robbens, R. Blust and K. De Wael, *Anal. Chem.*, 2012, **84**, 6753–6758.
210. J. Mehta, B. Van Dorst, E. Rouah-Martin, W. Herrebout, M.-L. Scippo, R. Blust and J. Robbens, *J. Biotechnol.*, 2011, **155**, 361–369.
211. S. R. Han, J. Yu and S.-W. Lee, *Biochem. Biophys. Res. Commun.*, 2014, **448**, 397–402.
212. X. Liu, L. Su, L. Zhu, X. Gao, Y. Wang, F. Bai, Y. Tang and J. Li, *Sens. Actuators, B*, 2016, **233**, 394–401.
213. H. Ni, S. Zhang, X. Ding, T. Mi, Z. Wang and M. Liu, *Anal. Lett.*, 2014, **47**, 2844–2856.
214. C. Wang, C. Liu, J. Luo, Y. Tian and N. Zhou, *Anal. Chim. Acta*, 2016, **936**, 75–82.
215. H. Wang, Y. Wang, S. Liu, J. Yu, Y. Guo, Y. Xu and J. Huang, *RSC Adv.*, 2016, **6**, 43501–43508.
216. X. Qin, Y. Yin, H. Yu, W. Guo and M. Pei, *Biosens. Bioelectron.*, 2016, **77**, 752–758.
217. Y. Xu, L. Sun, X. Huang, Y. Sun and C. Lu, *Anal. Methods*, 2016, **8**, 726–730.
218. Z. Liu, C. Tian, L. Lu and X. Su, *RSC Adv.*, 2016, **6**, 10205–10214.
219. H.-Y. Song, T.-F. Kang, N.-N. Li, L.-P. Lu and S.-Y. Cheng, *Anal. Methods*, 2016, **8**, 3366–3372.
220. M. Ramezani, N. M. Danesh, P. Lavaee, K. Abnous and S. M. Taghdisi, *Sens. Actuators, B*, 2016, **222**, 1–7.
221. R. Liu, Z. Yang, Q. Guo, J. Zhao, J. Ma, Q. Kang, Y. Tang, Y. Xue, X. Lou and M. He, *Electrochim. Acta*, 2015, **182**, 516–523.
222. L. S. Khabbaz, M. Hassanzadeh-Khayyat, P. Zaree, M. Ramezani, K. Abnous and S. M. Taghdisi, *Anal. Methods*, 2015, **7**, 8611–8616.

223. J. Chen, Z. Li, J. Ge, R. Yang, L. Zhang, L. Qu, H. Wang and L. Zhang, *Talanta*, 2015, **139**, 226–232.
224. W. Guo, N. Sun, X. Qin, M. Pei and L. Wang, *Biosens. Bioelectron.*, 2015, **74**, 691–697.
225. N. Zhou, J. Luo, J. Zhang, Y. You and Y. Tian, *Anal. Methods*, 2015, **7**, 1991–1996.
226. X. Qin, W. Guo, H. Yu, J. Zhao and M. Pei, *Anal. Methods*, 2015, **7**, 5419–5427.
227. C. Liu, C. Lu, Z. Tang, X. Chen, G. Wang and F. Sun, *Microchim. Acta*, 2015, **182**, 2567–2575.
228. N. Zhou, J. Zhang and Y. Tian, *Anal. Methods*, 2014, **6**, 1569.
229. T. K. Sharma, R. Ramanathan, P. Weerathunge, M. Mohammadtaheri, H. K. Daima, R. Shukla and V. Bansal, *Chem. Commun.*, 2014, **50**, 15856–15859.
230. X. Bai, H. Hou, B. Zhang and J. Tang, *Biosens. Bioelectron.*, 2014, **56**, 112–116.
231. F. Li, Y. Guo, X. Sun and X. Wang, *Eur. Food Res. Technol.*, 2014, **239**, 227–236.
232. W. Xu, Y. Wang, S. Liu, J. Yu, H. Wang and J. Huang, *New J. Chem.*, 2014, **38**, 4931–4937.
233. H. Li, D. Sun, Y. Liu and Z. Liu, *Biosens. Bioelectron.*, 2014, **55**, 149–156.
234. S. Niu, Z. Lv, J. Liu, W. Bai, S. Yang and A. Chen, *PLoS One*, 2014, **9**, e109263.
235. K.-H. Leung, H.-Z. He, D. S.-H. Chan, W.-C. Fu, C.-H. Leung and D.-L. Ma, *Sens. Actuators, B*, 2013, **177**, 487–492.
236. Y. Zhu, P. Chandra, K.-M. Song, C. Ban and Y.-B. Shim, *Biosens. Bioelectron.*, 2012, **36**, 29–34.
237. K.-M. Song, M. Cho, H. Jo, K. Min, S. H. Jeon, T. Kim, M. S. Han, J. K. Ku and C. Ban, *Anal. Biochem.*, 2011, **415**, 175–181.
238. C. Ocaña, A. Hayat, R. Mishra, A. Vasilescu, M. del Valle and J.-L. Marty, *Analyst*, 2015, **140**, 4148–4153.
239. C. Ocaña, A. Hayat, R. K. Mishra, A. Vasilescu, M. del Valle and J.-L. Marty, *Bioelectrochemistry*, 2015, **105**, 72–77.
240. Y. Xiao, Y. Wang, M. Wu, X. Ma and X. Yang, *J. Electroanal. Chem.*, 2013, **702**, 49–55.
241. K. Ling, H. Jiang, L. Zhang, Y. Li, L. Yang, C. Qiu and F.-R. Li, *Anal. Bioanal. Chem.*, 2016, **408**, 3593–3600.
242. Y. Zhang, Y. You, Z. Xia, X. Han, Y. Tian and N. Zhou, *RSC Adv.*, 2016, **6**, 99540–99545.
243. R. Su, J. Xu, Y. Luo, Y. Li, X. Liu, J. Bie and C. Sun, *Mater. Lett.*, 2016, **180**, 31–34.
244. M. Hosseini, F. Mehrabi, M. R. Ganjali and P. Norouzi, *Luminescence*, 2016, **31**, 1339–1343.
245. B. Tan, H. Zhao, L. Du, X. Gan and X. Quan, *Biosens. Bioelectron.*, 2016, **83**, 267–273.

246. C. Fang, S. Wu, N. Duan, S. Dai and Z. Wang, *Anal. Methods*, 2015, **7**, 2585–2593.
247. H. Zhang, C. Fang, S. Wu, N. Duan and Z. Wang, *Anal. Biochem.*, 2015, **489**, 44–49.
248. H. B. Seo, Y. S. Kwon, J. Lee, D. Cullen, H. (Moses) Noh and M. B. Gu, *Analyst*, 2015, **140**, 6671–6675.
249. C. Lu, Z. Tang, C. Liu, L. Kang and F. Sun, *Anal. Bioanal. Chem.*, 2015, **407**, 4155–4163.
250. C.-H. Kim, L.-P. Lee, J.-R. Min, M.-W. Lim and S.-H. Jeong, *Biosens. Bioelectron.*, 2014, **51**, 426–430.
251. H. Zhao, S. Gao, M. Liu, Y. Chang, X. Fan and X. Quan, *Microchim. Acta*, 2013, **180**, 829–835.
252. H. Hou, X. Bai, C. Xing, N. Gu, B. Zhang and J. Tang, *Anal. Chem.*, 2013, **85**, 2010–2014.
253. J. Mehta, E. Rouah-Martin, B. Van Dorst, B. Maes, W. Herrebout, M.-L. Scippo, F. Dardenne, R. Blust and J. Robbens, *Anal. Chem.*, 2012, **84**, 1669–1676.
254. J. Zhao, W. Guo, M. Pei and F. Ding, *Anal. Methods*, 2016, **8**, 4391–4397.
255. Y. Yin, X. Qin, Q. Wang and Y. Yin, *RSC Adv.*, 2016, **6**, 39401–39408.
256. A. S. Emrani, N. M. Danesh, P. Lavaee, M. Ramezani, K. Abnous and S. M. Taghdisi, *Food Chem.*, 2016, **190**, 115–121.
257. N. Mohammad Danesh, M. Ramezani, A. Sarreshtehdar Emrani, K. Abnous and S. M. Taghdisi, *Biosens. Bioelectron.*, 2016, **75**, 123–128.
258. V. Soheili, S. M. Taghdisi, M. Hassanzadeh Khayyat, B. S. Fazly Bazzaz, M. Ramezani and K. Abnous, *Microchim. Acta*, 2016, **183**, 1687–1697.
259. N. Zhou, J. Wang, J. Zhang, C. Li, Y. Tian and J. Wang, *Talanta*, 2013, **108**, 109–116.
260. J. Yan, Y. Huang, C. Zhang, Z. Fang, W. Bai, M. Yan, C. Zhu and A. Chen, *Microchim. Acta*, 2016, **184**, 59–63.
261. A. Chen, X. Jiang, W. Zhang, G. Chen, Y. Zhao, T. M. Tunio, J. Liu, Z. Lv, C. Li and S. Yang, *Biosens. Bioelectron.*, 2013, **42**, 419–425.
262. K.-M. Song, E. Jeong, W. Jeon, H. Jo and C. Ban, *Biosens. Bioelectron.*, 2012, **33**, 113–119.
263. S. Jahanbani and A. Benvidi, *Biosens. Bioelectron.*, 2016, **85**, 553–562.
264. T. H. Le, V. P. Pham, T. H. La, T. B. Phan and Q. H. Le, *Adv. Nat. Sci. Nanosci. Nanotechnol.*, 2016, **7**, 15008.
265. X. Zhan, G. Hu, T. Wagberg, S. Zhan, H. Xu and P. Zhou, *Microchim. Acta*, 2016, **183**, 723–729.
266. G. Zhao, *Int. J. Electrochem. Sci.*, 2016, **11**, 1699–1706.
267. A. Benvidi, M. D. Tezerjani, S. M. Moshtaghiun and M. Mazloum-Ardakani, *Microchim. Acta*, 2016, **183**, 1797–1804.
268. Y. Guo, Q. Zhang, Q. Xu, N. Ma, X. Sun and X. Y. Wang, *Int. J. Electrochem. Sci.*, 2016, **11**, 1691–1698.
269. S. M. Taghdisi, N. M. Danesh, M. Ramezani and K. Abnous, *Biosens. Bioelectron.*, 2016, **85**, 509–514.

270. Y. Guo, X. Y. Wang and X. Sun, *Int. J. Electrochem. Sci.*, 2015, **10**, 3668–3679.
271. S. Wang, J. Liu, Y. Dong, H. Su and T. Tan, *RSC Adv.*, 2015, **5**, 53796–53801.
272. M. Ramezani, N. Mohammad Danesh, P. Lavaee, K. Abnous and S. Mohammad Taghdisi, *Biosens. Bioelectron.*, 2015, **70**, 181–187.
273. Y. Luo, J. Xu, Y. Li, H. Gao, J. Guo, F. Shen and C. Sun, *Food Control*, 2015, **54**, 7–15.
274. S. H. Jalalian, S. M. Taghdisi, N. M. Danesh, H. Bakhtiari, P. Lavaee, M. Ramezani and K. Abnous, *Anal. Methods*, 2015, **7**, 2523–2528.
275. Y. Guo, G. Shen, X. Sun and X. Wang, *IEEE Sens. J.*, 2015, **15**, 1951–1958.
276. D. Chen, D. Yao, C. Xie and D. Liu, *Food Control*, 2014, **42**, 109–115.
277. Y. S. Kwon, N. H. Ahmad Raston and M. B. Gu, *Chem. Commun.*, 2014, **50**, 40–42.
278. L. He, Y. Luo, W. Zhi, Y. Wu and P. Zhou, *Aust. J. Chem.*, 2013, **66**, 485.
279. L. He, Y. Luo, W. Zhi and P. Zhou, *Food Anal. Methods*, 2013, **6**, 1704–1711.
280. S. Jeong and I. Rhee Paeng, *Sci. World J.*, 2012, **2012**, 1–10.
281. L. Zhou, D.-J. Li, L. Gai, J.-P. Wang and Y.-B. Li, *Sens. Actuators, B*, 2012, **162**, 201–208.
282. L. R. Schoukroun-Barnes, E. P. Glaser and R. J. White, *Langmuir*, 2015, **31**, 6563–6569.
283. M. Jokar, M. H. Safaralizadeh, F. Hadizadeh, F. Rahmani and M. R. Kalani, *J. Biomol. Struct. Dyn.*, 2016, **34**, 2505–2517.
284. B. Lin, Y. Yu, R. Li, Y. Cao and M. Guo, *Sens. Actuators, B*, 2016, **229**, 100–109.
285. Y. Qi, F.-R. Xiu, M. Zheng and B. Li, *Biosens. Bioelectron.*, 2016, **83**, 243–249.
286. Y. Tian, Y. Wang, Z. Sheng, T. Li and X. Li, *Anal. Biochem.*, 2016, **513**, 87–92.
287. J. Guo, Y. Li, L. Wang, J. Xu, Y. Huang, Y. Luo, F. Shen, C. Sun and R. Meng, *Anal. Bioanal. Chem.*, 2016, **408**, 557–566.
288. W. Hu, Q. Chen, H. Li, Q. Ouyang and J. Zhao, *Biosens. Bioelectron.*, 2016, **80**, 398–404.
289. X. Liu, Y. Li, J. Liang, W. Zhu, J. Xu, R. Su, L. Yuan and C. Sun, *Talanta*, 2016, **160**, 99–105.
290. C. Wang, D. Chen, Q. Wang and Q. Wang, *Anal. Sci.*, 2016, **32**, 757–762.
291. R. Rapini and G. Marrazza, *IEEE*, 2015, 1–4.
292. Z. Yang, J. Qian, X. Yang, D. Jiang, X. Du, K. Wang, H. Mao and K. Wang, *Biosens. Bioelectron.*, 2015, **65**, 39–46.
293. D. Jiang, X. Du, Q. Liu, L. Zhou, L. Dai, J. Qian and K. Wang, *Analyst*, 2015, **140**, 6404–6411.
294. P. Weerathunge, R. Ramanathan, R. Shukla, T. K. Sharma and V. Bansal, *Anal. Chem.*, 2014, **86**, 11937–11941.
295. H. Shi, G. Zhao, M. Liu, L. Fan and T. Cao, *J. Hazard. Mater.*, 2013, **260**, 754–761.

296. L. Fan, G. Zhao, H. Shi, M. Liu and Z. Li, *Biosens. Bioelectron.*, 2013, **43**, 12–18.
297. J. He, Y. Liu, M. Fan and X. Liu, *J. Agric. Food Chem.*, 2011, **59**, 1582–1586.
298. R. Williams, C. Crihfield, S. Gattu, L. Holland and L. Sooter, *Int. J. Mol. Sci.*, 2014, **15**, 14332–14347.
299. R. M. Williams, A. R. Kulick, S. Yedlapalli, L. Battistella, C. J. Hajiran and L. J. Sooter, *J. Nucleic Acids*, 2014, **2014**, 1–8.
300. S. Li, X. Wu, C. Liu, G. Yin, J. Luo and Z. Xu, *Anal. Chim. Acta*, 2016, **941**, 94–100.
301. M. Jokar, M. H. Safaralizadeh, F. Hadizadeh, F. Rahmani and M. R. Kalani, *J. Biomol. Struct. Dyn.*, 2016, 1–11.
302. F. Cao, X. Lu, X. Hu, Y. Zhang, L. Zeng, L. Chen and M. Sun, *Biosci. Biotechnol. Biochem.*, 2016, **80**, 823–832.
303. Y. S. Kwon, V.-T. Nguyen, J. G. Park and M. B. Gu, *Anal. Chim. Acta*, 2015, **868**, 60–66.
304. L. Wang, H. Ye, H.-Q. Sang and D.-D. Wang, *Chin. J. Anal. Chem.*, 2016, **44**, 799–803.
305. N. Prabhakar, H. Thakur, A. Bharti and N. Kaur, *Anal. Chim. Acta*, 2016, **939**, 108–116.
306. R. Bala, M. Kumar, K. Bansal, R. K. Sharma and N. Wangoo, *Biosens. Bioelectron.*, 2016, **85**, 445–449.
307. F. Barahona, C. L. Bardliving, A. Phifer, J. G. Bruno and C. A. Batt, *Ind. Biotechnol.*, 2013, **9**, 42–50.
308. P. Wang, Y. Wan, A. Ali, S. Deng, Y. Su, C. Fan and S. Yang, *Sci. Chin. Chem.*, 2016, **59**, 237–242.
309. T. Tang, J. Deng, M. Zhang, G. Shi and T. Zhou, *Talanta*, 2016, **146**, 55–61.
310. X. Dou, X. Chu, W. Kong, J. Luo and M. Yang, *Anal. Chim. Acta*, 2015, **891**, 291–297.
311. W. Bai, C. Zhu, J. Liu, M. Yan, S. Yang and A. Chen, *Environ. Toxicol. Chem.*, 2015, **34**, 2244–2249.
312. S. Pang, T. P. Labuza and L. He, *Analyst*, 2014, **139**, 1895.
313. C. Zhang, L. Wang, Z. Tu, X. Sun, Q. He, Z. Lei, C. Xu, Y. Liu, X. Zhang, J. Yang, X. Liu and Y. Xu, *Biosens. Bioelectron.*, 2014, **55**, 216–219.
314. L. Wang, X. Liu, Q. Zhang, C. Zhang, Y. Liu, K. Tu and J. Tu, *Biotechnol. Lett.*, 2012, **34**, 869–874.
315. R. Bala, R. K. Sharma and N. Wangoo, *Anal. Bioanal. Chem.*, 2016, **408**, 333–338.
316. O. A. Alsager, S. Kumar, B. Zhu, J. Travas-Sejdic, K. P. McNatty and J. M. Hodgkiss, *Anal. Chem.*, 2015, **87**, 4201–4209.
317. J.-J. Zhang, J.-T. Cao, G.-F. Shi, K.-J. Huang, Y.-M. Liu and Y.-H. Chen, *Anal. Methods*, 2014, **6**, 6796.
318. K.-J. Huang, Y.-J. Liu, G.-W. Shi, X.-R. Yang and Y.-M. Liu, *Sens. Actuators, B*, 2014, **201**, 579–585.

319. J. Liu, W. Bai, S. Niu, C. Zhu, S. Yang and A. Chen, *Sci. Rep.*, 2014, **4**, 7571.
320. F. Long, H. Shi and H. Wang, *RSC Adv.*, 2014, **4**, 2875–2881.
321. Z. Lin, L. Chen, G. Zhang, Q. Liu, B. Qiu, Z. Cai and G. Chen, *Analyst*, 2012, **137**, 819–822.
322. N. Yildirim, F. Long, C. Gao, M. He, H.-C. Shi and A. Z. Gu, *Environ. Sci. Technol.*, 2012, **46**, 3288–3294.
323. G. Dong Huy, N. Jin, B.-C. Yin and B.-C. Ye, *Bioprocess Biosyst. Eng.*, 2011, **34**, 189–195.
324. D. Chen, M. Yang, N. Zheng, N. Xie, D. Liu, C. Xie and D. Yao, *Biosens. Bioelectron.*, 2016, **80**, 525–531.
325. M. Roushani and F. Shahdost-fard, *Microchim. Acta*, 2016, **183**, 185–193.
326. K. Abnous, N. M. Danesh, M. Ramezani, S. M. Taghdisi and A. S. Emrani, *Sens. Actuators, B*, 2016, **224**, 351–355.
327. Z. Chen and M. Lu, *Talanta*, 2016, **160**, 444–448.
328. Z. Chen, Y. Tan, K. Xu, L. Zhang, B. Qiu, L. Guo, Z. Lin and G. Chen, *Biosens. Bioelectron.*, 2016, **75**, 8–14.
329. E. Guler, G. Bozokalfa, B. Demir, Z. P. Gumus, B. Guler, E. Aldemir, S. Timur and H. Coskunol, *Drug Test. Anal.*, 2016, **9**, 578–587.
330. F. Haddache, A. Le Goff, N. Spinelli, P. Gairola, K. Gorgy, C. Gondran, E. Defrancq and S. Cosnier, *Electrochim. Acta*, 2016, **219**, 82–87.
331. M. Oroval, M. Coronado-Puchau, J. Langer, M. N. Sanz-Ortiz, Á. Ribes, E. Aznar, C. Coll, M. D. Marcos, F. Sancenón, L. M. Liz-Marzán and R. Martínez-Máñez, *Chem. – Eur. J.*, 2016, **22**, 13488–13495.
332. S. Zhang, L. Wang, M. Liu, Y. Qiu, M. Wang, X. Liu, G. Shen and R. Yu, *Anal. Methods*, 2016, **8**, 3740–3746.
333. Y. Zhang, Z. Sun, L. Tang, H. Zhang and G.-J. Zhang, *Microchim. Acta*, 2016, **183**, 2791–2797.
334. F. Luo, L. Zheng, S. Chen, Q. Cai, Z. Lin, B. Qiu and G. Chen, *Chem. Commun.*, 2012, **48**, 6387.
335. C. Ma, W. Wang, Q. Yang, C. Shi and L. Cao, *Biosens. Bioelectron.*, 2011, **26**, 3309–3312.
336. Y. Wen, H. Pei, Y. Wan, Y. Su, Q. Huang, S. Song and C. Fan, *Anal. Chem.*, 2011, **83**, 7418–7423.
337. K. Kang, A. Sachan, M. Nilsen-Hamilton and P. Shrotriya, *Langmuir*, 2011, **27**, 14696–14702.
338. A. A. Ensafi, N. Kazemifard and B. Rezaei, *Biosens. Bioelectron.*, 2016, **77**, 499–504.
339. F. Divsar, K. Habibzadeh, S. Shariati and M. Shahriarinour, *Anal. Methods*, 2015, **7**, 4568–4576.
340. L. Song, K. Mao, X. Zhou and J. Hu, *Talanta*, 2016, **146**, 285–290.
341. B. Yang, X. Chen, R. Liu, B. Liu and C. Jiang, *RSC Adv.*, 2015, **5**, 77755–77759.
342. M. Tang, G. Wen, A. Liang and Z. Jiang, *Luminescence*, 2014, **29**, 603–608.

343. L. Ye, G. Wen, J. Dong, Y. Luo, Q. Liu, A. Liang and Z. Jiang, *RSC Adv.*, 2014, **4**, 32960.
344. M. Yu, *Aust. J. Chem.*, 2014, **67**, 813.
345. Y. Wu, S. Zhan, H. Xing, L. He, L. Xu and P. Zhou, *Nanoscale*, 2012, **4**, 6841.
346. Z. Chen, L. Li, X. Mu, H. Zhao and L. Guo, *Talanta*, 2011, **85**, 730–735.
347. C. Wang, X. Cui, Y. Li, H. Li, L. Huang, J. Bi, J. Luo, L. Q. Ma, W. Zhou, Y. Cao, B. Wang and F. Miao, *Sci. Rep.*, 2016, **6**, 21711.
348. W. Deng, L.-R. Hong, M. Zhao, Y. Zhuo and M. Gao, *Analyst*, 2015, **140**, 4206–4211.
349. M. Park, H. D. Ha, Y. T. Kim, J. H. Jung, S.-H. Kim, D. H. Kim and T. S. Seo, *Anal. Chem.*, 2015, **87**, 10969–10975.
350. Z. S. Qian, X. Y. Shan, L. J. Chai, J. R. Chen and H. Feng, *Biosens. Bioelectron.*, 2015, **68**, 225–231.
351. B.-F. Ye, Y.-J. Zhao, Y. Cheng, T.-T. Li, Z.-Y. Xie, X.-W. Zhao and Z.-Z. Gu, *Nanoscale*, 2012, **4**, 5998.
352. L. Ma, H. Liu, G. Wu, N. Sun, L. Meng, Y. Li, Z. Liu and A. Diao, *Analyst*, 2016, **141**, 3997–4000.
353. Y.-Q. Huang, J.-C. Yin, Y.-S. Wang, X.-L. Xiao, B. Zhou, J.-H. Xue, X. Tang, X.-F. Wang, Y.-F. Zhu and S.-H. Chen, *Sens. Actuators, B*, 2016, **235**, 507–514.
354. H. Xuan, J. Ren, Y. Zhu, B. Zhao and L. Ge, *RSC Adv.*, 2016, **6**, 36827–36833.
355. Q. Wei, R. Nagi, K. Sadeghi, S. Feng, E. Yan, S. J. Ki, R. Caire, D. Tseng and A. Ozcan, *ACS Nano*, 2014, **8**, 1121–1129.
356. L. Zhang, H. Chang, A. Hirata, H. Wu, Q.-K. Xue and M. Chen, *ACS Nano*, 2013, **7**, 4595–4600.
357. Y. Helwa, N. Dave, R. Froidevaux, A. Samadi and J. Liu, *ACS Appl. Mater. Interfaces*, 2012, **4**, 2228–2233.
358. R. Freeman, X. Liu and I. Willner, *J. Am. Chem. Soc.*, 2011, **133**, 11597–11604.
359. F. Long, C. Gao, H. C. Shi, M. He, A. N. Zhu, A. M. Klibanov and A. Z. Gu, *Biosens. Bioelectron.*, 2011, **26**, 4018–4023.
360. S.-F. Torabi and Y. Lu, *Faraday Discuss.*, 2011, **149**, 125–135.
361. X. Chen, H. Zhou, N. Zhai, P. Liu, Q. Chen, L. Jin and Q. Zheng, *Anal. Lett.*, 2015, **48**, 1892–1906.
362. S. H. Suh, H. P. Dwivedi and L.-A. Jaykus, *LWT – Food Sci. Technol.*, 2014, **56**, 256–260.
363. S. L. Burrs, M. Bhargava, R. Sidhu, J. Kiernan-Lewis, C. Gomes, J. C. Claussen and E. S. McLamore, *Biosens. Bioelectron.*, 2016, **85**, 479–487.
364. D. O. Demirkol and S. Timur, *Int. J. Polym. Mater. Polym. Biomater.*, 2016, **65**, 85–90.
365. Y. Guo, Y. Wang, S. Liu, J. Yu, H. Wang, Y. Wang and J. Huang, *Biosens. Bioelectron.*, 2016, **75**, 315–319.

366. M. Ling, Z. Peng, L. Cheng and L. Deng, *Appl. Spectrosc.*, 2015, **69**, 1175–1181.
367. W. Wu, S. Zhao, Y. Mao, Z. Fang, X. Lu and L. Zeng, *Anal. Chim. Acta*, 2015, **861**, 62–68.
368. C. Zhang, X. Lv, X. Han, Y. Man, Y. Saeed, H. Qing and Y. Deng, *Anal. Methods*, 2015, **7**, 6339–6345.
369. Y. S. Kim, M. Y. Song, J. Jurng and B. C. Kim, *Anal. Biochem.*, 2013, **436**, 22–28.
370. E. A. Lamont, P. Wang, S. Enomoto, K. Borewicz, A. Abdallah, R. E. Isaacson and S. Sreevatsan, *PLoS One*, 2014, **9**, e114622.
371. K. Urmann, S. Arshavsky-Graham, J. G. Walter, T. Scheper and E. Segal, *Analyst*, 2016, **141**, 5432–5440.
372. C. L. A. Hamula, H. Peng, Z. Wang, A. M. Newbigging, G. J. Tyrrell, X.-F. Li and X. C. Le, *J. Mol. Evol.*, 2015, **81**, 194–209.
373. L. Zhang, R. Huang, W. Liu, H. Liu, X. Zhou and D. Xing, *Biosens. Bioelectron.*, 2016, **86**, 1–7.
374. S.-H. Lee, J.-Y. Ahn, K.-A. Lee, H.-J. Um, S. S. Sekhon, T. Sun Park, J. Min and Y.-H. Kim, *Biosens. Bioelectron.*, 2015, **68**, 272–280.
375. S. H. Suh, H. P. Dwivedi, S. J. Choi and L.-A. Jaykus, *Anal. Biochem.*, 2014, **459**, 39–45.
376. G. Liu, Y. Lian, C. Gao, X. Yu, M. Zhu, K. Zong, X. Chen and Y. Yan, *J. Integr. Agric.*, 2014, **13**, 1121–1129.
377. F. He, Y. Xiong, J. Liu, F. Tong and D. Yan, *Biosens. Bioelectron.*, 2016, **77**, 799–804.
378. F. Jia, N. Duan, S. Wu, R. Dai, Z. Wang and X. Li, *Microchim. Acta*, 2016, **183**, 337–344.
379. G. Liu, X. Yu, F. Xue, W. Chen, Y. Ye, X. Yang, Y. Lian, Y. Yan and K. Zong, *Microchim. Acta*, 2012, **178**, 237–244.
380. V. C. Ozalp, G. Bayramoglu, Z. Erdem and M. Y. Arica, *Anal. Chim. Acta*, 2015, **853**, 533–540.
381. X. Ma, Y. Jiang, F. Jia, Y. Yu, J. Chen and Z. Wang, *J. Microbiol. Methods*, 2014, **98**, 94–98.
382. P. Zhang, H. Liu, S. Ma, S. Men, Q. Li, X. Yang, H. Wang and A. Zhang, *Biosens. Bioelectron.*, 2016, **80**, 538–542.
383. Z. Fang, W. Wu, X. Lu and L. Zeng, *Biosens. Bioelectron.*, 2014, **56**, 192–197.
384. W. Wu, Z. Fang, S. Zhao, X. Lu, L. Yu, T. Mei and L. Zeng, *RSC Adv.*, 2014, **4**, 22009.
385. Y. Ning, L. Cheng, M. Ling, X. Feng, L. Chen, M. Wu and L. Deng, *Pathog. Dis.*, 2015, **73**, ftv034.
386. X. Yan, W. Li, K. Liu and L. Deng, *Anal. Methods*, 2015, **7**, 10243–10250.
387. Y. Ning, W. Li, Y. Duan, M. Yang and L. Deng, *J. Biomol. Screen.*, 2014, 1087057114528538.
388. Z. Bagheryan, J.-B. Raoof, M. Golabi, A. P. F. Turner and V. Beni, *Biosens. Bioelectron.*, 2016, **80**, 566–573.

389. N. Duan, B. Chang, H. Zhang, Z. Wang and S. Wu, *Int. J. Food Microbiol.*, 2016, **218**, 38–43.
390. H. Kurt, M. Yüce, B. Hussain and H. Budak, *Biosens. Bioelectron.*, 2016, **81**, 280–286.
391. C. Zhu, Y. Hong, Z. Xiao, Y. Zhou, Y. Jiang, M. Huang, X. Xu and G. Zhou, *Anal. Methods*, 2016, **8**, 6560–6565.
392. X. Wang, Y. Huang, S. Wu, N. Duan, B. Xu and Z. Wang, *Int. J. Food Microbiol.*, 2016, **237**, 172–179.
393. T. Qiu, Y. Wang, J. Yu, S. Liu, H. Wang, Y. Guo and J. Huang, *RSC Adv.*, 2016, **6**, 62031–62037.
394. P. Singh, R. Gupta, M. Sinha, R. Kumar and V. Bhalla, *Microchim. Acta*, 2016, **183**, 1501–1506.
395. Q.-Y. Wang and Y.-J. Kang, *Nanoscale Res. Lett.*, 2016, **11**, 150.
396. R. Wang, Y. Xu, T. Zhang and Y. Jiang, *Anal. Methods*, 2015, **7**, 1701–1706.
397. E. Sheikhzadeh, M. Chamsaz, A. P. F. Turner, E. W. H. Jager and V. Beni, *Biosens. Bioelectron.*, 2016, **80**, 194–200.
398. J. Y. Park, H. Y. Jeong, M. I. Kim and T. J. Park, *J. Nanomater.*, 2015, **2015**, 1–9.
399. J. Moon, G. Kim and S. Park, *Anal. Methods*, 2014, **6**, 7442.
400. N. Duan, W. Gong, Z. Wang and S. Wu, *Anal. Methods*, 2016, **8**, 1390–1395.
401. N. Duan, S. Wu, S. Dai, T. Miao, J. Chen and Z. Wang, *Microchim. Acta*, 2015, **182**, 917–923.
402. S. M. Yoo, D.-K. Kim and S. Y. Lee, *Talanta*, 2015, **132**, 112–117.
403. H. Zhang, X. Ma, Y. Liu, N. Duan, S. Wu, Z. Wang and B. Xu, *Biosens. Bioelectron.*, 2015, **74**, 872–877.
404. V. C. Ozalp, G. Bayramoglu, M. Kavruk, B. B. Keskin, H. A. Oktem and M. Y. Arica, *Anal. Biochem.*, 2014, **447**, 119–125.
405. H.-C. Park, I. A. Baig, S.-C. Lee, J.-Y. Moon and M.-Y. Yoon, *Appl. Biochem. Biotechnol.*, 2014, **174**, 793–802.
406. N. Duan, S. Wu, X. Chen, Y. Huang, Y. Xia, X. Ma and Z. Wang, *J. Agric. Food Chem.*, 2013, **61**, 3229–3234.
407. H. P. Dwivedi, R. D. Smiley and L.-A. Jaykus, *Appl. Microbiol. Biotechnol.*, 2013, **97**, 3677–3686.
408. J. Yuan, Z. Tao, Y. Yu, X. Ma, Y. Xia, L. Wang and Z. Wang, *Food Control*, 2014, **37**, 188–192.
409. G. Singh, P. Vajpayee, N. Rani, A. Jyoti, K. C. Gupta and R. Shanker, *Ecotoxicol. Environ. Saf.*, 2012, **78**, 320–326.
410. W. Zhu, Z. Li, X. Liu, X. Yan and L. Deng, *Anal. Lett.*, 2015, **48**, 2870–2881.
411. W. Gong, N. Duan, S. Wu, Y. Huang, X. Chen and Z. Wang, *Anal. Methods*, 2015, **7**, 3625–3631.
412. Y. Wan, P. Qi, Y. Zeng, Y. Sun and D. Zhang, *Sens. Actuators, B*, 2016, **233**, 454–458.
413. J. Wang, X. Wu, C. Wang, N. Shao, P. Dong, R. Xiao and S. Wang, *ACS Appl. Mater. Interfaces*, 2015, **7**, 20919–20929.

414. M. R. T. Rahman, Z. Lou, H. Wang and L. Ai, *Anal. Lett.*, 2015, **48**, 2414–2422.
415. J. Moon, G. Kim, S. Park, J. Lim and C. Mo, *Sensors*, 2015, **15**, 8884–8897.
416. Y. Lian, F. He, H. Wang and F. Tong, *Biosens. Bioelectron.*, 2015, **65**, 314–319.
417. A. D. Friedman, D. Kim and R. Liu, *Biomaterials*, 2015, **36**, 110–123.
418. A. Abbaspour, F. Norouz-Sarvestani, A. Noori and N. Soltani, *Biosens. Bioelectron.*, 2015, **68**, 149–155.
419. D. Cheng, M. Yu, F. Fu, W. Han, G. Li, J. Xie, Y. Song, M. T. Swihart and E. Song, *Anal. Chem.*, 2016, **88**, 820–825.
420. A. Baumstummler, D. Lehmann, N. Janjic and U. A. Ochsner, *Lett. Appl. Microbiol.*, 2014, **59**, 422–431.
421. Y.-C. Chang, C.-Y. Yang, R.-L. Sun, Y.-F. Cheng, W.-C. Kao and P.-C. Yang, *Sci. Rep.*, 2013, **3**, 1863.
422. W. Chen, Y. Yan, Y. Zhang, X. Zhang, Y. Yin and S. Ding, *Sci. Rep.*, 2015, **5**, 11190.
423. C. L. A. Hamula, H. Peng, Z. Wang, G. J. Tyrrell, X.-F. Li and X. C. Le, *Methods*, 2016, **97**, 51–57.
424. N. Duan, Y. Yan, S. Wu and Z. Wang, *Food Control*, 2016, **63**, 122–127.
425. N. Duan, S. Wu, X. Chen, Y. Huang and Z. Wang, *J. Agric. Food Chem.*, 2012, **60**, 4034–4038.
426. N.-R. Ha, S.-C. Lee, J.-W. Hyun and M.-Y. Yoon, *Process Biochem.*, 2016, **51**, 24–33.
427. Z. Xi, R. Huang, Z. Li, N. He, T. Wang, E. Su and Y. Deng, *ACS Appl. Mater. Interfaces*, 2015, **7**, 11215–11223.
428. C. Roh, S. E. Kim and S.-K. Jo, *Sensors*, 2011, **11**, 6685–6696.
429. S. Karash, R. Wang, L. Kelso, H. Lu, T. J. Huang and Y. Li, *J. Virol. Methods*, 2016, **236**, 147–156.
430. L. Xu, R. Wang, L. C. Kelso, Y. Ying and Y. Li, *Sens. Actuators, B*, 2016, **234**, 98–108.
431. Y. Pang, Z. Rong, J. Wang, R. Xiao and S. Wang, *Biosens. Bioelectron.*, 2015, **66**, 527–532.
432. J. Lum, R. Wang, B. Hargis, S. Tung, W. Bottje, H. Lu and Y. Li, *Sensors*, 2015, **15**, 18565–18578.
433. T. T. Le, B. Adamiak, D. J. Benton, C. J. Johnson, S. Sharma, R. Fenton, J. W. McCauley, M. Iqbal and A. E. G. Cass, *Chem. Commun.*, 2014, **50**, 15533–15536.
434. H. Bai, R. Wang, B. Hargis, H. Lu and Y. Li, *Sensors*, 2012, **12**, 12506–12518.
435. K. Kiilerich-Pedersen, J. Daprà, S. Cherré and N. Rozlosnik, *Biosens. Bioelectron.*, 2013, **49**, 374–379.
436. B. I. Escudero-Abarca, S. H. Suh, M. D. Moore, H. P. Dwivedi and L.-A. Jaykus, *PLoS One*, 2014, **9**, e106805.

CHAPTER 9

Carbon Nanotube Sensing in Food Safety and Quality Analysis

RAGHID NAJJAR, JOSEPH R. NALBACH AND WEI XUE*

Rowan University, Department of Mechanical Engineering, 201 Mullica Hill Road, Glassboro, NJ 08028, USA
*Email: xuew@rowan.edu

9.1 Introduction

Carbon nanotubes (CNTs) have attracted tremendous attention during the past two decades due to their exceptional properties. They represent one of the most studied novel nanomaterials and have been considered a promising candidate for future nanoelectronics and high-performance sensors. Their unique molecular structure grants them remarkable electrical and electrochemical properties. The intrinsic mobility of an individual semiconducting carbon nanotube can exceed 100 000 $cm^2 Vs^{-1}$ at room temperature, greater than any other known semiconductors.[1,2] Field-effect transistors (FETs) based on CNTs have demonstrated high carrier mobility of 3000 $cm^2 Vs^{-1}$.[3] In principle, CNTs play the same role in nanoelectronics as silicon in traditional electronic devices. Because of their unique nanostructures and inherent electric properties, CNTs can continue functioning at the molecular/atomic scale where silicon and other standard semiconductors cease to work.[4,5] However, it is extremely difficult to control the rolling orientation, called chirality, of CNTs in the material preparation process. Because the electrical properties of CNTs are determined by

Food Chemistry, Function and Analysis No. 2
Sensing Techniques for Food Safety and Quality Control
Edited by Xiaonan Lu

Published by the Royal Society of Chemistry, www.rsc.org

chirality, the resulting CNTs are typically a mixture of semiconducting and metallic nanotubes. The lack of precise control of chirality limits the potential of CNTs in nanoelectronics research and applications. Interestingly, the impact of the chirality-induced material property variation for sensor applications is relatively small for the CNTs. Both semiconducting and metallic CNTs have been successfully used in high-performance sensors.

It has long been recognized that CNTs are a powerful sensing material as a result of their large aspect ratio and high surface-to-volume ratio. The CNTs are a near-ideal one-dimensional material and are highly sensitive to their surrounding environment. They are commonly used as the sensing elements and conducting components in sensors. CNTs exhibit excellent adsorption properties because of their large specific surface area and nanoscale structure. The honeycomb lattice structures provide a large number of sites on their tubular surfaces where gaseous and aqueous molecules can react. The CNTs, pristine or functionalized, have shown high sensitivity towards chemicals, biomolecules and gas molecules.[6–8] Researchers have demonstrated a broad variety of sensing mechanisms including mechanical, electrical, optical and electrochemical principles using CNTs.[9] Both individual CNTs and CNT thin films have been used in the construction of different sensors. Various designs including conductimetric devices, potentiometric electrodes, amperometric sensors and ion-sensitive field-effect transistors (ISFETs) have been implemented.[10]

The US Centers for Disease Control and Prevention (CDC) estimates that 1 in 6 Americans, or 48 million people, get sick due to foodborne diseases each year with 3000 deaths.[11] Food supply contaminated by pathogenic bacteria often results in severe, sometimes life-threatening, conditions. The concern over unknown links between some of the severe diseases and the food supply has grown dramatically. Some concerns are controversial, but others are well founded, such as the elevated lead concentration in water or the presence of *Escherichia coli* (*E. coli*) in agri-food products. Nonetheless, the truth remains that monitoring food quality is challenging due to the wide variety of threats from various sources. Hence, sensors need to be developed to target many potential contaminants in the food supply. Different sensing principles such as amperometry, impedance, fluorescence and resonance have been successfully used in the development of sensors for food quality control.[12,13] However, many of the existing sensors suffer from low sensitivity, poor specificity, low detection limit, long detection time and high prices for testing equipment. In addition, some of them are bulky and can only be operated by skilled users. In comparison, CNTs are biocompatible, highly reactive, small, relatively inexpensive and can be functionalized by other molecules.[8] These merits make them a powerful material for sensing applications in the food industry. There has been an increasing number of research articles on this topic during the past ten years.

This chapter will review CNT-based sensors, their specific applications and performance for the detection of foodborne threats. The basic concepts of CNTs and their characteristics will be summarized. Sensor designs and their corresponding transduction principles will be described and explained. Due to the wide variety of substances that may cause food safety concerns, creative designs have been used in CNT-based sensors. A wide range of sensing methodologies, such as measuring pH changes in a solvent, using vibration to detect mass changes on CNTs, monitoring the bacteria-antibody binding events and pairing specific DNA strands in pathogenic *E. coli* and *Salmonella* bacteria, has been investigated and validated by the recent publications. In addition, we will review the latest progress in CNT sensors for specific substances, such as allergens, that are known to cause major illnesses in a selective group of the population. With their unique properties and sensing capabilities, CNT-based sensors have demonstrated great potential in food safety control and quality analysis. Further research and development in this field is currently underway. That will result in smaller and cheaper sensors with higher performance in the near future.

9.2 Materials

9.2.1 Carbon Nanotubes

There are two types of CNTs, single-walled carbon nanotubes (SWNTs) and multi-walled carbon nanotubes (MWNTs). An SWNT can be conceptualized as a nanoscale cylinder rolled up by a single-layer graphene sheet. The two edges of the sheet are seamlessly joined together and the ends of the resulting cylinders are closed to the hemispheres of the fullerene structures. SWNTs consist of only surface carbon atoms. Their length can reach up to several centimetres while the diameter is only 1–2 nm.[14] An MWNT is made of multiple coaxial SWNTs of varying diameters. The overall diameter of an MWNT is typically in the range of 10–50 nm. Although the two groups of materials are similar, SWNTs have more confined dimensions with less variation. This leads to more predictable material properties and often devices with higher performance. Therefore, there have been significantly more research activities on the devices, especially on electronics and sensors, by using SWNTs than MWNTs. On the other hand, MWNTs are much cheaper and their performance is usually sufficient for many low-cost applications. As a result, they have been widely used for the development and construction of thin film-based devices or coatings.

The base graphene lattice and the resulting SWNT are illustrated in Figure 9.1. The orientation of an SWNT (called "chirality") can be depicted by using the chiral vector C_h and the chiral angle θ. The chiral vector is defined in eqn (9.1):[15]

$$C_h = na_1 + ma_2 \tag{9.1}$$

where a_1 and a_2 are the unit vectors in a two-dimensional graphene lattice, and n and m are integers representing the numbers of unit vectors along the two directions. The chiral vector, often shortened to (n, m), connects two crystallographically equivalent sites on the graphene lattice. An SWNT is

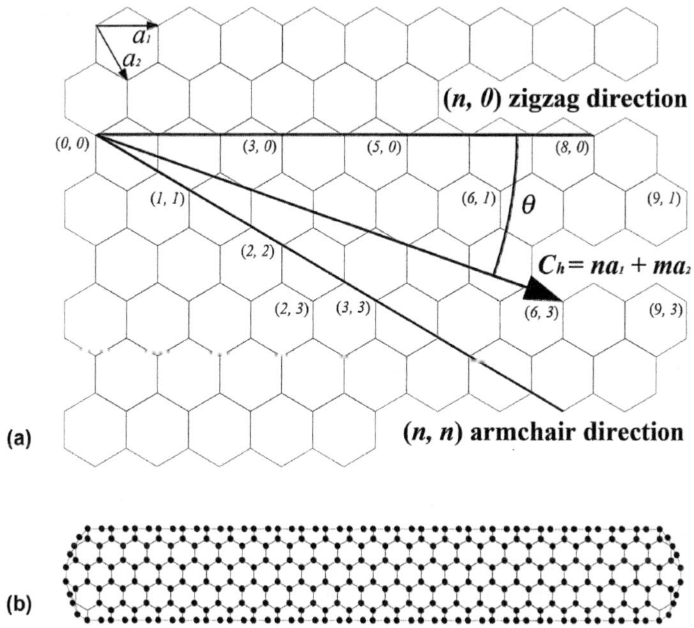

Figure 9.1 (a) 2D lattice of a graphene sheet illustrating chiral vector C_h and chiral angle θ. (b) Schematic model for an armchair SWNT.

called a zigzag nanotube when $n \neq 0$ and $m = 0$, and an armchair nanotube when $n = m \neq 0$. When a graphene sheet is rolled into a tube, the two points connected by the chiral vector coincide. The chiral vector therefore becomes the circumference of the tubular structure. The chiral angle between the chiral vector and the zigzag direction can be expressed as:

$$\theta = \arctan\left(\frac{\sqrt{3}m}{2n + m}\right) \tag{9.2}$$

The diameter of the nanotube can be expressed as:

$$d = \frac{C_h}{\pi} = \frac{\sqrt{3}a\sqrt{m^2 + mn + n^2}}{\pi} \tag{9.3}$$

where a is the in-plane nearest carbon-carbon bond length of 1.42 Å.

The geometric structure of an SWNT determines its electrical properties. For an SWNT with chirality of (n, m), it is a metallic nanotube if $n - m \neq 3i$ where i is an integer; otherwise, the nanotube is semiconducting. Therefore, one-third of nanotubes are predicted to be metallic and the other two-thirds are semiconducting.

The bonding structure among carbon nanotubes of SWNTs is a planar sp^2 bond, which is stronger than the tetrahedral sp^3 bond identified in diamond. As a result, SWNTs have outstanding mechanical strength with an estimated Young's modulus great than 1 TPa.[16–18] The chemical reactivity of the SWNTs is directly related to the local curvature engendered by the misalignment of the π-orbital.[19] The SWNTs are more chemically reactive than flat graphene sheets due to the curvature of the tube surface. For the same reason, SWNTs with smaller diameters often have higher reactivity than large-diameter nanotubes. Nonetheless, for the purpose of simplicity in this chapter, the term "CNTs" will be used in the following sections.

9.2.2 Sensing Properties of Carbon Nanotubes

The CNTs have great potential in sensing applications due to their inherent properties, such as small diameters, large surface areas and high conductivity. Compared with the bulky sensors, the nanomaterial-based sensors have much higher surface-to-volume ratios, which permit rapid accumulation and depletion of charge carriers.[20–24] Previous experimental studies have shown that nanosensors can achieve detection limits in the order of femtomoles (fM),[20,25,26] which has been echoed by numerical analyses.[27,28] Furthermore, theoretical studies have validated that the nanosensors are more sensitive than the corresponding planar devices.[29–32]

The adsorption of the target molecules onto the surface of a CNT alters its electrical properties, resulting in a conductance change. Pristine semiconducting CNTs are sensitive to electron-donating (*e.g.*, NH_3) and electron-accepting (*e.g.*, NO_2) gas molecules.[33] With appropriate chemical, physical and/or biological modifications, the CNTs are capable of detecting more types of gas molecules.[34–37] In an aqueous environment, the CNTs demonstrate a similar electrical property change when exposed to small traces of chemicals or biomolecules.[38,39] For example, the binding of proteins onto the surface of a CNT can alter its conductance.

Figure 9.2 shows various functionalization mechanisms of CNTs.[6] The functionalization can be achieved through defect groups, covalent bonded molecules, π–π interactions, non-covalent aggregates with surfactants, non-covalent wrapping with polymers and/or DNA strands and potentially inner-tubular insertion of C_{60}. Many of these mechanisms have been utilized by researchers for sensor development.[40]

9.3 Carbon Nanotube Sensors in Food Safety and Quality Analysis

CNT-based sensors can be categorized into different groups depending upon their signal transduction principles. For food safety applications, the sensors typically convert the concentration of a target into a signal measurable by other equipment. The three commonly used methods include electrochemical, electromechanical and optical sensing. The electrochemical sensors convert chemical concentrations into electrical output signals in forms of voltage, current or resistance. The electromechanical sensors produce

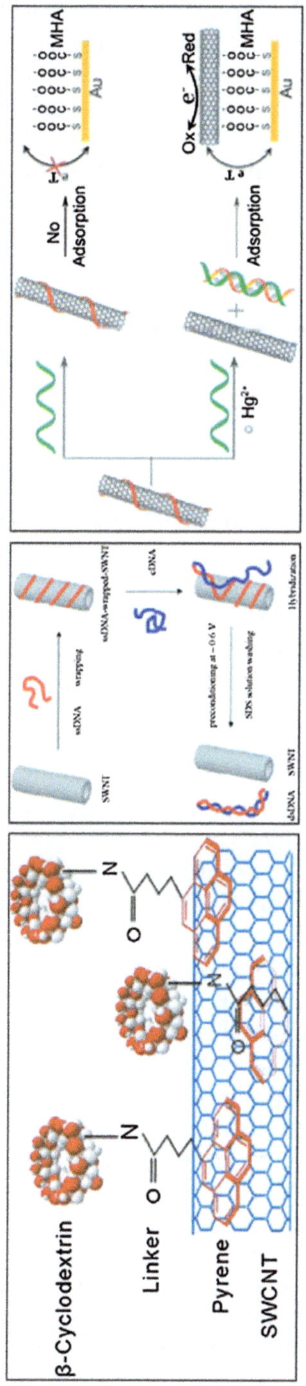

Figure 9.2 Various functionalization mechanisms of CNTs.
(Reproduced from ref. 6 with permission from the Royal Society of Chemistry.)

mechanical deflection or vibration with its values directly affected by the chemical concentration. The optical sensors display chemical concentrations as light signal changes in fluorescence, colour or intensity.

Another common method used to categorize sensors is based upon the applications or targeted chemicals. For example, the sensors could be used in an aqueous environment for solvent-based detection or in a gaseous environment as electronic noses. Within each environment, the sensors can be configured to detect pH values, bacteria, viruses, DNA and/or proteins. Because the tested samples are typically mixtures of various materials, the critical task for sensor development is to prepare the sensor that can specifically respond to only one particular target in the mixture. This section reviews key aspects such as device design, material preparation and detection performance of CNT sensors for food safety applications. The sensors are primarily grouped based upon the detected targets while the transduction principles are also considered.

9.3.1 Sensors in Food Safety and Quality Analysis

9.3.1.1 pH Sensors

The pH value is an indicator of hydronium ion concentration in an aqueous solution. It is measured on a scale of 0–14 with 7 being neutral. The pH value should remain unchanged in a stable sample. Any change to this value indicates the shifting of the hydronium ion concentration, very likely caused by a chemical change. For pH sensing in foods, if a pH change is not caused by the intended processing (*e.g.*, cooking, fermentation, *etc.*), it can be the indication of pathogens or unwanted ions being present. This may cause concerns over the food quality.

One of the earlier types of pH sensors are ion-selective electrodes.[41] These devices rely on potentiometric measurements by converting ion concentrations into measurable electrical potentials. A membrane electrode, or ion-selective membrane, is used to ensure the selectivity of specific ions. The most common membrane material is glass, but other materials in solid and liquid forms have also been used. Recent research efforts on ion-selective electrodes have been focused on improving their detection limit.[42] However, the typical limit is still above the nanomolar (nM) range, which is significantly higher than many nanomaterial-based sensors (detection limit in femtomolar to picomolar range, fM to pM). In addition, ion-selective electrodes are bulky and are difficult to integrate with other portable measurement systems.

ISFET-based sensors have also attracted considerable attention for pH sensing. They are highly compatible with the current IC microfabrication technology and can be easily integrated into the control circuits.[30,31] However, most ISFET sensors are planar devices fabricated on bulk silicon wafers. The accumulation or depletion of charge carriers, affected by the ion concentrations in the testing samples, only occurs in the surface region of the device. By contrast, CNT-modified ISFET sensors provide porous, 3D sensing elements that can "amplify" the electrical response of the ion concentration.[43,44] The CNTs pose no additional size limit to the ISFETs and the overall sensor dimensions can be controlled for portable sensing application.

The operating principles of CNT-based pH sensors can be either electrochemical or optical, depending upon the device design and the sensing elements.[45–47] A simple but unique sensor that combined both principles into one device was reported by Kaempgen and Roth.[45] Figure 9.3 shows the fabricated sensor as well as the pH sensing results using both electrochemical and optical principles. A polymer substrate is used for its flexibility and transparency. The sensing material is a mixture of CNTs and polyaniline, prepared through electro-polymerization. The pH dependence of the CNT-polyaniline network shows high linearity between pH values 1 and 13. The sensitivity is calculated as 58 mV decade^{-1}. The CNT-polyaniline composite outperforms pure CNT thin films as the sensing element in terms of

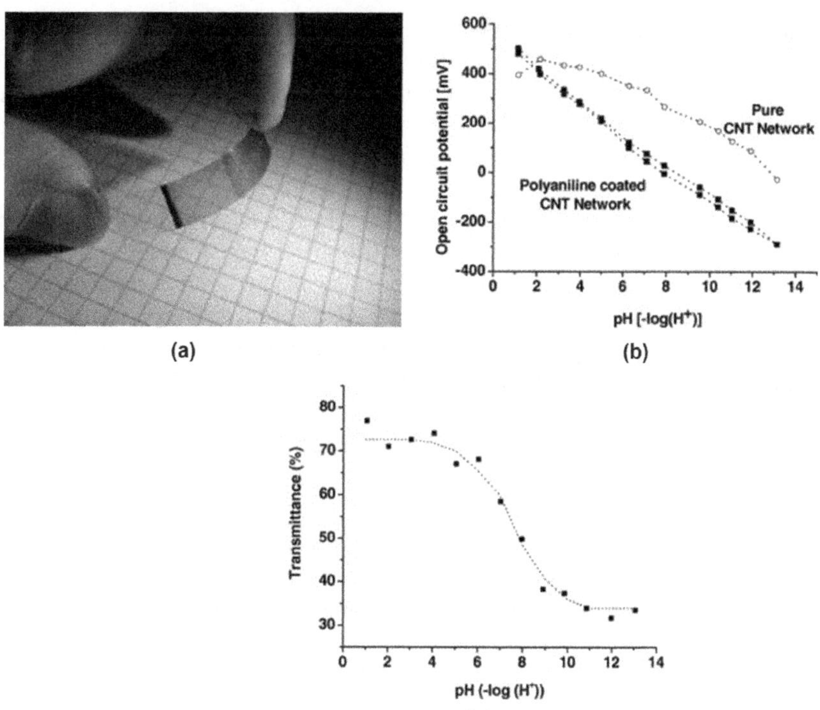

Figure 9.3 (a) A flexible CNT pH sensor. (b) Electrochemical response of the sensor at different pH solutions. (c) Optical transmittance of the sensor after immersing in different pH buffer solutions.
Reprinted from *the Journal of Electroanalytical Chemistry*, 586, M. Kaempgen and S. Roth, Transparent and flexible carbon nanotube/polyaniline pH sensors, 72–76, Copyright 2006, with permission from Elsevier.

Figure 9.4 (a) A wireless RFID pH sensor using a CNT/PAA chemiresistor. (b) Scanning electron microscopy (SEM) image of oxidized CNTs functionalized with PAA.
Reprinted with permission from Macmillan Publishers Ltd: *Scientific Reports*, 4, P. Gou, N. D. Kraut, I. M. Feigel, H. Bai, G. J. Morgan, Y. Chen, et al., Carbon nanotube chemiresistor for wireless pH sensing, copyright 2014.

linearity and sensitivity. The optical response of the same sensor is determined by measuring the light (wavelength of 575 nm) transmittance through the CNT-polyaniline film. A linear range can be observed in the pH range between 4 and 10. Although the performance is slightly reduced, the simple testing of the optical setup enables fast and low-cost measurement that can be used for a wide range of applications.

Efforts have been made to integrate CNT pH sensors into on-board control circuits as standalone devices. This eliminates external and often complicated measurement systems for data collection and analysis. Figure 9.4 illustrates a wirelessly powered, implantable radio-frequency identification (RFID) tag prototype using a CNT chemiresistor-based pH sensor.[48] The chemiresistor configuration is selected over the ISFET due to its simpler device structure without the need for both the reference electrode and the gate electrode. The sensing element is oxidized CNTs functionalized with a conductive polymer poly(1-aminoanthracene) (PAA). The CNT/PAA network, as shown in Figure 9.4b, can be deposited onto a substrate using either dielectrophoresis or a drop-cast method. The sensor shows a linear response to a wide range of pH values between 2 and 12. The integrated RFID tag can power the sensor, record the pH signals and transmit the data through a 1 cm tissue phantom. Although the device is still a prototype and its functions are relatively limited, it shows great possibilities of creating batteryless sensors for remote sensing. In the long term, sensors like this can potentially change the ways that foodborne materials are sensed in the food industry.

9.3.1.2 Gas Sensors as Electronic Noses

Another viable method of detecting potentially harmful substances in foods is to analyze the gases that ar emitted from agri-food products. Exposure to certain gases can cause detectable changes to properties of CNTs.

By analyzing the transformation of these properties, useful information about the gases can be ascertained.[49,50] Because many biological specimens emit gases as they decompose, sensors can be developed to monitor the spoiling stage of the foods besides detecting harmful microbes and/or pathogens. These sensors, often called electronic noses, are adept at selectively sensing aerosolized chemical compounds that may or may not cause odours from foods. They have become a powerful tool in recent years for monitoring food and beverage safety.[51,52]

An electronic nose is typically made of a gas sensor or an array of gas sensors that can simultaneously detect multiple compounds. In the multiplexed setting, each sensor can be treated with different sensing elements that can respond to a particular compound in a gas mixture. An electronic nose can monitor the concentration of a critical compound in real time. This direct measurement compares the recorded data with a pre-set threshold value to determine the food quality. An indirect method for food quality control is to use sensory descriptors (*e.g.*, sour odour, meat odour, *etc.*) to determine the elapsed storage time.[53] The results obtained from electronic noses can predict the stage (early *vs.* late) of the foods in storage. Both direct and indirect methods have validated that electronic noses are useful for on-line implementation in food quality control.

Figure 9.5 shows a portable electronic nose that can detect methanol contaminants in whiskeys.[51] The sensor uses interdigitated Cr/Au electrodes on an alumina substrate. The CNT-SiO_2 thin film is coated on top of the

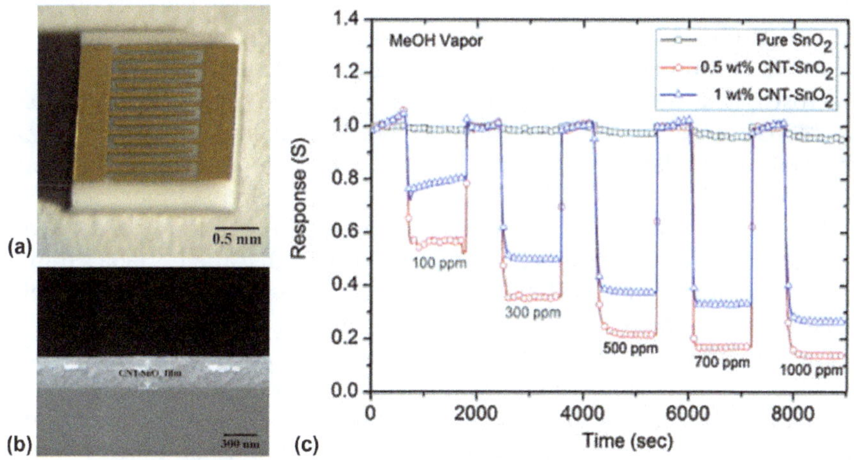

Figure 9.5 (a) An electronic nose using a CNT-SiO_2 sensor. (b) Cross-sectional SEM image of the CNT-SiO_2 thin film. (c) Response of three sensors to different concentrations of methanol.
Reprinted from *Sensors and Actuators B: Chemical*, Vol 147, C. Wongchoosuk, A. Wistsoraat, A. Tuantranont and T. Kerdcharoen, Portable electronic nose based on carbon nanotube–SO_2 gas sensors and its application for detection of methanol contamination in whiskeys, 392–399, Copyright 2010, with permission from Elsevier.

electrodes using electron-beam evaporation with CNT-SiO_2 mixed pallets as the target. The evaporated film is approximately 270 nm in thickness. Two different weight ratios, 0.5 wt% and 1 wt%, of the CNT-SiO_2 mixture are used for comparison. Pure SiO_2 films are also used as a controlled sensing element. The CNT-SiO_2 films behave as n-type semiconductors due to the decreased resistance in deoxidizing gases. Figure 9.5c shows the response of the three sensors to methanol vapours at different concentrations. The sensor with a pure SiO_2 film shows low sensitivity while both CNT-SiO_2 sensors demonstrate significantly enhanced sensitivity towards methanol. Furthermore, the weight ratio of the CNT-SiO_2 mixture can be changed to tune the sensitivity and selectivity to the desired gases. The 1 wt% CNT-SiO_2 sensor shows higher performance for methanol sensing while the 0.5 wt% CNT-SiO_2 performs better for ethanol sensing. A small device like this could be a simple yet useful tool for quick screening of various contaminants in the food industry. Sensors generated in this manner take advantage of both CNTs and other sensing materials. The hybrid of the sensing elements represents an important aspect of research as it seeks to combine the benefits of multiple materials for higher sensitivity.

9.3.1.3 Bacteria Sensors

CNTs can be used to directly sense bacteria using mechanical principles. The CNTs can be configured as resonators with known initial resonant frequencies. Any mass change caused by attached molecules on the CNTs induces a shift in resonant frequency. The frequency shift of the CNT vibration can be correlated to the added mass as well as a quantified value for the molecules. Mathematically, many vibration calculations used in the macroscopic structures can be applied to CNT-based sensors in most cases. These nanoscale sensors typically behave like beams supported by two common methods: cantilevers and bridges. Mathematical models on both cases can be built on well-founded engineering analysis of idealized beam deflection with necessary modifications and/or constraints. Recent theoretical and modelling investigations have indicated that the CNT-based mass sensors can reach a resolution of 10^{-24} kg (or zepto-grams).[54–56] This level of sensitivity has also been experimentally verified by other groups.[57]

The mechanical behaviour of a CNT mass sensor can be modelled using either continuum or atomistic approaches. The continuum modelling approach is straightforward, simple to use and can provide sufficient accuracy when used to predict sensor performance. It has become more powerful based upon modern mathematical and computational tools. Figure 9.6 shows the two continuum models used for vibrating CNT-based bacteria sensors.[54] The CNT sensor can be mathematically idealized to either a cantilever or a bridge. For the cantilever sensor, the mass is attached to the free end, which causes beam deflection and resonant frequency shift. Using the idealized beam model, the relationship between the resonant frequency and the mass can be established. The bridge sensor model works in a similar

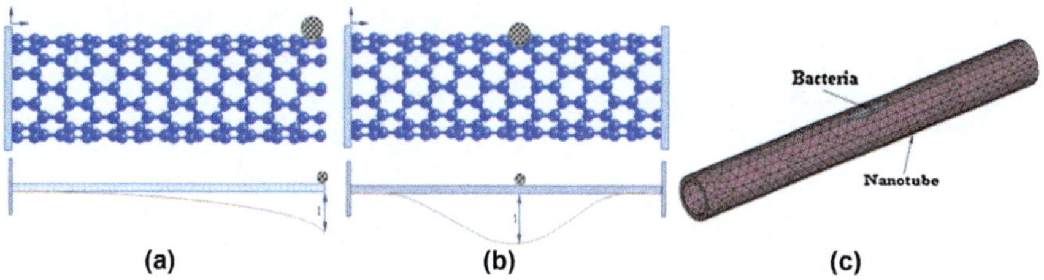

Figure 9.6 (a) Cantilever-based CNT resonator and its mathematically idealized model for bacteria sensing. (b) Bridge-based CNT resonator and its idealized model. (c) Finite element model of a CNT-based bacteria sensor.
Reprinted from *Physica E: Low dimensional Systems and Nanostructures*, Vol 42, R. Chowdhury, S. Adhikari and J. Mitchell, Vibrating carbon nanotube based biosensors, 104–109, Copyright 2009, with permission from Elsevier.

manner, with the exception that the mass lands in the middle of the sensor. These continuum models can be extended beyond the realm of point masses as bacteria are large objectives and may not have evenly distributed mass. Figure 9.6c shows a finite element model of a CNT-based bacteria sensor. The nanotube and bacteria (*E. coli*) are modelled as a circular cylinder and an ellipsoid, respectively. The simulation results match closely with the analytical results using beam deflection equations. The theoretical advancement on this topic indicates that CNT mass sensors have great potential to become high-sensitivity devices for bacteria sensing.

E. coli is a type of bacterium that exists in the intestines of humans and animals. There are many types of *E. coli*, most of which are harmless and even have a symbiotic relationship with humans. They can help digest food,[58] produce vitamin K[59] and create an anaerobic environment that is needed for the rest of the microbiome in the intestine to survive. However, some other types of *E. coli* bacteria are pathogenic. Some symptoms of *E. coli* infections are haemorrhagic diarrhoea, abdominal cramps and in some cases acute kidney failure. Pathogenic *E. coli* bacteria can be transmitted *via* food or drinks. In the United States alone, there were at least two *E. coli*-related outbreaks per year between 2010 and 2016. Identifying the presence of pathogenic *E. coli* in agri-food products is therefore critical. However, as the most commonly used methods for the detection of *E. coli*, enzyme linked immunosorbent assay (ELISA) and polymerase chain reaction (PCR), are time consuming, labour intensive, costly and require specialized laboratories.[60]

CNT-based *E. coli* sensors have attracted considerable attention in recent years.[60–62] They can simplify the testing procedure and provide timely results for *E. coli* sensing. Figure 9.7 shows an example of an electronic *E. coli* sensor using a CNT transistor.[63] The sensor fabrication starts with patterning electrodes onto a silicon wafer. A solution containing CNTs and 1-pyrenebutanoic acid succinimidyl ester (PASE) is placed on top of the device

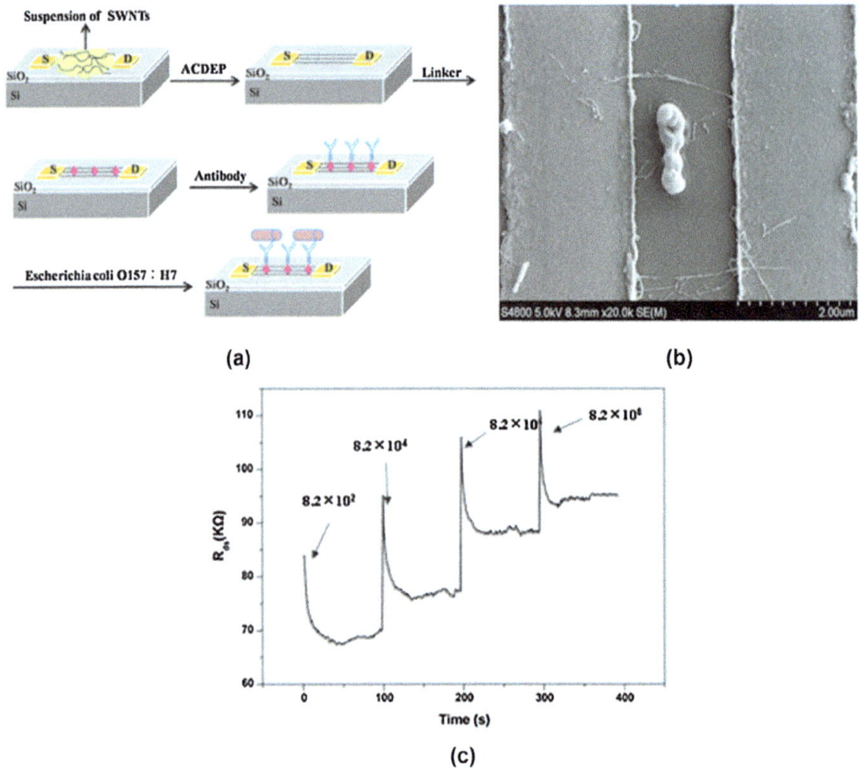

Figure 9.7 (a) The fabrication process of the CNT transistor biosensor for the detection of *E. coli*. (b) SEM image of the CNTs and pathogenic strain *E. coli* O157:H7. (c) Sensor response as the concentration of *E. coli* increases over time.
Reproduced from *Scientific Research*, Vol 4, X. Zhang, D. Wang, D. Yang, S. Li and Z. Shen, Electronic Detection of Escherichia coli O157 : H7 Using Single-Walled Carbon Nanotubes Field-Effect Transistor Biosensor, 2012, under Creative Commons License CC BY 4.0.

overnight, followed by rinsing with deionized water and drying with nitrogen. The PASE molecules function as the linkers that promote the attachment between CNTs and antibodies. Next, the device is submerged overnight in a phosphate-buffered saline (PBS) solution that contains *E. coli* O157:H7 antibodies. The sensor is then exposed to *E. coli* solutions with different concentrations. Due to antibody-bacteria binding, its electrical response is consequently recorded using a semiconductor parameter analyzer. SEM inspection shows that the CNTs are aligned between the electrodes. With the PACE modified CNTs, the antibodies are attached to the surface of the nanotube, resulting in easy attachment of *E. coli* during the sensing experiment. In comparison, the *E. coli* bacteria do not attach to the CNTs without the PACE linkers, which validates the poor adhesion between the CNTs and the antibodies in the preparation process. The sensor is exposed to solutions of *E. coli* with concentrations of 8.2×10^2, 8.2×10^4, 8.2×10^6 and finally 8.2×10^8 CFU mL^{-1}. The concentration is increased every 100 s to test the time dependence of the sensor. The results demonstrate a real-time response of the sensor with its resistance values increasing at a higher concentration. This research work shows a positive correlation between the *E. coli* concentration and the electrical properties of pre-treated CNTs.

The detection method of other pathogenic bacteria, such as *Salmonella*, is similar to that of *E. coli*. There has been increasing effort in the research community in using CNTs to detect *Salmonella*.[64–66] Although materials, preparation and sensing processes vary depending upon the scope of each project, most sensors for the detection of *Salmonella* rely on the specific interactions between antibodies and *Salmonella*. In most reported procedures, specific antibodies are selected to conjugate with the CNTs in order to bind to *Salmonella*. These devices have demonstrated low detection limits in the range of 1000 CFU ml^{-1} and can measure high-concentrations of *Salmonella* over 10^8 CFU ml^{-1}.

9.3.1.4 *Virus Sensors*

Viruses, the simplest form of life, are parasitic entities that use other organisms to replicate. They have an outer shell called a capsid and nucleic acids (DNA or RNA) inside the shell. When a virus infects an organism, it injects genetic materials into the body of the host cell and uses the host's organelles to make new copies of itself. The virus keeps replicating inside the host cell until finally the cell bursts and releases newly generated viruses. One of the common ways viruses can transmit to humans is through food. A European Union Summary Report published in 2014 reports that approximately 14.5% of all reported foodborne outbreaks are caused by viruses, mainly by norovirus, flavivirus and hepatitis virus.[67] However, there are no broad spectrum antiviral medications available. It is therefore necessary to identify the specific virus to issue the corresponding antiviral treatment. Traditional virus detection methods using ELISA and PCR are time consuming and labour intensive. Nanoscale immunosensors using CNTs provide an effective, low-cost, simple and quick method for accurate detection of different viruses.[68,69]

In a typical CNT immunosensor, the CNTs are placed between two metal electrodes. Anti-virus antibodies that can respond to a specific type of virus are attached onto the surface of the CNTs. When the viruses interact with the antibodies, measurable electrical signals in the form of resistance or conductance resulting from the interaction can be sensed by the CNTs. The changes are proportional to the concentration of the viruses present in the sample. Figure 9.8 shows the design, device and other key elements of an immunosensor for the detection of swine influenza virus (SIV) H1N1.[70] The sensor has a two-terminal structure with a CNT thin film bridging the gap between the electrodes. The CNTs are deposited onto a silicon substrate using an alternating coating method called "layer-by-layer self-assembly". Polyelectrolytes are used as the intermediate layers to provide the proper

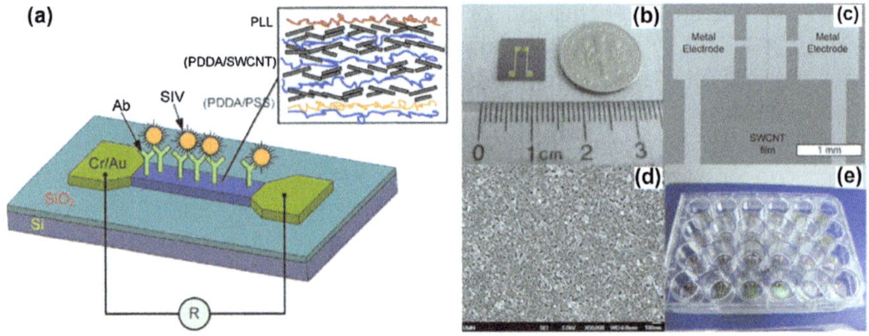

Figure 9.8 (a) Schematic illustration of a CNT immunosensor and its sensing principles. (b) A fabricated virus sensor on silicon. (c) SEM image of the electrodes and the CNT film. (d) The coated CNTs. (e) Immunoassay in a 24-well plate using sorted CNT sensors.
Reprinted from *Biosensors and Bioelectronics*, Vol 26, D. Lee, Y. Chander, S. M. Goyal and T. Cui, Carbon nanotube electric immunoassay for the detection of swine influenza virus H1N1, 3482–3487, Copyright 2011, with permission from Elsevier.

charging properties that can create a multilayer structure. The top surface of the multi-layer is poly-L-lysine (PLL) that is used to immobilize the anti-SIV antibodies based upon electrostatic attraction forces. The anti-SIV antibodies can bind with SIV strains in a liquid sample. The device has small dimensions of 1 cm×1 cm and can easily form an array for immunoassay. The viruses in the sample can bind with the antibodies and in turn change the electrical properties of the CNT film. The sensor resistance increases upon surface adsorption of SIV viruses. The reported detection limit is 180 $TCID_{50}$ ml^{-1} ($TCID_{50}$: 50% tissue culture infective dose) for SIV sensing and shows high selectivity against other types of viruses, such as transmissible gastroenteritis virus (TGEV) and feline calicivirus (FCV). The sensor performance is higher than other non-CNT-based immunoassay methods. Given the desired characteristics, the CNT immunosensors can potentially be used as a point-of-care sensing platform for rapid and reliable identification of different types of viruses.

Another recent paper on using CNT immunosensors for the detection of H1N1 virus uses a similar approach, as illustrated in Figure 9.9.[71] The virus detection is based upon the sensor resistance change caused by binding events between antibodies and viruses. The CNTs are coated between two parallel electrodes to form a uniformly distributed and aligned nanotube film. They are deposited with a solution-processed method called dielectrophoresis. The dielectrophoretic forces from an alternating electrical field stretch the dispersed CNTs and coat them on a pre-treated substrate. The orientation and density of the CNT film can be precisely controlled. Biotinylated monoclonal antibodies are used as the specific molecules to bind with the H1N1 viruses. The sensors show linear response to the viruses from 1 to 10 000 PFU ml^{-1}. A higher resistance change can be observed when more

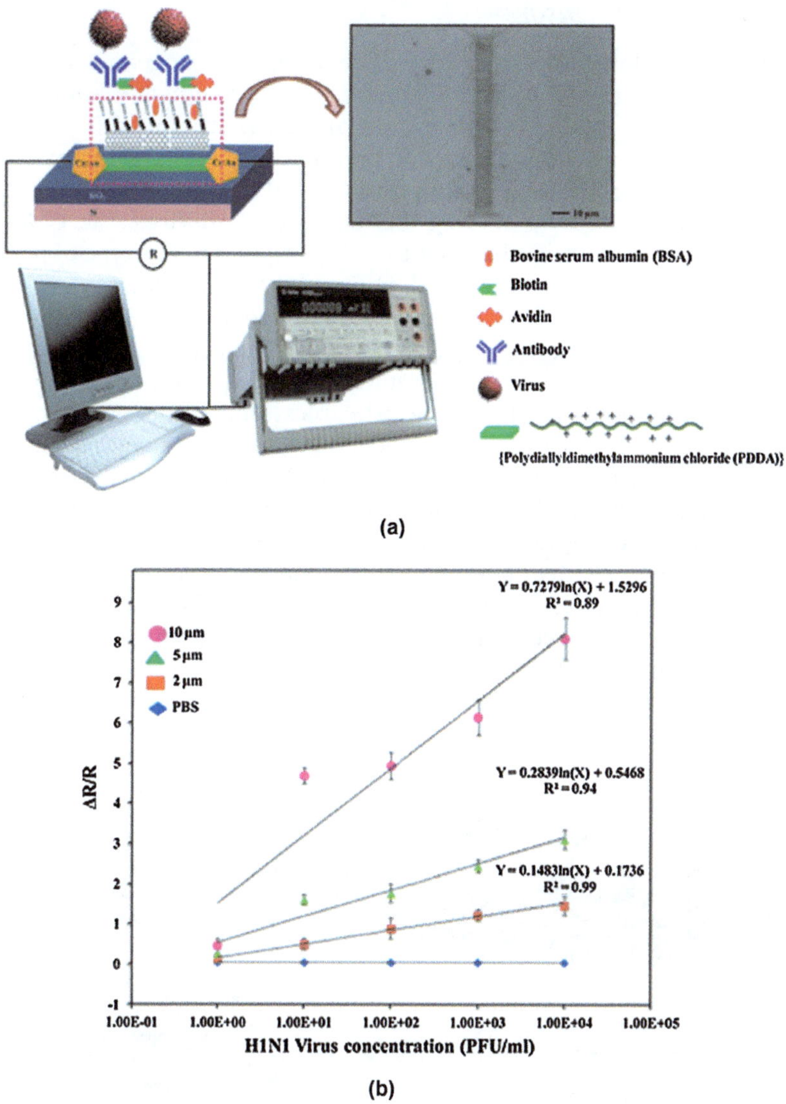

Figure 9.9 (a) Schematic illustration of a CNT immunosensor for the detection of H1N1 virus and the sensing system. (b) Sensing performance of immunosensors with different channel lengths responding to various concentrations of H1N1 virus samples.
Reproduced from ref. 71 with permission from the Royal Society of Chemistry.

viruses are absorbed on the sensor surface. The sensor geometry is also an important factor that determines the sensor performance. The sensor with the shortest channel shows the lowest detection limit of 1 PFU ml^{-1}; the device with the longest channel has a higher limit of 100 PFU ml^{-1} but with a greater slope for sensor response.

9.3.1.5 Protein Sensors

Proteins can be indicators of the presence of many pathogens and harmful substances, such as prions and allergens. Some proteins are known to be unique to certain strains of bacteria or viruses. Detection of proteins can therefore serve as an indicator for the presence of these targeted materials. Most CNT-based protein sensors rely on electrical signals that are induced by specific antigen–antibody binding events. In these devices, the CNTs are treated with specific antibodies that can only respond to one type of protein. The sensors then become unique identifiers of that protein.

The protein sensors have expanded their applications from the early emphasis on cancer markers to other types of molecules, including food-borne substances in recent years. Despite the wide spectrum of applications and target molecules, their sensing principles remain almost unchanged. The most common design for the detection of protein is an electrochemical sensor using field-effect transistors. Figure 9.10 illustrates the device design, material preparation and performance of a protein sensor used for the detection of prostate-specific antigen (PSA).[72] The sensor uses a back-gate transistor structure with either n-type In_2O_3 nanowires or p-type CNTs as the semiconductor material. The nanotube surface is first functionalized with 1-pyrenebutanoic acid succinimidyl ester. Next, the functionalized CNTs are treated with an anti-PSA monoclonal antibody buffer solution. The antibodies are anchored on the CNTs and become the receptors for PSA antigen.

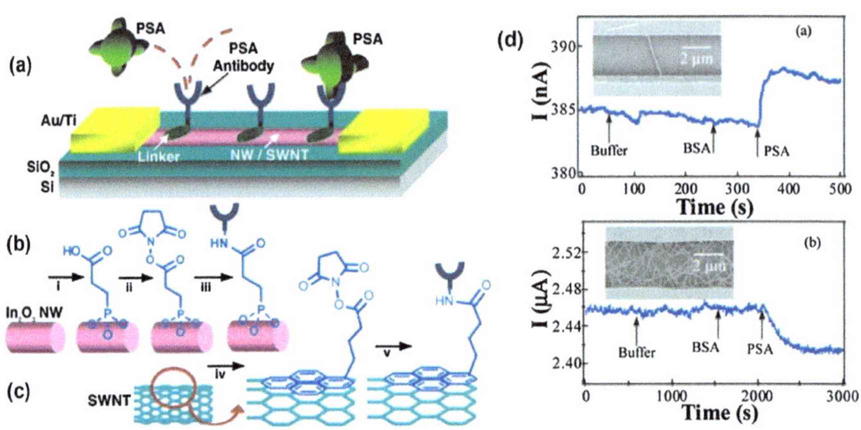

Figure 9.10 (a) Protein sensor structure and its sensing principle based upon antigen–antibody binding. Material preparation of (b) In_2O_3 nanowires and (c) CNTs to anchor antibodies on their surfaces. (d) Sensor response curves of the In_2O_3 and CNT sensors.
Reprinted with permission from C. Li, M. Curreli, H. Lin, B. Lei, F. N. Ishikawa, R. Datar, *et al.*, Complementary detection of prostate-specific antigen using In_2O_3 nanowires and carbon nanotubes, *Journal of the American Chemical Society*, vol. 127, pp. 12484–12485, Copyright 2005, American Chemical Society.

The CNT sensor responds differently to various solutions. It shows little change when it is exposed to a PBS solution. In comparison, the sensor shows a clear decrease in its electrical current when it is exposed to a PSA solution. The current reduction is due to the specific PSA antigen–antibody binding. Non-specific binding between the anchored anti-PSA antibodies and other types of antigen is suppressed, as the device shows no apparent response to a bovine serum albumin (BSA) solution. Similar behaviours have been observed in other CNT-based PSA sensors.[73,74] Exposure to the PSA solutions tends to decrease the electrical conductivity of the functionalized CNTs. In conclusion, these results validated that the CNT transistors can be used as effective protein sensors.

Another common approach to using CNTs in protein sensing is based upon bio-fluorescence. This relies on the principle that CNTs can be considered as quasi 1D quantum wires and have unique energy band structures and densities of electronic states.[75] CNTs have distinctive optical properties, such as strong optical absorption in the near-IR range and enhanced Raman spectroscopic signatures. In addition, the chirality of each CNT affects the emitted light wavelength. Based upon these reasons, CNTs have been successfully used in biomedical imaging and detection applications. Figure 9.11 illustrates a multiplexed protein microarray using isotopically labelled CNTs (^{12}C and ^{13}C isotopes).[76] The multiplexed detection is based upon the binding between the proteins on the microarray and the conjugated antibodies on the CNTs. The microarray is fabricated on a gold-coated glass slide. Its surface is carefully treated with two types of proteins, namely human immunoglobulin G (IgG) and mouse IgG, at different locations using a robotic array printer. The CNTs are synthesized with chemical vapour deposition (CVD) and then conjugated with two types of antibodies: anti-human IgG and anti-mouse IgG. When a mixture of the two conjugates is applied to the microarray, specific binding occurs at different protein spots. The Raman scattering spectra show clear differential detection of human IgG (green) and mouse IgG (red) on the same colour map. The CNT Raman tags are effective in targeting and detecting specific proteins. The "colour-coding" approach of using CNT tags has already been used for *in vitro* and *in vivo* applications in biology and medicine.[75] It can be easily adopted by the field of food safety control and potentially become a powerful method, as multiplexed detection of pathogens in food sources is critical.

9.3.1.6 DNA Sensors

DNA and RNA are the molecules that hold all the genetic information of all organisms and viruses. They are made of nucleic acid sequences. These strands of sequences contain four nucleobases: cytosine (C), guanine (G), adenine (A) and thymine (T). The nucleobases can pair together to form the two strands of DNA. When a single strand of DNA is available, it will only pair with its complementary strand that has the exact sequence needed to satisfy the pairing rule: only A-T and C-G can pair together. Furthermore,

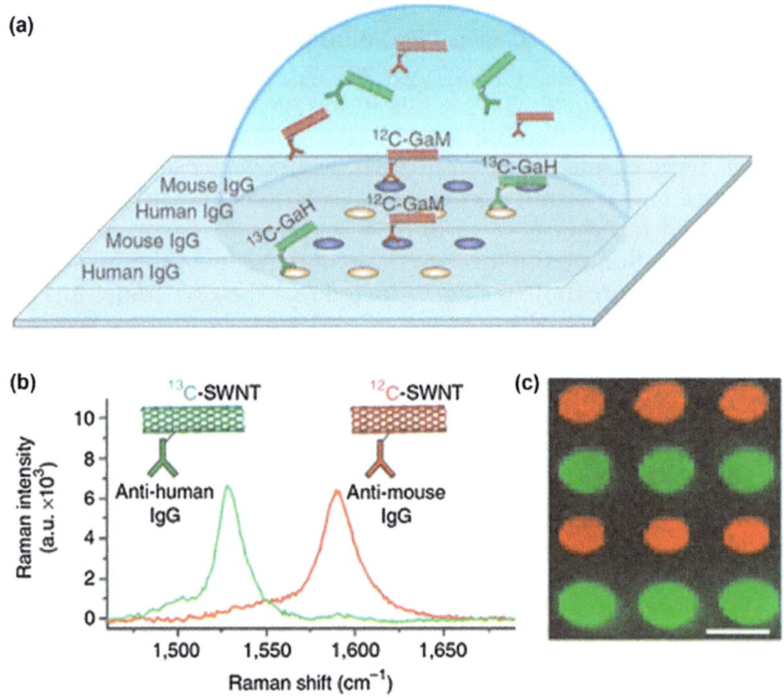

Figure 9.11 A microarray for multiplexed detection of proteins. CNTs are functionalized with specific antibodies to detect multiple proteins. The microarray uses a multicolour Raman scattering map.
Reprinted with permission from Macmillan Publishers Ltd: *Nature Biotechnology* (ref. 76). Copyright 2008.

some DNA sequences can be unique to a species. Using DNA as a sensing mechanism is therefore an accurate method of detecting the presence of specific pathogens. A single strand of DNA on a sensor can be used to detect the complementary strand that is known to belong to a pathogen.[77]

CNTs and DNA can be combined together to create sensors with high performance and high specificity. In a typical device, the CNTs are deposited onto a silicon wafer to form a back-gate transistor. Single strands of DNA (called probe DNA) are immobilized onto the surface of the nanotubes. They can only bind with target DNA strands with complementary sequences. Non-specific binding of DNA strands is therefore supressed. The binding between the probe and target DNA causes an electrical behaviour change in the CNTs, measurable by external electrical systems.

It has been shown that the sensor performance can be enhanced when the electrostatic charges from the target DNA increase.[78] DNA has a negative charge and attaching more DNA strands to the target DNA can amplify the overall sensitivity of the sensor. The additional DNA strands, called "reporter DNA", can therefore be used. This principle has been experimentally validated in a CNT transistor-based DNA sensor.[79] Gold nanoparticles are used as the carrying medium for DNA. The reporter DNA strands are firstly attached to the particles. Next, the target DNA strands are partially bound with the reporter DNA strands. This partial bond has no overlap with the location where the probe DNA and target DNA need to bind. Multiple reporter DNA strands are used to bind with the gold nanoparticles. Only one of the strands is used to bind with the target DNA, but the sensor uses all the electrostatic charges. The sensor is based upon an FET using CNT thin films. The probe DNA is immobilized onto the CNT surface as the receptors. A polyethylene glycol (PEG) solution is used to treat the sensor surface as a blocking agent to reduce non-specific binding. Four sensing experiments are carried out using the following samples: gold nanoparticles with both target DNA and reporter DNA, target DNA, gold nanoparticles with mismatched DNA and gold nanoparticles with reporter DNA. The sensors' drain current decrease percentages are obtained and compared. The statistical results show that the current reduction is significantly higher when all the three components (*i.e.*, gold nanoparticles, reporter DNA and target DNA) are present. Two types of reporter DNA are used, both of which demonstrate high performance for specific DNA sensing. This technique has been validated to be effective in enhancing the performance of the sensor with a reported detection limit down to 100 fM.

Allergens are various types of antigens that can cause allergic reactions for a collection of the population. These reactions are based upon abnormal immune responses from the immune system. Many of the allergic reactions are caused by ingestion of food and therefore the detection of allergens becomes critical for the selected populations. The detection of allergen can be based upon either protein sensing using antigen–antibody binding or indirectly by DNA sensing.

Ara h1, a protein from peanuts, is responsible for most of the cases of severe peanut allergic reactions. Currently, testing for the presence of Ara h1 is performed in an indirect manner, in which the presence of other peanut proteins is detected and the Ara h1 concentration is calculated based upon the results. Therefore, there is a need for the construction of direct detection systems for the presence of Ara h1 in foods to minimize errors. CNT sensors using a stem-loop DNA technique show great potential for accurate detection of the Ara h1 allergen. The fabrication process and the sensing principle of the sensor are illustrated in Figure 9.12.[80] The device is based upon a glassy carbon electrode with an additional coating of chitosan-treated CNTs. A film of spongy gold is then deposited onto the surface using electrodeposition. A single strand DNA complementary to the Ara h1 DNA is looped around and attached to a biotin molecule from one end. The other end of the DNA strand is fixed onto the gold sponge surface. This process keeps the DNA in a stable shape until the next step of the process and also deactivates the sensor.[81] When a solution with the Ara h1 DNA is added to the sensor surface, the complementary DNA strands join together. This re-shapes the DNA strands in the typical straight double helix formation and releases the biotin away from the gold sponge layer. Streptavidin-horseradish peroxidises (HRP-SA)

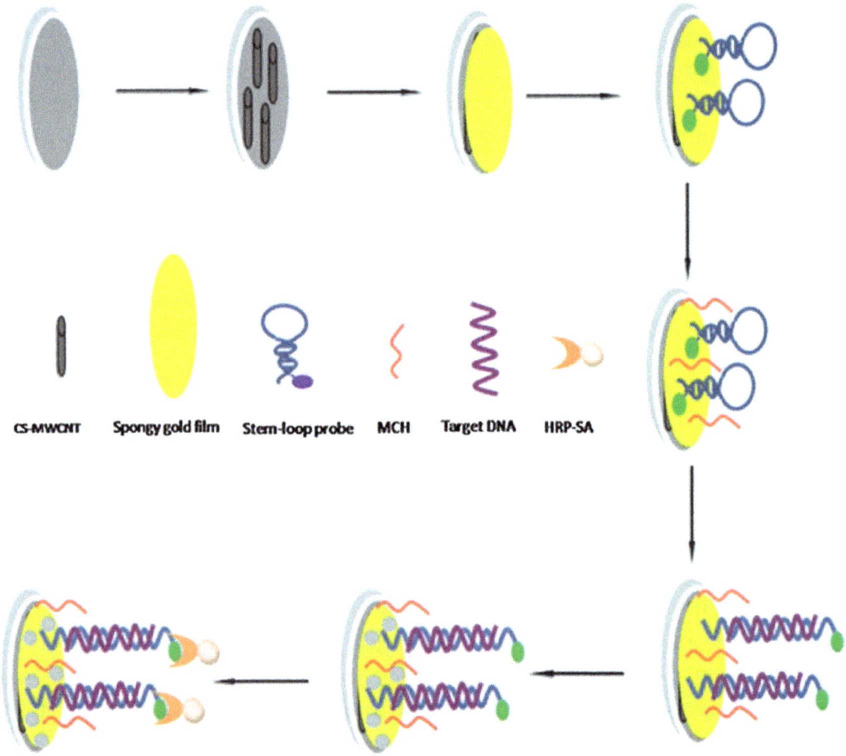

Figure 9.12 The fabrication and sensing mechanisms of the Ara h1 allergen using CNT-based stem-loop DNA biosensors.
Reprinted from *Talanta*, Vol 131, X. Sun, M. Jia, J. Ji, L. Guan, Y. Zhang, L. Tang and Z. Li, Enzymatic amplification detection of peanut allergen Ara h1 using a stem-loop DNA biosensor modified with a chitosan-mutiwalled carbon nanotube nanocomposite and spongy gold film, 521–527, Copyright 2015, with permission from Elsevier.

molecules are then added to the mix. The HRP-SA molecules are attracted to the biotin. As the HRP-SA attach to the biotin, a redox reaction takes place, which can be measured. The recorded signals from the electrochemical measurement system can therefore indicate the presence of Ara h1 DNA.

9.3.2 Summary of CNT Sensors in Recent Literature

For the past two decades, researchers have shown great interest in the unique properties of CNTs and have explored various applications for this material. One of the emerging fields for CNT research is in the construction of biosensors used for the food industry. There have been increasing numbers of publications each year with the emphasis on CNT sensors for food safety monitoring and quality control. Table 9.1 is a representative sample of

Table 9.1 Summary of the recent CNT-based biosensors developed and used for food safety related sensing applications.

Sensor	Technique	Analyte	Limit	Range	Ref.
pH	Electrochemical	pH		1 to 13	45
pH	Electrical	pH		2 to 12	48
Pesticide	Electrochemical	Paraoxon & malaoxon	2 ppb	2 to 50 ppb	82
Pesticide	Electrochemical	Paraoxon & methyl parathion	0.15 μM	0.15 to 10 μM	83
Metal ion	Electrochemical	Cadmium & lead	0.04 $\mu g\,L^{-1}$		84
Metal ion	Electrochemical	Mercury ion	0.5 nM	0.5 nM to 0.7 mM	37
Gas	Electrochemical	Organic vapour	44 ppb		85
Gas	Electrochemical	Methanol	1% by volume		51
Bacteria	Electrical	*E. coli*	10^2 CFL ml^{-1}	10^2 to 10^5 CFL ml^{-1}	60
Bacteria	Electromechanical	*E. coli*	10^{-24} kg		54
Bacteria	Electrochemical	*E. coli*	0.2 CFU ml^{-1}	0.2 to 10^3 CFU ml^{-1}	86
Bacteria	Electrochemical	*E. coli*	0.15 $\mu g\,ml^{-1}$	0.15 to 25 $\mu g\,ml^{-1}$	61
Bacteria	Electrical	*E. coli*	8.2×10^2 CFL ml^{-1}	8.2×10^2 to 8.2×10^8 CFL ml^{-1}	63
Bacteria	Electrical	*Salmonella*	10^3 CFU ml^{-1}	10^2 to 10^8 CFL ml^{-1}	64
Bacteria	Electrochemical	*Salmonella*	10^3 CFU ml^{-1}	10^2 to 10^7 CFL ml^{-1}	66
Virus	Electrical	H1N1	1 PFU ml^{-1}	1 to 10^4 PFU ml^{-1}	71
Virus	Electrical	H1N1	1.77×10^2 $TCID_{50}\,ml^{-1}$	10^3 to 10^5 $TCID_{50}\,ml^{-1}$	70
Virus	Electrochemical	Hepatitis C	0.1 fM	0.1 fM to 1 pM	87
Protein	Electrochemical	Human cellular prions	0.5 pM	0.5 pM to 10 μM	88
Protein	Fluorescence	Protein	1 fM		76
DNA	Electrical	DNA	10^{-13} M	10^{-6} to 10^{-13} M	78
DNA	Electromechanical	DNA	50 nM	50 to 5×10^3 nM	89
DNA	Electrochemical	Ara h1 peanut allergen	1.3×10^{-17} $mol\,l^{-1}$	3.91×10^{-17} to 1.25×10^{-15} $mol\,l^{-1}$	80
DNA	Electrochemical	p53 gene	0.32 μM	0.32 to 0.96 μM	90
DNA	Electrochemical	DNA	10 pM	10 pM to 0.1 μM	91

these sensors. It summarizes some of the existing research on CNT sensors in the recent literature. The table compares the sensor types, the sensing techniques used by the sensor, the target analytes, the detection limit and the detection range of the sensors. These results have greatly improved the technologies and methods used in the food industry. They are not only proofs of the current progress, but also indicators that CNT-based biosensors will provide a great technological leap in the near future.

9.4 Conclusion

The inclusion of CNTs in sensors has been proven to be effective in enhancing the sensor performance. CNTs have been studied and utilized in a wide spectrum of sensors that can be used for the detection of foodborne materials. The increased accuracy and effectiveness have made CNTs a promising sensing material for low-cost, portable and powerful sensors. A great number of techniques have been employed in developing CNT-based biosensors. In many of these cases, the use of CNTs in sensors has overcome the bottleneck of many existing detection methods being only accessible to specialized medical laboratories with highly skilled technicians. The unique physical and structural properties of CNTs have enabled high performance in sensors, in terms of detection limit, sensitivity, sensing range and response time. Many of the existing sensing principles used in other fields, such as electrical characteristic changes caused by additional materials, specific binding between various biomolecules, mechanical vibration analysis to determine mass and fluorescence mapping to determine various molecules, have adopted CNTs in their designs and shown significant performance enhancement.

New applications of CNT-based devices have been continuously investigated, showing great and broad impact on many scientific fields. For sensing applications, CNT sensors have expanded from the early fields of gas sensing and chemical sensing to other critical areas including the food industry. Many types of illness-causing substances in foods can be detected using CNT-based sensors. These substances may come from simple chemicals that are related to pH changes in solvents, gases that indicate the elapsed storage time of stored foods, bacteria that cause life-threatening illnesses, viruses that lead to stomach flu and allergens that can result in severe allergic reactions. In addition, new research efforts are constantly finding innovative ways to use CNT-based sensors in food safety and quality analysis. For example, new metal ion sensors can detect specific metals in foods, which allow better control over the heavy metal concentrations in foods.[84,92] Powerful sensors that can distinguish between metals are also needed. At present, two different sensing techniques that can selectively sense mercury by CNTs have been developed.[37,93] Furthermore, as the detection of pesticide is becoming increasingly important due to the fact that farming relies more on pesticides to maximize crop yields, CNT-based sensors that are capable of detecting multiple pesticides are needed.[82,83]

These foodborne substances vary in many ways: from sub-nm to mm scales in size, from ions to molecules, from protein or DNA to the entire organisms. The sensors need to be designed and prepared accordingly to meet these requirements. Because of the versatility and customizability of CNT sensors, using CNTs in food safety and quality control can potentially revolutionize the food industry. The combinations of CNTs and smart sensor designs have proven to be highly effective and are expected to be a major technological leap in food sensing in the near future. The enhancement in accuracy, ease of use and low cost of these sensors will have significant impact on portable systems. With the extensive research on CNTs and sensors in recent years, it is anticipated that new knowledge will be continuously generated and CNT sensors will become more useful in the food industry.

References

1. T. Dürkop, S. A. Getty, E. Cobas and M. S. Fuhrer, *Nano Lett.*, 2004, **4**, 35–39.
2. T. Dürkop, B. M. Kim and M. S. Fuhrer, *J. Phys.: Condens. Matter*, 2004, **16**, R553–R580.
3. A. Javey, H. Kim, M. Brink, Q. Wang, A. Ural, J. Guo, P. McIntyre, P. McEuen, M. Lundstrom and H. Dai, *Nat. Mater.*, 2002, **1**, 241–246.
4. P. G. Collins and P. Avouris, *Sci. Am.*, 2000, 62–69.
5. P. G. Collins, M. S. Arnold and P. Avouris, *Science*, 2001, **292**, 706–709.
6. C. Gao, Z. Guo, J.-H. Liu and X.-J. Huang, *Nanoscale*, 2012, **4**, 1948–1963.
7. W. Yang, K. R. Ratinac, S. P. Ringer, P. Thordarson, J. J. Gooding and F. Braet, *Angew. Chem. Int. Ed.*, 2010, **49**, 2114–2138.
8. C. B. Jacobs, M. J. Peairs and B. J. Venton, *Anal. Chim. Acta*, 2010, **662**, 105–127.
9. Q. Wang and B. Arash, *Comput. Mater. Sci.*, 2014, **82**, 350–360.
10. J. Wang, *Electroanalysis*, 2005, **17**, 7–14.
11. Centers for Disease Control and Prevention (CDC), *Burden of Foodborne Illness: Findings*, 2011.
12. S. Li, A. Simonian and B. A. Chin, *Electrochem. Soc. Interface*, 2010, **19**, 41–46.
13. N. Wang, N. Zhang and M. Wang, *Comput. Electron. Agric.*, 2006, **50**, 1–14.
14. H. W. Zhu, C. L. Xu, D. H. Wu, B. Q. Wei, R. Vajtai and P. M. Ajayan, *Science*, 2002, **296**, 884–886.
15. C. N. R. Rao, B. C. Satishkumar, A. Govindaraj and M. Nath, *ChemPhysChem*, 2001, **2**, 78–105.
16. Y. Wu, M. Huang, F. Wang, X. M. H. Huang, S. Rosenblatt, L. Huang, H. Yan, S. P. O'Brien, J. Hone and T. F. Heinz, *Nano Lett.*, 2008, **8**, 4158–4161.
17. S. Gupta, K. Dharamvir and V. K. Jindal, *Phys. Rev. B*, 2005, **72**, 165428.

18. L. Deng, S. J. Eichhorn, C.-C. Kao and R. J. Young, *ACS Appl. Mater. Interfaces*, 2011, **3**, 433–440.
19. S. Niyogi, M. A. Hamon, H. Hu, B. Zhao, P. Bhowmik, R. Sen, M. E. Itkis and R. C. Haddon, *Acc. Chem. Res.*, 2002, **35**, 1105–1113.
20. F. Patolsky and C. M. Lieber, *Mater. Today*, 2005, **8**, 20–28.
21. J.-i. Hahm and C. M. Lieber, *Nano Lett.*, 2003, **4**, 51–54.
22. Y. Cui, Q. Wei, H. Park and C. M. Lieber, *Science*, 2001, **293**, 1289–1292.
23. A. Star, J.-C. P. Gabriel, K. Bradley and G. Grüner, *Nano Lett.*, 2003, **3**, 459–463.
24. K. Bradley, J.-C. P. Gabriel, A. Star and G. Grüner, *Appl. Phys. Lett.*, 2003, **83**, 3821–3823.
25. F. Patolsky, G. Zheng and C. M. Lieber, *Anal. Chem.*, 2006, **78**, 4260–4269.
26. E. Stern, J. F. Klemic, D. A. Routenberg, P. N. Wyrembak, D. B. Turner-Evans, A. D. Hamilton, D. A. LaVan, T. M. Fahmy and M. A. Reed, *Nature*, 2007, **445**, 519–522.
27. P. R. Nair and M. A. Alam, *Nano Lett.*, 2008, **8**, 1281–1285.
28. P. E. Sheehan and L. J. Whitman, *Nano Lett.*, 2005, **5**, 803–807.
29. S. V. Dzyadevych, A. P. Soldatkin, Y. I. Korpan, V. N. Arkhypova, A. V. El'skaya, J.-M. Chovelon, C. Martelet and N. Jaffrezic-Renault, *Anal. Bioanal. Chem.*, 2003, **377**, 496–506.
30. Y. Miao, J. Guan and J. Chen, *Biotechnol. Adv.*, 2003, **21**, 527–534.
31. P. Bergveld, *Sens. Actuators, B*, 2003, **88**, 1–20.
32. A. B. Kharitonov, M. Zayats, A. Lichtenstein, E. Katz and I. Willner, *Sens. Actuators, B*, 2000, **70**, 222–231.
33. J. Kong, N. R. Franklin, C. Zhou, M. G. Chapline, S. Peng, K. Cho and H. Dai, *Science*, 2000, **287**, 622–625.
34. Y. Wong, *Sens. Actuators B*, 2003, **93**, 327–332.
35. T. Xu, N. Scafa, L.-P. Xu, L. Su, C. Li, S. Zhou, Y. Liu and X. Zhang, *Electroanalysis*, 2014, **26**, 449–468.
36. M. Mittal and A. Kumar, *Sens. Actuators, B*, 2014, **203**, 349–362.
37. H. R. Rajabi, M. Roushani and M. Shamsipur, *J. Electroanal. Chem.*, 2013, **693**, 16–22.
38. E. Katz and I. Willner, *ChemPhysChem*, 2004, **5**, 1084–1104.
39. D. Cai, L. Ren, H. Zhao, C. Xu, L. Zhang, Y. Yu, H. Wang, Y. Lan, M. F. Roberts, J. H. Chuang, M. J. Naughton, Z. Ren and T. C. Chiles, *Nat Nano*, 2010, **5**, 597–601.
40. A. Hirsch, *Angew. Chem., Int. Ed.*, 2002, **41**, 1853–1859.
41. E. Pungor and K. Toth, *Analyst*, 1970, **95**, 625–648.
42. E. Bakker and E. Pretsch, *Anal. Chem.*, 2002, **74**, 420 A–426 A.
43. W. Xue and T. Cui, *Sens. Actuators, B*, 2008, **134**, 981–987.
44. D. Lee and T. Cui, *Biosens. Bioelectron.*, 2010, **25**, 2259–2264.
45. M. Kaempgen and S. Roth, *J. Electroanal. Chem.*, 2006, **586**, 72–76.
46. J. L. Kenneth, K. Junhee, P. L. Jerome, K. Nadine Wong Shi and A. K. Nicholas, *Smart Mater. Struct.*, 2007, **16**, 429.
47. H. Kwon, M. Kim, B. Meany, Y. Piao, L. R. Powell and Y. Wang, *J. Phys. Chem. C*, 2015, **119**, 3733–3739.

48. P. Gou, N. D. Kraut, I. M. Feigel, H. Bai, G. J. Morgan, Y. Chen, Y. Tang, K. Bocan, J. Stachel, L. Berger, M. Mickle, E. Sejdic and A. Star, *Sci. Rep.*, 2014, **4**, 4468.
49. Y. Wang and J. T. W. Yeow, *J. Sens.*, 2009, **2009**, 24.
50. N. Van Hieu, L. T. B. Thuy and N. D. Chien, *Sens. Actuators, B*, 2008, **129**, 888–895.
51. C. Wongchoosuk, A. Wisitsoraat, A. Tuantranont and T. Kerdcharoen, *Sens. Actuators, B*, 2010, **147**, 392–399.
52. H. Yu and J. Wang, *Sens. Actuators, B*, 2007, **122**, 134–140.
53. J. S. Vestergaard, M. Martens and P. Turkki, *LWT - Food Sci. Technol.*, 2007, **40**, 1095–1101.
54. R. Chowdhury, S. Adhikari and J. Mitchell, *Phys. E*, 2009, **42**, 104–109.
55. A. Y. Joshi, S. C. Sharma and S. P. Harsha, *Sensor. Actuators, A*, 2011, **168**, 275–280.
56. H.-L. Lee, J.-C. Hsu and W.-J. Chang, *Nanoscale Res. Lett.*, 2010, **5**, 1774–1778.
57. J. Moser, J. Guttinger, A. Eichler, M. J. Esplandiu, D. E. Liu, M. I. Dykman and A. Bachtold, *Nat. Nano*, 2013, **8**, 493–496.
58. A. Wagner and D. A. Fell, *Proc. R. Soc. London, Ser. B*, 2001, **268**, 1803–1810.
59. K. Ramotar, J. Conly, H. Chubb and T. Louie, *J. Infect. Dis.*, 1984, **150**, 213–218.
60. K. Yamada, C.-T. Kim, J.-H. Kim, J.-H. Chung, H. G. Lee and S. Jun, *PLoS One*, 2014, **9**, e105767.
61. K. Kerman, Y. Morita, Y. Takamura and E. Tamiya, *Anal. Bioanal. Chem.*, 2005, **381**, 1114–1121.
62. H.-M. So, D.-W. Park, E.-K. Jeon, Y.-H. Kim, B. S. Kim, C.-K. Lee, S. Y. Choi, S. C. Kim, H. Chang and J.-O. Lee, *Small*, 2008, **4**, 197–201.
63. X. Zhang, D. Wang, D. Yang, S. Li and Z. Shen, *Engineering*, 2013, **4**, 94.
64. M. B. Lerner, B. R. Goldsmith, R. McMillon, J. Dailey, S. Pillai, S. R. Singh and A. T. C. Johnson, *AIP Adv.*, 2011, **1**, 042127.
65. B. Stephen Inbaraj and B. H. Chen, *J. Food Drug Anal.*, 2016, **24**, 15–28.
66. M. Yang, Z. Peng, Y. Ning, Y. Chen, Q. Zhou and L. Deng, *Sensors*, 2013, **13**, 6865.
67. European Food Safety Authority and European Centre for Disease Prevention and Control, *EFSA J.*, 2014, **12**, 3547.
68. M. Lu, D. Lee, W. Xue and T. Cui, *Sens. Actuators, A*, 2009, **150**, 280–285.
69. J. V. Veetil and K. Ye, *Biotechnol. Prog.*, 2007, **23**, 517–531.
70. D. Lee, Y. Chander, S. M. Goyal and T. Cui, *Biosens. Bioelectron.*, 2011, **26**, 3482–3487.
71. R. Singh, A. Sharma, S. Hong and J. Jang, *Analyst*, 2014, **139**, 5415–5421.
72. C. Li, M. Curreli, H. Lin, B. Lei, F. N. Ishikawa, R. Datar, R. J. Cote, M. E. Thompson and C. Zhou, *J. Am. Chem. Soc.*, 2005, **127**, 12484–12485.
73. B. L. Allen, P. D. Kichambare and A. Star, *Adv. Mater.*, 2007, **19**, 1439–1451.
74. K. Atsuhiko, H. Chan Kyeong, K. Takafumi, M. Masatoshi and M. Kazuhiko, *Jpn. J. Appl. Phys.*, 2005, **44**, 1596.

75. Z. Liu, S. Tabakman, K. Welsher and H. Dai, *Nano Res.*, 2009, **2**, 85–120.
76. Z. Chen, S. M. Tabakman, A. P. Goodwin, M. G. Kattah, D. Daranciang, X. Wang, G. Zhang, X. Li, Z. Liu, P. J. Utz, K. Jiang, S. Fan and H. Dai, *Nat. Biotechnol.*, 2008, **26**, 1285–1292.
77. E. Paleček, *Electroanalysis*, 2009, **21**, 239–251.
78. X. Dong, C. M. Lau, A. Lohani, S. G. Mhaisalkar, J. Kasim, Z. Shen, X. Ho, J. A. Rogers and L.-J. Li, *Adv. Mater*, 2008, **20**, 2389–2393.
79. D. Fu and L.-J. Li, *Nano Rev.*, 2010, **1**, 5354.
80. X. Sun, M. Jia, J. Ji, L. Guan, Y. Zhang, L. Tang and Z. Li, *Talanta*, 2015, **131**, 521-527.
81. X. Mao, J. Jiang, X. Xu, X. Chu, Y. Luo, G. Shen and R. Yu, *Biosens. Bioelectron.*, 2008, **23**, 1555–1561.
82. A. N. Ivanov, R. R. Younusov, G. A. Evtugyn, F. Arduini, D. Moscone and G. Palleschi, *Talanta*, 2011, **85**, 216–221.
83. R. P. Deo, J. Wang, I. Block, A. Mulchandani, K. A. Joshi, M. Trojanowicz, F. Scholz, W. Chen and Y. Lin, *Anal. Chim. Acta*, 2005, **530**, 185–189.
84. G. Liu, Y. Lin, Y. Tu and Z. Ren, *Analyst*, 2005, **130**, 1098–1101.
85. J. Li, Y. Lu, Q. Ye, M. Cinke, J. Han and M. Meyyappan, *Nano Lett.*, 2003, **3**, 929–933.
86. G. A. Zelada-Guillén, J. Riu, A. Düzgün and F. X. Rius, *Angew. Chem., Int. Ed.*, 2009, **48**, 7334–7337.
87. B. Zribi, E. Roy, A. Pallandre, S. Chebil, M. Koubaa, N. Mejri, H. Magdinier Gomez, C. Sola, H. Korri-Youssoufi and A.-M. Haghiri-Gosnet, *Biomicrofluidics*, 2016, **10**, 014115.
88. A. Miodek, G. Castillo, T. Hianik and H. Korri-Youssoufi, *Anal. Chem.*, 2013, **85**, 7704–7712.
89. T. Wang, Z.-Z. Zheng, X.-X. Feng, H. Chen and M. Lu, *Biomed. Microdevices*, 2015, **17**, 1–7.
90. G. Congur, M. Plucnara, A. Erdem and M. Fojta, *Electroanalysis*, 2015, **27**, 1579–1586.
91. S. Shahrokhian, R. Salimian and H. R. Kalhor, *RSC Adv.*, 2016, **6**, 15592–15598.
92. J. Morton, N. Havens, A. Mugweru and A. K. Wanekaya, *Electroanalysis*, 2009, **21**, 1597–1603.
93. L. R. Pokhrel, N. Ettore, Z. L. Jacobs, A. Zarr, M. H. Weir, P. R. Scheuerman, S. R. Kanel and B. Dubey, *Sci. Total Environ.*, 2017, **574**, 1379–1388.

CHAPTER 10

Graphene-electrochemical Sensing in Food Safety and Quality Analysis

XIAN ZHANG

Department of Materials Design and Innovation, School of Engineering and Applied Sciences, University at Buffalo, 311 Bell Hall, Buffalo, NY 14260-5030, USA
Email: xazhang@buffalo.edu

10.1 Introduction

The food industry is important because it is considered as one of the most essential components to our daily life and is worth trillions of dollars according to the world population. It is also related to a collective of diverse businesses, such as agriculture, manufacturing, marketing, public health and others. Although significant attention has been paid to food safety regulation and inspection, foodborne illness is a major problem for public health and a burden that can lead to significant costs of health care. According to the World Health Organization (WHO) and Centers for Disease Control and Prevention (CDC), there are over 76 million cases of foodborne illnesses in the United States alone, which result in around 325 000 cases of hospitalizations and 5000 deaths per year. In the United States, foodborne illness results in a $77.7 billion economic burden every year for medical costs and productivity losses.

There are several important and indispensable steps for the food industry, including processing of raw food products, nutritional values, labelling and

Food Chemistry, Function and Analysis No. 2
Sensing Techniques for Food Safety and Quality Control
Edited by Xiaonan Lu

Published by the Royal Society of Chemistry, www.rsc.org

packaging, integrated resource management, from the farm to the table, the customer service chain and food preparation at home. It is essential to pay attention to food safety and quality in all of these steps because they are the priorities for all consumers.

Among all these steps in the food industry, agriculture and food processing are the two most important pillars for food safety and quality. Agriculture is the very first step of the food industry and it is the process that produces agri-food products. It covers research works in the areas of plant genetics, plant physiology, meteorology and soil science. It is sensitive to agricultural environment, such as soil, water and air, and extremely easy to be affected by these factors. For example, on-farm contamination can threaten the safety and quality of crops. Moreover, pesticides, herbicides, insecticides and fungicides are commonly used to improve agricultural productivity, but this can generate significant toxicity if their residues remain in the food chain.[1] Therefore, it is critical to evaluate the safety and quality of agri-food products before they are delivered to markets and consumers.

Food processing techniques are those that can be used to transform raw crops into food products for human consumption. By performing food processing, potentially hazardous chemicals and contamination may be introduced into the final food products. For example, preservatives, colorants and other additives are intentionally supplemented into agri-food products to enhance the appearance to consumers. Unfortunately, overdosing some of these chemicals may lead to health risks to consumers. In January 2013, the US Food and Drug Administration (FDA) published a proposed rule on the preventive control of hazards and contaminants in agri-food products. This plan includes hazard analysis, preventive control, monitoring procedure and other items. Among them, hazard analysis is the crucial and the most challenging section to be handled. In addition, direct detection of chemical and microbiological hazards is still complicated. Recently, standard food safety analysis detection has gradually caused problems due to its limited capability to determine all types of food contaminations, especially emerging food contamination and adulteration.[2]

Therefore, it is highly important to employ accurate and reliable sensing and analytical methods to detect chemical and biological hazards in agricultural and food products. Many detection methods have been taken into consideration or developed, such as optical-based methods, chromatographic-based methods and spectroscopic-based techniques. Most analyses of food contaminants are performed in laboratories with a limited number of samples that can be tested. Therefore, more effective and high-throughput methods are highly in demand. Several enzyme-based and bio-affinity assays have been reported.[3–6] Although there has been significant progress in these assays during the past several decades, these methods are still complicated and costly, and thus not applicable for wide use in industry.

Due to their popularity in various research areas, nanomaterials have been increasingly used in sensing a variety of substances in recent years.[7,8] They can increase functionality, effectiveness, portability, stability, selectivity and sensitivity of the constructed sensors and analytical measurement technologies. Nanomaterials have been widely used in electronics, sensing, biomaterials and catalysis.[9–11] Recently, their application has been expanded to the agri-food industry, especially for food analysis.[12]

Applications of nanomaterials in the food industry are highly diversified and dynamic. For example, titanium dioxide (TiO_2) and silver nanoparticles (AgNPs) are used as antimicrobial agents to build storage containers for food and beverages. In addition, gold nanoparticles (AuNPs),[13,14] Au nanorods[15,16] and carbon nanotubes (CNTs)[17] have been widely applied to detect the presence of gases, aromas, chemical contaminants and pathogens. Nanomaterial-based encapsulation and delivery systems have also been developed to increase the flavour and shelf-life of food products while other nanotechnologies (*e.g.*, radio frequency identification (RFID)) are applied for tracking, tracing and brand protection.[12,18] Nano-biosensors have the potential to improve the speed as well as the sensitivity in detecting various food contaminants in a trace amount scale.[19] The increasing development of the synthesis and control of sizes, surface properties and textures of nanomaterials aids in the development of sensing systems using various types of nanotechnology. For example, the electrochemical method[20] and the colorimetric method[21] are now integrated into low-cost platforms. Because of their portability, low cost and user-friendliness, they have advantages over the traditional methods in the areas of biomedical diagnosis,[22] environmental monitoring and food safety and quality control.[23] For example, using screen-printed carbon electrodes (SPCE) as the portable platform has been reported to sense environmental pollutants and food quality factors by using electrochemical methods.[24]

However, there is a potential side effect from the increased use of these nanomaterials and nanoparticles because they have potential toxicity and generate a negative impact on human health. For example, TiO_2 nanoparticles used in chewing gums have a toxicity effect.[25] Several studies suggested that these nanoparticles could pass through the barriers of the gastrointestinal tract and gradually accumulate in the organs.[26] In addition, wide use of nanomaterials and nanoparticles may enhance their distribution to the environment, soil and finally agri-food products. Due to the technical challenges in detection and readiness to distribute and accumulate, it is important to confirm the safety of these nanomaterials before they are applied in sensing agri-food products.

Graphene is one of the most popular nanomaterials used in engineering and chemistry. Because of its simple and pure composition of carbon, there is no chemical or biological toxicity derived from graphene. Graphene, together with graphite, fullerenes, carbon nanotubes and amorphous carbon, are all isotopes of carbon and have the same "non-toxic" properties. This group of materials is thus extensively used as biosensors, immobilization matrices, stabilizers and mediators. Therefore, there are no side effects from using it as a sensor.

Graphene is an isolated single atomic layer of graphite, with a hexagonal structure of carbon atoms. Ever since its discovery in 2004 by Andre Geim and Konstantin Novoselov at the University of Manchester,[27] graphene has been widely used in almost every aspect in academic research, such as physics, chemistry, biology, electro-chemical analysis and engineering, in terms of thermal, optical, electrical and mechanical aspects. The extraordinary electronic transport properties of graphene have been reported in numerous studies.[28,29] It possesses high electron mobility (250 000 $cm^2 V s^{-1}$) and remarkable thermal conductivity (5000 $W m^{-1} K^{-1}$), both of which are exceptional properties with values significantly higher than the most conductible metals. It also possesses large specific surface area (>100 $m^2 g^{-1}$), excellent electron transfer ability and good biocompatibility.[30] In the past decade, there were numerous publications about the application of graphene in science, engineering and technology.[31,32] At the same time, graphene has been studied to ensure food quality and safety.[33,34] In brief, it is a new sorbent for biological and food samples.[35] It is a better option than other nanomaterials because they are easily synthesized, cheap and possess a high recovery.

Electrochemistry is the study of chemical processes that cause electrons to move. Its mechanism is shown in Figure 10.1.

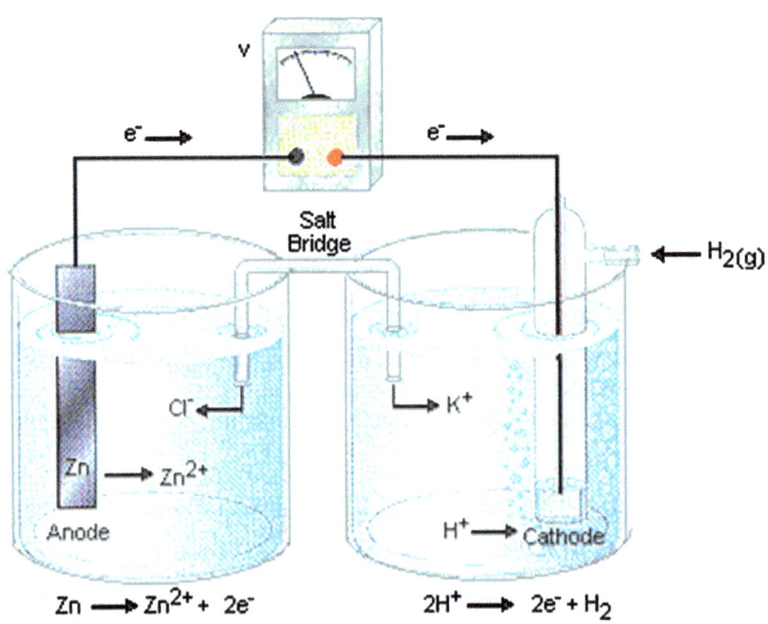

Figure 10.1 Electrochemical reaction, with an example of electricity being generated by Zn metal reacted with KCl solution.
Reproduced with permission from Professor George Bodner at Purdue University.

Specifically, an oxidation–reduction reaction generates moving electrons with the following formula:

$$Zn(s) + 1H^{+}(aq) \rightarrow Zn^{2+}(aq) + H_2(g)$$

Electrochemistry has been used as an effective sensing technique in the area of food safety and quality. An electrochemical sensor consists of a diffusion barrier, a sensing electrode (also called "working electrode", "measuring electrode" or "anode"), a counter electrode (also called "cathode") and an electrolyte. The structure of an electrochemical sensor is shown in Figure 10.2.

The reactions include electric charges moving between two electrodes and an electrolyte. It is the interaction between electrical energy and chemical change. A chemical reaction is called an electrochemical reaction when it is caused by an externally supplied current, and in electrolysis as in a battery. The electrons are transferred directly between molecules or atoms, from one electrode to another electrode, and these phenomena are called "oxidation–reduction reactions". For example, when an atom of sodium reacts with an atom of chlorine, sodium gives one electron and attains an oxidation state of +1. Chlorine accepts the electron and its oxidation state is then reduced to −1. The sign of the oxidation state (positive or negative) actually corresponds to the value of each ion's electronic charge. The attraction of the differently charged sodium and chlorine ions is the reason that they can form an ionic bond from this process.

There are various important electrochemical processes in both research and industry, such as coating of objects with metals through electrodeposition. Moreover, the generation of chemical energy through photosynthesis is also an electrochemical process. There is also a wide range of new emerging technologies, such as fuel cells, the new generation of lithium

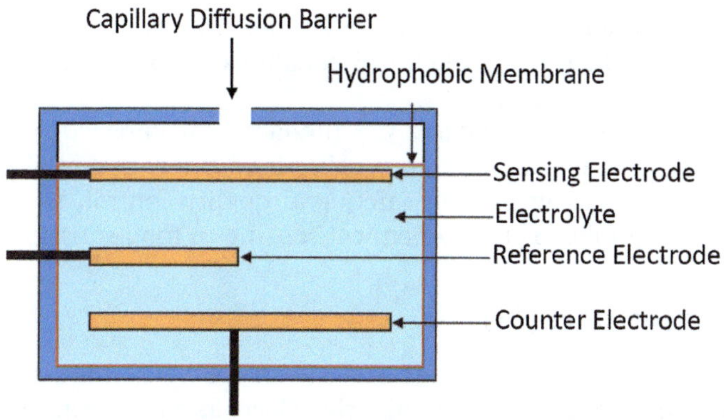

Figure 10.2 Schematic illustration of a basic electrochemical sensing system, including sensing electrode, reference electrode, counter electrode, electrolyte, hydrophobic membrane and capillary diffusion barrier.

batteries, electrochemical reactors and super-capacitors, all of which draw research attention and have become increasingly commercialized. In addition, electrochemistry has important applications in the food industry, such as the assessment of the interaction between food products and packaging materials,[36] the analysis of milk composition,[37] the characterization and determination of the freezing end-point of ice-cream mixture and the determination of free acidity in olive oil.[38]

Electrochemical sensors have advantages over other sensors in the aspects of high sensitivity, rapidity, ease of use, easy miniaturization (*e.g.*, microelectrodes) and low detection limit. They can be designed as compact and portable units for on-line and in-field uses.[39,40] They can be applied over a large temperature range (from −30 °C to 1600 °C) to gain *in situ* information in real sample matrices, such as cell cultures, body fluids, food samples, pharmaceutical samples and environmental samples. For example, Guth and coworkers[41] reported a conventional sensor with liquid electrolyte at 140 °C and a solid-electrolyte-assisted sensor withstanding up to 500 °C. Compared to most other analytical techniques, equipment for electrochemical analysis is relatively simple in its setup and has low cost because of its simple production (*e.g.*, microelectronic circuits), straightforward electronic read-out and ease of data processing. Liu and coauthors[42] reported a developed technique with reusable electrodes for multiple tests by storing at −4 °C in buffer for about two weeks and by easily regenerating with sorbitol or acidic buffers. Screen-printed electrodes that combine working, reference and counter electrodes in one set have been commercially available because of their reusable and portable applications. Various electrochemical sensors for multiplex detection were also developed and reported in several other publications.[43,44,107] In general, electrochemical sensors have fast sensing responses, good stability, portability and low cost. Electrochemical biosensors can also minimize or eliminate the need for tedious sample pre-treatments. Specifically, liquid samples can be either directly tested or diluted accordingly while solid samples are either solubilized or suspended in solutions. Moreover, it is user friendly. Taken together, using graphene materials to conduct electrochemical sensing of food safety and quality control is highly promising.

This chapter aims to overview the fundamental knowledge about graphene, application of graphene in sensing food safety and quality control, electrochemical sensing in food safety and quality control, as well as the prospective of graphene-electrochemical sensing in food safety and quality.

10.2 Nanomaterials

In general, nanomaterials are materials with a single unit sized between 1 and 100 nm. This size is within the electron microscope range, and nanomaterials are just slightly larger than atoms (Figure 10.3).

Nanomaterials can exist everywhere in our daily life. Common examples include the gecko's foot, the microscopic image of a butterfly wing, viral

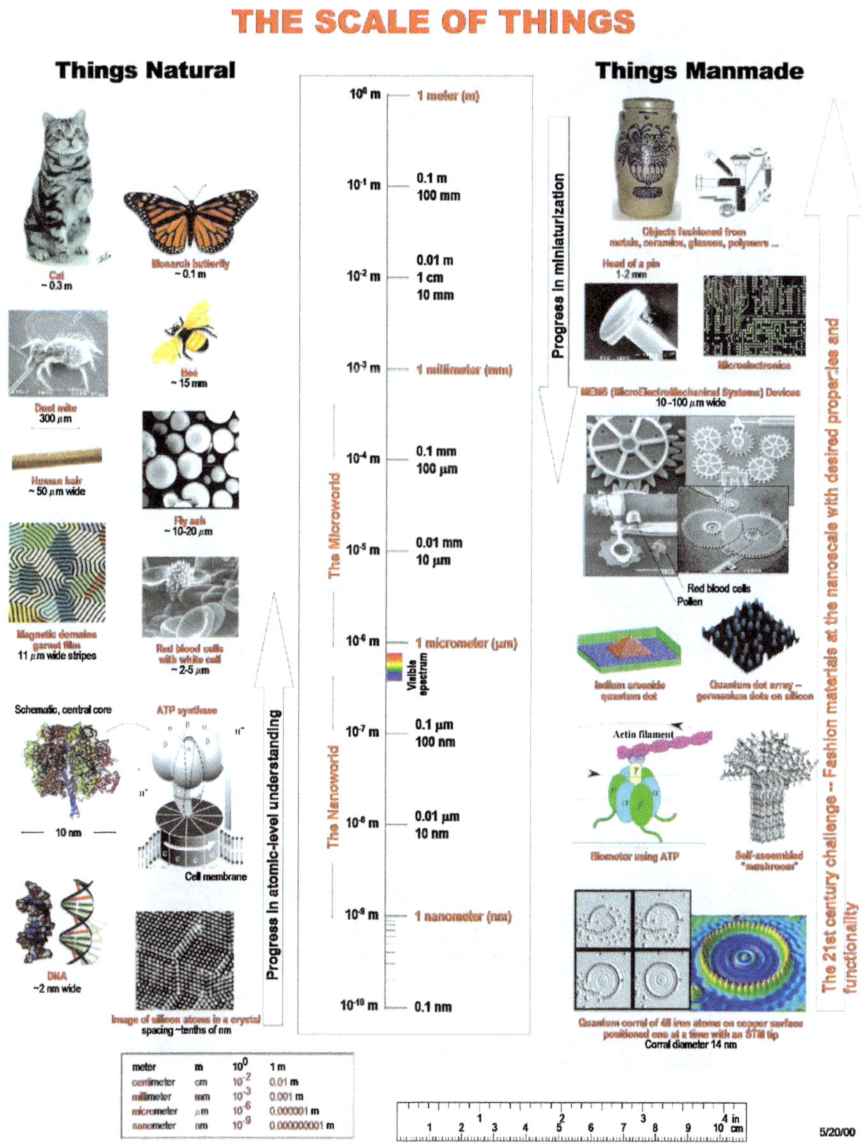

Figure 10.3 Size scale of different materials.
Reproduced with permission from Professor Ray Carpenter and Professor Andrew Chizmeshya at Arizona State University.

capsid with a size of a few nanometres (Figure 10.4) and the surface of a lotus leaf, Brazilian crystal opal, the Lycurgus cup in the 4th century. The colour of Brazilian crystal opal is caused by the interference and diffraction of light between silica spheres, which are a few nanometres in diameter. The Lycurgus cup contains gold and silver particles of 70 nm and presents the dichroic effect.

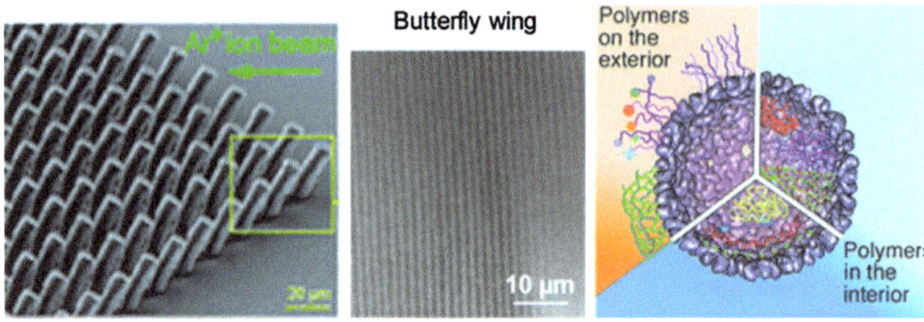

Figure 10.4 Examples of nano-scaled materials in nature.
Left: Reproduced from M.-W. Moon, T.-G. Cha, K.-R. Lee, A. Vaziri and H.-Y. Kim, *Soft Matter*, 2010, **6**, 3924–3929, with permission from the Royal Society of Chemistry. Middle: Reproduced from M. Rother, M. G. Nussbaumer, K. Renggli and N. Bruns, *Chem. Soc. Revs*, 2016, **45**, 6213–6249, with permission from the Royal Society of Chemistry. Right: Reproduced from G. D. Bixler and B. Bhushan, *Soft Matter*, 2012, **8**, 11271–11284, with permission from the Royal Society of Chemistry.

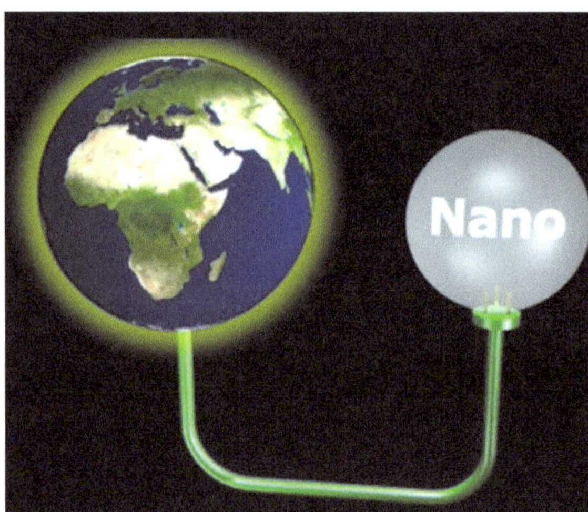

Figure 10.5 Powering the planet with energy nanomaterials.
Reproduced with permission from Professor Geoffrey Ozin at the University of Toronto.

Nanomaterials are popular in current research. Nanotechnology-related research takes place in the fields of material science, physics, chemistry, biology, electrical engineering, mechanical engineering, *etc*. There are usually well-developed methods for the synthesis, fabrication and characterization of nanomaterials. Nanotechnology has developed so rapidly that sometimes interesting and even crazy ideas are generated from using nanomaterials. For example, "Powering the planet with energy nanomaterials!" was proposed by Professor Geoffrey Ozin at the University of Toronto in 2011 (Figure 10.5).

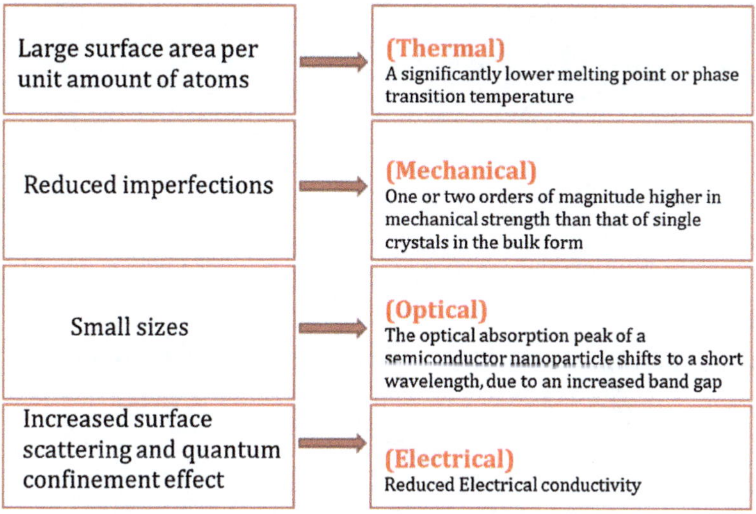

Figure 10.6 Overview of physical properties of nanomaterials, which are superior to conventional materials.

The unique structure of nanomaterials gives them very unique properties (Figure 10.6). For example, the large surface area per unit amount of atoms provides them with a significantly lower melting point or phase transition temperature. In addition, the reduced imperfections of nanomaterials make them one or two orders of magnitude higher in mechanical strength than that of the single crystals in bulk form. Moreover, the small sizes of nanomaterials make the optical absorption peak of a semiconductor nanoparticle shift to a short wavelength, which is due to an increased band gap. The increased surface scattering and quantum confinement effect of the nanomaterials can result in a reduced electrical conductivity as well.

The forms of nanomaterials include nanoparticles, nano-wires and two-dimensional materials (with one side extremely small and two other sides with normal lengths). For convenience, they are also called 0-dimensional, 1-dimensional and 2-dimensional materials. Currently carbon-based nanomaterials are popular for research because of their ease of synthesis, characterization and measurement, as well as excellent physical properties (*e.g.*, mechanical, electrical, optical and thermal properties). Common carbon-based nanomaterials for research include fullerene, carbon nanotubes and graphene.

10.3 Graphene

Carbon is one of the most abundant elements on Earth and it is vital in chemical reactions, biological systems and ecosystems. There is an increasing number of studies based upon carbon-derived materials in the applications of electrochemical sensors.

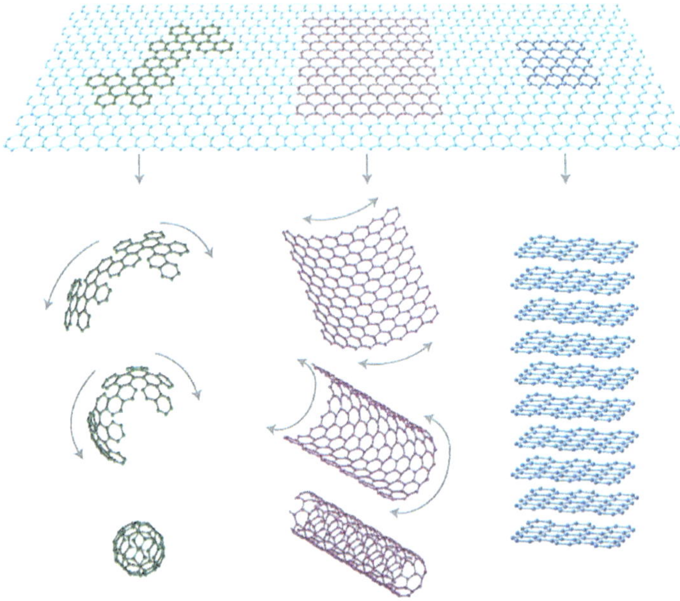

Figure 10.7 Structures of fullerene, carbon nanotubes and graphite, all of which can be viewed as formed by a single sheet of graphite – graphene. Reproduced with permission from Professor Andre Geim from the University of Manchester.

Figure 10.7 shows the typical structure of carbon-based materials include graphene, fullerene carbon nanotubes and graphite. Carbon, as an array of atoms in certain structural forms, exists in the forms of diamond, graphite, graphene, fullerenes, carbon nanotubes (CNTs) and amorphous carbon with a significant number of appealing properties for various purposes.

Before 1985, pure solid carbon was believed to exist in only two forms: diamond (Figure 10.8) and graphite. Diamond is formed by carbon atoms in a compact structure, thus possessing the robust mechanical property. Diamond has sp^3 hybridization with small size, close packing of atoms and shorter and stronger covalent bonding, giving it superior mechanical, optical and thermal properties. Formed by stacks of graphene sheets layered together, graphite, however, has sp^2 hybridization, resulting in excellent electrical conductivity. CNTs have a particular seamless structure from the sp^2 carbon units with hexagonal lattices, which is the same as graphene. They have closed topology and tubular structure typically in nano-scaled diameter and micron-scaled length.

In 1985, an interesting discovery of a new pure solid carbon, C_{60}, was achieved by a group of researchers: Richard Smalley and Robert Curl at Rice University in Houston and Harry Kroto at the University of Sussex in England. They used an intense pulse of laser light to vaporize a sample of graphite and brought the vaporized carbon by a stream of helium gas into

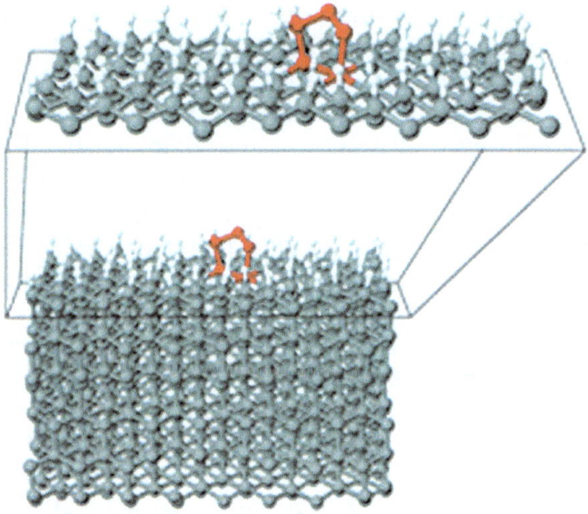

Figure 10.8 Diamond in compact structure.
Reproduced from ref. 45 with permission from the Royal Society of Chemistry.

the mass spectrometer, thus making the C_{60} molecules. Of its 32 faces, 12 were pentagons and 20 were hexagons, which are exactly like a soccer ball.

It is named after an architect, Buckminster Fuller, who designed the first geodomes, as shown in the picture of "spaceship earth" at the Epcot Center at Disney World, Florida. The C_{60} molecule was named "buckerminsterfullerene" or "buckyball" for short. After this discovery, the soccer-shaped carbon allotropes were discovered, such as C_{36}, C_{70}, C_{76} and C_{84}.

In 1991, a new shape of carbon allotrope, namely carbon nanotubes, with a shape of long cylindrical tubes, was discovered by Sumio Iijima in Japan. This is a brand new shape of nanomaterial, with a cylindrical diameter of a few nanometres and a length-to-diameter ratio up to over 1 million to 1, significantly larger than any other material. CNTs are one of the most interesting nanomaterials for sensor applications due to their unique properties, such as high electrical conductivity, chemical stability, large surface area, high surface/volume ratio, high mechanical strength and chemically modifiable surfaces. These interesting properties are related to their quasi-1D shape and sp^2 and π-bonding C atoms. There are two types of CNTs, namely multi-walled CNTs (MWCNTs) and single-walled CNTs (SWCNTs). MWCNTs are composed of concentric and closed graphite tubules, each of which is made of a rolled graphite sheet respectively with interspacing distances in between. In contrast, SWCNTs are made of a single graphite sheet rolled seamlessly, and have a tendency to aggregate in parallel to each other if not properly functionalized. Usually the diameter of a SWCNT is around 1 nanometre, while a MWCNT has a diameter of 2–5 nanometres. CNTs have been electrically used as semiconductors for a certain form of chirality and widely applied in electrochemical sensors to decrease over potential and improve sensitivity by enhancing electro-reactivity and promoting electron transfers. Therefore, CNT-based sensors generally have not only high sensitivities, but also lower detection limits and faster electron-transfer kinetics. Besides, CNT-modified electrodes show better bio-compatibility, such as coating biomolecules and alleviating surface fouling effects. Each individual CNT can act as a nano-electrode that facilitates electron transfers without any mediators. Further modification of vertically aligned CNTs (VACNTs) with nanoparticles (NPs) can not only increase the effective surface area and electrocatalytic activities but also provide continuous conducting pathways for the transportation of electrons, which is beneficial for the applications of electrochemical sensors and has a synergistic effect on sensing performances. The tube-like structure of carbon nanotubes can also be viewed as a rolled up graphene as shown in Figure 10.7.

Graphene has been extensively researched since 2004. It is a single-atom-thick planar sheet of hexagonally-arrayed sp^2-bonded carbon atoms packed in a 2D honeycomb crystal lattice. It is a two-dimensional building material for carbon materials of all other dimensionalities. Graphene can be wrapped up into 0D buckyballs, rolled into a 1D nanotube or stacked into 3D graphite. In general, graphene is a strong, thin and 2-dimensional material, with a one-atom-thick layer of carbon atoms in a hexagonal lattice. It is the allotrope of carbon. This unique structure offers graphene physical and chemical properties that are outstanding from other materials, and many of its interesting properties have already been revealed in spite of its relatively short history. For example, its strength is 100 times more than the strongest metal. It also has excellent thermal conductivity and electrical conductivity. It also has high surface-to-volume ratio, large surface area, high electro-catalytic activity, fast electron transfers, low cost, robust mechanical properties and superior thermal conductivity, making it a promising material for application in electronics, optoelectronics, sensors, composites, batteries and supercapacitors. Its accessible surface area can be as large as 5 cm and is favourable for further modification and bio-applications.

Graphene-based sensors have been reported to have higher sensitivity due to its low electronic noise from thermal effects, fast electron transfer kinetics and optical effects. Moreover, due to its simple chemical synthesis from the inexpensive graphite, it serves as both a superior carbon support for the growth of nanoparticles and an electrical contact, providing electrode micro-environments as high-performance biosensors and promoting the performance of electrocatalysts. There are numerous applications of graphene in various research fields, such as transparent conductors, flexible electronics, field-effect transistors, fuel cells, batteries, solar cells, biomaterials, biosensors and water purifiers.

10.3.1 Discovery and Synthesis

Graphene was first isolated and discovered in 2004 by Andre Geim and Konstantin Novoselov at the University of Manchester. The high-quality graphene was also validated to be surprisingly easy to isolate. Numerous research works have been published since then. This work has been recognized by awarding the two scientists the Nobel Prize in Physics in 2010 for "groundbreaking experiments regarding the two-dimensional material graphene". This mechanical exfoliation method is achieved by placing a piece of graphite onto scotch tape. By folding and separating the scotch tape several times, the graphite becomes extremely thin. The final step is to press the graphite onto the target substrate by scotch tape, which leaves a significant amount of single-atomic layer of graphite on the substrate.

The chemical vapour deposition (CVD) method is another widely used method to produce graphene, especially when there is a need for a large area. It uses high-temperature (>800 °C) gas mixtures (CH_4 and alcoholic vapor) to pass through a catalytic material (*e.g.*, copper), which is also a place where graphene grows. It produces a large area of graphene (which can be as large as 5 cm), but is not as precise as that produced by the exfoliation method.

The global market for graphene was reported to have reached $9 million by 2012, with the most sales in the semiconductor, electronics, battery energy and composites industries.

10.3.2 Physical Properties of Graphene

Before monolayer graphene was isolated in 2004, it was theoretically believed that two-dimensional compounds could not exist due to thermal instability when they are separated. However, once graphene was isolated, it was clear that this was completely possible. After suspended graphene sheets were studied by transmission electron microscopy, scientists assumed that it is due to slight rippling in the graphene and subsequently modifying the structure of the material. However, later research suggested that it is actually due to the fact that the carbon-to-carbon bonds in graphene are so small and strong that they prevent thermal fluctuations from destabilizing it.

One of the most useful properties of graphene is that it is a zero-overlap semi-metal (with both holes and electrons as charge carriers) with very high electrical conductivity. Carbon atoms have a total of six electrons, with two in the inner shell and four in the outer shell. The four outer shell electrons in an individual carbon atom are available for chemical bonding but, in graphene, each atom is connected to three other carbon atoms on the two-dimensional plane, leaving one electron freely available in the third dimension for electronic conduction. These highly mobile electrons are called pi (π) electrons and are located above and below the graphene sheet. These pi orbitals overlap and aid in enhancing the carbon-to-carbon bonds in graphene. Fundamentally, the electronic properties of graphene are dictated by the bonding and anti-bonding (the valance and conduction bands) of these pi orbitals.

Another outstanding feature of graphene is its inherent strength. Due to the strength of its 0.142 nm-long carbon bonds, graphene is the strongest material ever discovered, with an ultimate tensile strength of 130 000 000 000 pascals (or 130 gigapascals), compared to 400 000 000 pascals for A36 structural steel and 375 700 000 pascals for aramid (Kevlar). Not only is graphene extraordinarily strong, it is also very light at 0.77 mg m^{-2} (for comparison purposes, 1 m^2 of paper is roughly 1000 times heavier). It is often said that a single sheet of graphene (being only 1 atom thick) large enough to cover an entire football field would weigh less than a single gram.

Graphene also has elastic properties and is able to retain its initial size after strain. Atomic force microscopic (AFM) tests were carried out on graphene sheets that were suspended over silicone dioxide cavities. These tests showed that graphene sheets (with thicknesses of between 2 and 8 nm) had spring constants in the region of 1–5 N m^{-1} and a Young's modulus (different from that of three-dimensional graphite) of 0.5 TPa. Again, these superlative figures are based upon a theoretical prospective using graphene that is unflawed, containing no imperfections. This is currently very expensive and difficult to reproduce artificially, although production techniques are steadily improving with reduced costs and complexity.

Graphene can absorb a rather large 2.3% of white light, which is also a unique and interesting property, especially considering that it is only 1 atom thick. This is due to its aforementioned electronic properties. The electrons act like massless charge carriers with very high mobility. A few years ago, it was validated that the amount of white light absorbed is based upon the Fine Structure Constant rather than being dictated by material specifics. Adding another layer of graphene increases the amount of white light absorbed by approximately the same value (2.3%). Graphene has an opacity value of $\pi\alpha \approx 2.3\%$, which equates to a universal dynamic conductivity value of $G = e^2/4\hbar$ ($\pm$2–3%) over the visible frequency range.

Due to these impressive characteristics, it has been observed that once optical intensity reaches a certain threshold (known as the saturation fluence) saturable absorption takes place and that very high intensity light causes a reduction in absorption. This is an important characteristic with regards to the mode-locking of fibre lasers. Due to the properties of wavelength-insensitive ultrafast saturable absorption, full-band mode locking has been achieved using an erbium-doped dissipative soliton fibre laser capable of obtaining wavelength tuning as large as 30 nm.

10.4 Application of Graphene in Sensing Food Safety and Quality

The application of graphene in sensing food safety and quality has been widely applied in recent years.[46,47] In the following section, its application to detect and characterize different targeted analytes in agricultural and food products will be introduced and discussed.

10.4.1 Detection of Chemical Contaminants in Agri-food Products

Surface-enhanced Raman spectroscopy (SERS) is a technique that has been widely used to detect hazardous biomolecules and chemicals in food products.[48] For details about Raman spectroscopy and SERS, please refer to Chapter 1. In particular, the intensity of Raman scattering signals could be significantly enhanced by placing the prepared sample onto the substrates, such as graphene nano-mesh, as reported in several recent publications.[49,50] In other words, graphene can be used as a Raman active substrate.

Liu and colleagues[49] fabricated a graphene nano-mesh to detect rhodamine B (RhB) by the SERS technique. The authors fabricated the graphene nano-mesh by depositing a Cu film on graphene and then removing it by annealing, which left holes around the Cu nanoparticles. This method could ensure controllable size and density of the holes. The graphene nano-meshes can produce a higher intensity of Raman spectral signal due to the ability of edges in graphene nano-meshes to increase the localized charge transfer and the ability to absorb RhB molecules. The Raman spectra of different samples are shown in Figure 10.9. The samples would have higher Raman signal intensities when placed onto single-layer graphene nano-meshes than those placed onto bare single-layer graphene, which validated the effectiveness of the SERS effect.

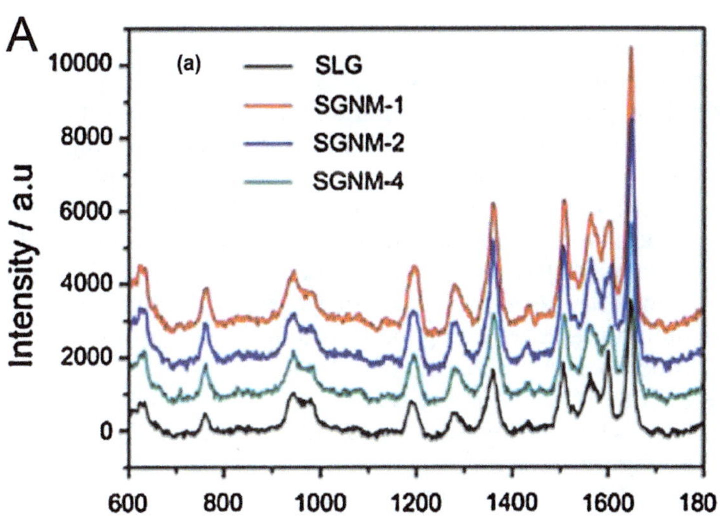

Figure 10.9 Raman spectra of different substrates with 0.5 nm thick RhB deposited – single layer graphene, and single layer graphene nano-mesh (sample 1, sample 2 and sample 3).
Reproduced from J. Liu, H. Cai, X. Yu, *et al.*, *J. Phys. Chem. C*, 2012, **116**(29), 15741–15746. Copyright 2012 American Chemical Society.

In another study, a SERS sensor was constructed to detect prohibited colorants and illegal dyes in food products by using graphene/Ag substrate. Xie and coauthors[51] used graphene/Ag nanoparticles as the substrates to detect prohibited colorants in foods. Due to the stronger absorbance ability of graphene/Ag nanoparticles than Ag nanoparticles alone, the signal intensities of the Raman spectra were greatly enhanced, leading to a lower limit of detection of food colorants.

Oxalic acid (OA) is a natural element in plants, fruits, vegetables and nuts, and it could form a less soluble complex with cations in the form of Ca^{2+}.[52] OA has the potential risk of causing irritation of the digestive system, especially when it has a relatively high level in the diet.[53] Chen and colleagues[54] reported a method of using a platinum nanoparticles/graphene nanosheet electrode to detect OA with the concentration range of 0.1–50 mM. The platinum nanoparticles/graphene nanosheet film had a high electrochemical activity in the detection of OA.

Hydrogen peroxide (H_2O_2) is a by-product in food processing and is also a mediator in food, pharmaceutical and clinical analysis.[55] A palladium nanoparticles/graphene nanosheet film-coated glassy carbon electrode presents an increased electrochemical reduction of H_2O_2. This palladium nanoparticles/graphene nanosheet film-coated glassy carbon electrode could therefore in turn act as a sensor to detect H_2O_2 in the presence of ascorbic acid, glucose and dopamine.

Gan and colleagues[56] prepared graphene layer-wrapped phosphotungstic acid (PTA) hybrid film that aided in detecting synthetic aromatic azo dyes. Synthetic aromatic azo dyes are harmful because they cause allergies, migraines, eczema, anxiety, diarrhoea and cancer when an excessive amount is consumed by humans.[57,58] This graphene/PTA hybrid film greatly enhanced the detection efficiency of azo dyes.

In addition, graphene-based materials can be used as the separation elements towards various chemical hazards in foods. The high extraction efficiency of graphene nano-mesh as adsorbent in solid-phase micro extraction (SPME) is summarized in Table 10.1.[64] SPME can be integrated into a sensing platform for automated determination of food safety and quality.

Graphene dots (GQDs) are graphene sheets with lateral dimensions less than 100 nm. They have very low toxicity, high stability, high solubility, stable photoluminescence and good bio-compatibility. A previous study reported a fluorescence "off-to-on" mechanism of bright blue fluorescent glutathione (GSH)-functionalized GQDs for the determination of adenosine triphosphate (ATP).[67] This strategy could quantify ATP in the range of 25–250 μM (Figure 10.10).

Moreover, GQDs possess more efficient peroxidase-like catalytic activity than graphene oxide (GO) with large sheet size, as reported by Zheng and coworkers.[68] GQDs catalyzed the oxidation of peroxidase substrate 3,3,5,5-tetramethylbenzidine (TMB) in the presence of H_2O_2 and produced a blue product (Figure 10.11), which was used for the detection of H_2O_2 by monitoring the changes in absorbance.[69]

Table 10.1 Graphene-based extraction and chromatographic analysis of pesticides, toxins and oestrogens.

Substrate and separation technique	Extraction efficiency (%)	Sample matrix	Ref.
NH_2-Graphene–LC–MS/MS analysis	70.5–100.0	Rapeseed, peanut, soybean and sesame seeds	59
Magnetic graphene nanocomposite(G-Fe_3O_4) -GC-ECD	79.2–102.4	Tap water, reservoir water and grape juice	60
Graphene nanosheet-coated stainless steel fibre–GC-ECD	80.1–101.1	River water sample	61
Graphene-coated fibre–HPLC-DAD	83.8–95.4	Water samples (sea, lake and tap water)	62
Graphene-coated Fe_3O_4 magnetic NPs-HPLC-DAD	89.0–96.2	Water samples (lake, river and reservoir)	63
Graphene-based sol–gel coating as sorbent coupled with GC-MS	99–114	River water samples	64
GO as adsorbent-HPLC	85.1	Peanuts	43
Fe_3O_4@TiO_2/graphene oxide magnetic microspheres-HPLC	70.6	Milk powder	65
Graphene as an SPE adsorbent-IC	83.3	Dairy products	28
Graphene-based SPE coupled with UPLC	81.7	Fish tissues	29
GO/fused-silica fibre-GC	84.58	Yellow river and local waterworks	66

10.4.2 Detection and Characterization of Food Compositions

Cholesterol content in blood is one of the major risk factors for heart diseases. However, many common foods contain cholesterol, such as egg yolk, meats, seafood products and dairy products. To facilitate the determination of cholesterol content in agricultural and food products, biosensors have been developed and used. Among these biosensors, graphene-based biosensors are highly favoured. Cao and colleagues developed a biosensor by using TiO_2-graphene-Pt–Pd nano-composite coated with gold nanoparticles and applied it to achieve accurate detection of cholesterol in foods.[70]

Glucose is another important food component, which is directly associated with diabetes.[71] It is important to develop a fast and accurate method to quantify glucose content in biological and food samples. Three-dimensional graphene foam (3DGF) with highly crystalline manganese oxide (Mn_3O_4) was used to determine the content of glucose in foods.[72] It ensured a flexible, free-standing platform. Glucose was also determined in the range of 0.02–2.26 $mmol\,L^{-1}$ using a graphene-Au biosensor with a relatively high sensitivity (Figure 10.12).[73]

Moreover, graphene-based sensors were developed and used to detect β-lactoglobulin (BLG),[74] which composed up to 10% of the proteins in cow's

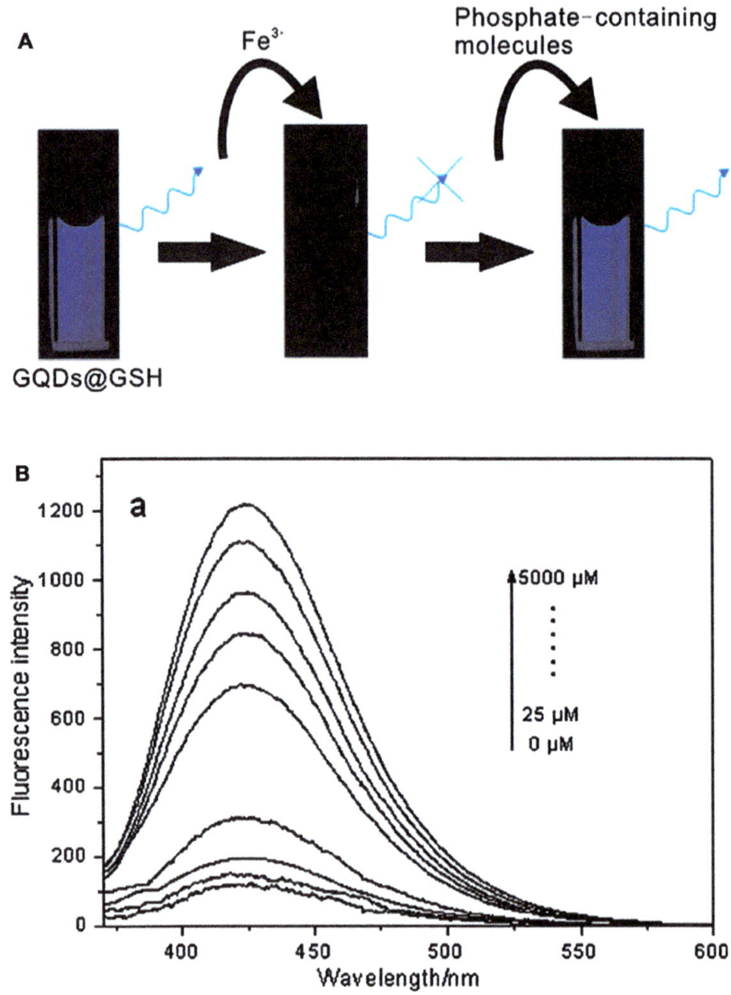

Figure 10.10 (A) Sensing process for phosphate-containing molecules based upon a GQD@GSH–Fe^{3+} probe. (B) Fluorescence emission spectra of GQD@GSH–Fe^{3+} (2 $\mu g\,mL^{-1}$) in the presence of different concentrations of ATP (0, 25, 50, 100, 250, 500, 1000, 2500 and 5000 μM). Reproduced from ref. 67 with permission from the Royal Society of Chemistry.

milk. This detection method was achieved by linking graphene to the anti-BLG antibodies using glutaraldehyde. A hybrid film composed of gold nanoparticles/poly/graphene was also synthesized for labelling horseradish peroxidase (HRP).[75]

One interesting application is a molecular beacon biosensor that uses GO as the "nanoquencher" (Figure 10.13). A strong adsorption of the hairpin-structured oligonucleotide on the GO was observed and the exceptional fluorescence-quenching effect of GO was developed.[76,77]

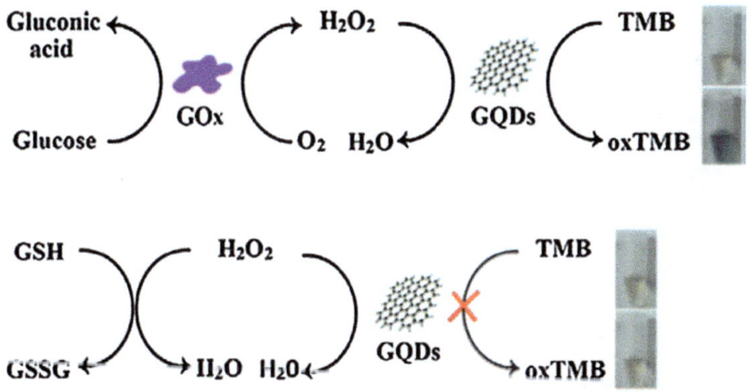

Figure 10.11 The highly efficient peroxidase-like activity of graphene dots (GQDs) for the detection of H_2O_2, glucose and reduced glutathione (GSH). Y. Song, K. Qu, C. Zhao, J. Ren, X. Qu, Graphene oxide: intrinsic peroxidase catalytic activity and its application to glucose detection, *Adv. Mater.*, 2010, **22**, 2206–2210. Copyright Wiley-VCH Verlag GmbH & Co. KGaA. Reproduced with permission.

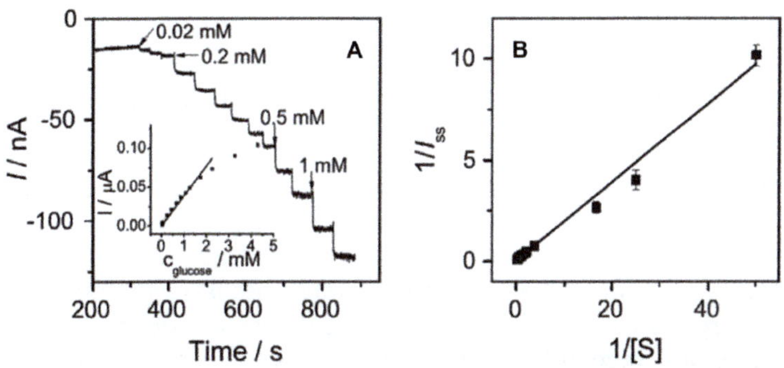

Figure 10.12 (A) Amperometric responses of the (GOD–graphene–AuNPs)5/GCE at 0.6 V upon successive additions of glucose to air-saturated PBS (pH 7.0) with stirring. Inset: calibration curve between current and the concentration of glucose. (B) Lineweaver–Burk plots for the (GOD–graphene–AuNPs)5/GCE with glucose as a substrate.
Reprinted from *Journal of Electroanalytical Chemistry*, Volume 697, Xiaodong Cao, Yongkang Ye, Ying Li, Xuan Xu, Jiachao Yu, Songqin Liu, Self-assembled glucose oxidase/graphene/gold ternary nanocomposites for direct electrochemistry and electrocatalysis, 10–14. Copyright 2013. With permission from Elsevier.

One unique detection method is electrochemical impedance spectroscopy. It is widely used to study biological events occurring at the electrode surface, such as cell growth, protein binding, bacterial growth, DNA hybridization and antigen–antibody reaction.[78] Another category of detection methods is graphene-based ion-selective electrode (ISE), which is a transducer that converts the activity of a specific ion dissolved in a solution into an electrical

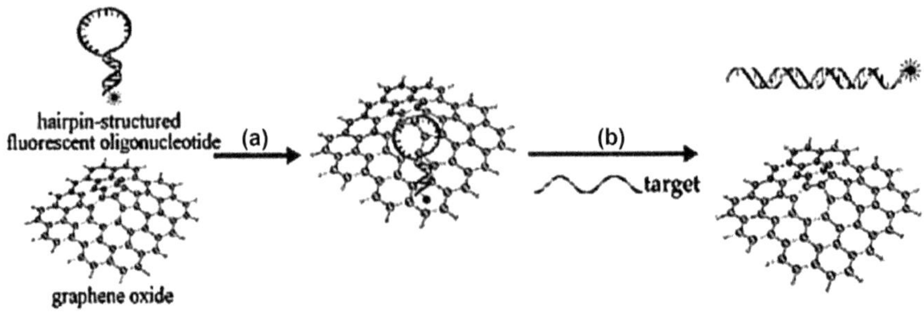

Figure 10.13 Target-induced fluorescence change of GO-quenched molecular beacon. Reprinted from C.-H. Lu, J. Li, J.-J. Liu, H.-H. Yang, X. Chen, G.-N. Chen, Increasing the sensitivity and single-base mismatch selectivity of the molecular beacon using graphene oxide as the "nanoquencher", *Chem. Eur. J.*, 2010, **16**, 4889–4894. Copyright Wiley-VCH Verlag GmbH & Co. KGaA. Reproduced with permission.

signal. Figure 10.14 depicts a graphene-based method that uses vacuum filtration of graphene suspension and directly uses it as a free-standing electrode substrate in all solid-state ISEs. All the signals are stable, thus proving it to be a reliable method.[79]

10.4.3 Detection of Volatile Organic Compounds

Graphene-based sensors were developed and used to detect volatile organic compounds released from food products, such as methanol, ethanol, acetone and methylacetate,[80–82] which can be regarded as the major chemical-based markers to determine food deterioration (Figure 10.15).

Methanol, ethanol, acetone, methylacetate, dimethyl sulfide and toluene vapours were tested because these chemicals are frequently released from meat, fruits and vegetables and considered as the representative chemical markers of food deterioration. The resistance relative amplitudes or relative differential resistance responses (Ar) of sensor devices were recorded and defined as Ar = (R–Ro)/Ro, where R is the resistance of the sensing materials in the presence of exposed vapours and Ro is the resistance of the sensing materials under a nitrogen flow. The vapour content was decreased from 1000 parts-per-million (ppm) to 2.5 ppm for methanol. Based upon the first two principal components (PC1 and PC2) that account for 98.25% of the total variance, six clusters corresponding to different volatile markers were well separated from each other (Figure 10.15), confirming that rGO-based sensors could simultaneously detect and differentiate the selective chemical vapours.

10.4.4 Detection of Toxins in Agricultural Food Products

In one study, Shen and coauthors reported the potential of using graphene as an adsorbent for the separation and determination of lipophilic marine

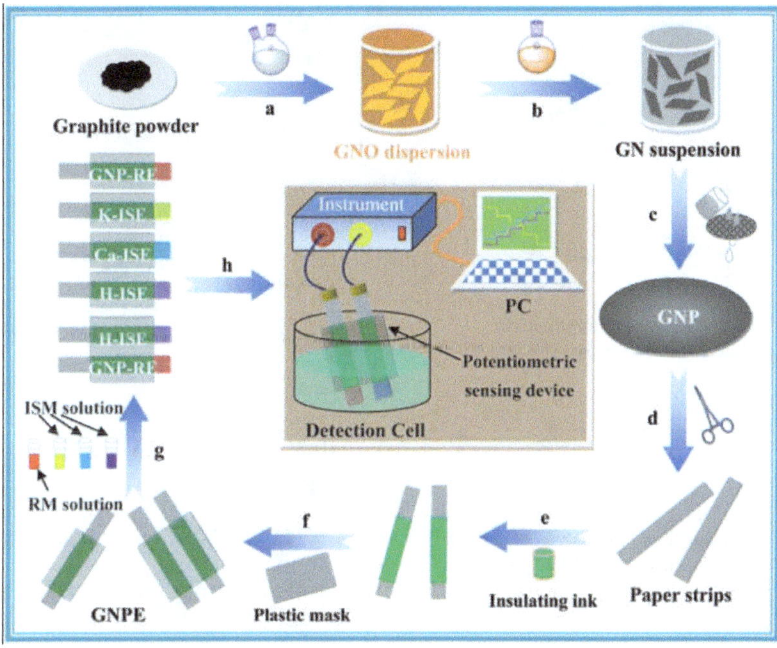

Figure 10.14 The fabrication process of a graphene paper (GNP)-based potentiometric sensing device: (a) oxidation and exfoliation, (b) reduction, (c) filtration, (d) cutting, (e) coating, (f) assembling, (g) dip coating and (h) detecting.
Reprinted from *Biosensors and Bioelectronics*, Volume 27, Qunfang Li, Dianping Tang, Juan Tang, Biling Su, Guonan Chen, Mingdeng Wei, Magneto-controlled electrochemical immunosensor for direct detection of squamous cell carcinoma antigen by using serum as supporting electrolyte. Copyright 2011, with permission from Elsevier.

toxins (LMTs) produced by aquatic food products, such as mussels, oysters and clams.[84] It was identified that graphene was superior compared to other materials in terms of serving as the PT-SPE sorbent. This graphene-based adsorbent showed excellent specificity, linearity, reproducibility and high extraction efficiency.

In summary, several graphene-based analytical devices with a similar working principle were developed for the detection of various types of biomolecules (Table 10.2).

Wu and coauthors developed a multiplexed fluorescence resonance energy transfer (FRET)-based turn-on assay for simultaneous detection of two types of mycotoxins, namely ochratoxin A (OTA) and fumonisin B1 (FB1).[100] This aptasensor provided a linear detection range of 0.05–100 ng mL^{-1} for OTA and 0.1–500 ng mL^{-1} for FB1, respectively. The LODs for OTA and FB1 were 0.02 ng mL^{-1} and 0.1 ng mL^{-1}, respectively. When contaminated maize samples were tested, the results obtained by using this sensor were

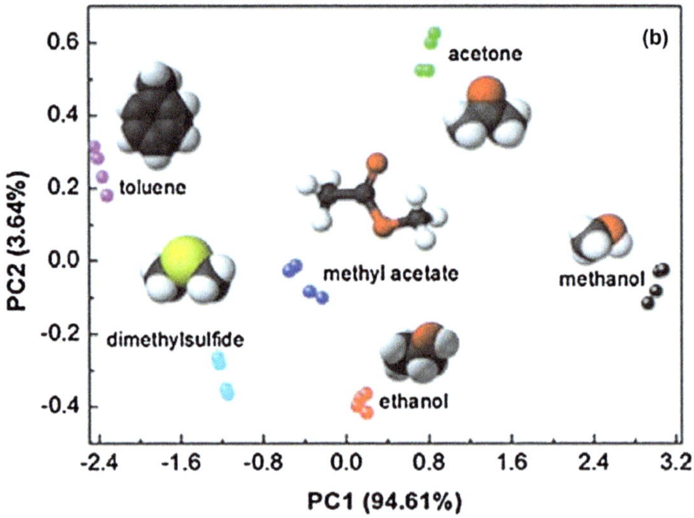

Figure 10.15 Pattern recognition of the e-nose exposed to some meat volatile markers after principal component analysis.
Reproduced from ref. 83 with permission from the Royal Society of Chemistry.

consistent with those determined by using an enzyme-linked immune sorbent assay (ELISA). In summary, graphene-based analytical sensing methods have the potential to provide outstanding sensitivity for various biomolecules in agri-food products.

10.4.5 Detection of Pesticides in Agricultural and Food Products

Pesticides are commonly used in the agricultural and crop industry to eliminate diseases. However, pesticides have strong side effects on human health due to their high biological activity and inherent toxicity. Conventional methods to detect trace levels of pesticides in food products require liquid chromatography coupled with mass spectrometry (LC-MS) techniques. With a relatively complicated sample preparation procedure and clean-up step, LC-MS can achieve extremely accurate and sensitive quantification of pesticide residues in agri-food products. However, the entire detection is time consuming and requires highly trained personnel. Therefore, innovative biosensors are needed to improve the efficiency of detection.

Wang and colleagues reported a graphene-based magnetic nanocomposite as an adsorbent for the separation and detection of pesticides in cucumber, cabbage and tomato samples.[101] In other studies, graphene was used as reversed-dispersive solid phase extraction material to combine with LC-MS for the detection of pesticide residues in oil crops[102] and fungicides in juice samples.[60]

Table 10.2 Detection of biomolecules in foods using graphene-based nanomaterials.

Graphene and its composite	Linear range	Sample matrix	Ref.
Graphene-Nafion film	5 $nmol\,L^{-1}$ $\mu mol\,L^{-1}$	Natural aloe extracts and human urine	85
4,5-dihydroxy-3-[(2-hydroxy-5-sulfophenyl)azo]-2,7-naphthalenedisulfonic acid/GO-Nafion film	1.0–36.0 $ng\,mL^{-1}$	Pork	86
Fe_3O_4-functionalized graphene nanoribbons	1–30 μM	—	87
rGO/AuNPs	10 to 1000 μM	Meat	88
Nickel hexacyanoferrate/rGO	1.0×10^{-6}–1.7×10^{-2} M	Human blood serum	89
Mn_3O_4/3D-graphene foam	0.1–8 mM	Serum, milk	72
Chitosan-rGO–NiNPs	0.2 to 9.0 mM	Urine samples	90
GOD-Au-Ag hollow microspheres/Prussian blue NPs/graphene	0.005–50 $ng\,mL^{-1}$	Clinical serum specimens	91
Graphene-Nafion/thionine/Pt-NPs + anti-kanamycin antibody	0.01–12.0 $ng\,mL^{-1}$	Chicken liver	92
Graphene oxide (GO)	25 $\mu g\,L^{-1}$ $mg\,L^{-1}$	Pork samples	93
GOD-rGO–AuPd alloy NP composites	0.5–3.5 mM	Human blood serum	94
Thionine-Nanogold-magnetic beads (signal tags)/nanogold–graphene + HRP-anti-TSH conjugates	0.01–20 $\mu IU\,mL^{-1}$	Human serum specimens	95
Nanogold-magnetic graphene nanosheet + HRP-anti-AFP conjugates	0.01–200 $ng\,mL^{-1}$	Human serum	96
Graphene/Prussian blue-chitosan/nanoporous gold + Kanamycin antibody	0.02–14 $ng\,mL^{-1}$	Pork	97
Carbon nanospheres-graphene-platinum nanosphere + labelled HRP-anti-CEA conjugates	0.001–100 $ng\,mL^{-1}$	Clinical serum samples	98
Graphene-Anti-CEA antibodies-Pt@Aunano labels + GOD + glucose–hydroquinone system	0.001–120 $ng\,mL^{-1}$	Serum specimens	99

10.5 Electrochemical Sensing in Foods

An electrochemical sensor is a self-contained integrated device capable of generating analytical signals based upon a recognition element with an electrochemical transduction element. Electrochemical analytical sensors have been used in a wide range of industries. The food and soft drink industries are in need of reliable detection technologies to assure product quality, and to monitor hazardous ingredients, contaminants and process control. Electrochemical sensors in food processing and quality control have several advantages, such as simplicity, rapidity, affordability and miniaturization for on-site detection. Specifically, different carbon and carbon hybrid-based electrochemical biosensors have been developed for the detection of food ingredients and additives, such as sugars.[103] Sugars are involved in the production of numerous food products and they are directly related to human health issues, such as obesity, glycemic index, metabolism and diseases. The most known health problem is diabetes, which is associated with elevated glucose level in blood and sometimes in urine. Several enzyme-free electrochemical sensors were developed for the detection of glucose based on different carbon hybrid nanocomposite materials, such as carbon nanotubes. The CVD-synthesized MWCNTs (multi-walled carbon nanotubes) were modified with copper nanoparticles or cobalt oxide/hydroxide nanoparticles, and exhibited much higher electrocatalytic activity towards glucose than before modification (Figure 10.16).[104] This innovative device also had a high sensitivity, fast response time, wide linear range, low detection limit and long-term stability over several weeks.[104]

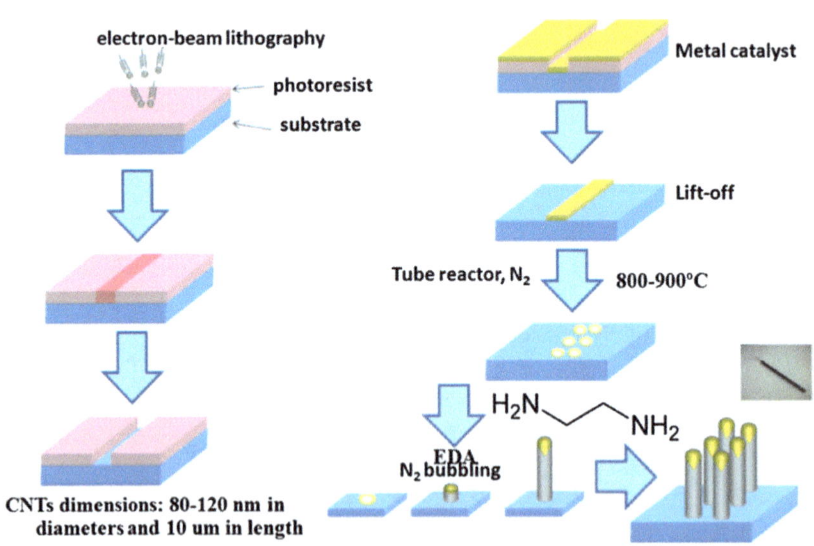

Figure 10.16 A schematic diagram of the controlled MWCNTs array CVD growth using EDA precursors and cobalt catalyst on Ta substrate. Reproduced with permission from the Ph.D. thesis of Dr Jiang Yang titled "Electrochemical sensing for food quality and safety".

Chapter 9 focuses on another type of carbon-based advanced material, namely carbon nanotubes. These advanced nano-carbon materials have numerous applications as electrochemical sensing platforms for food safety and quality control.

10.6 Application of Graphene in Detecting Food Safety and Quality by Electrochemical Methods

Two research works have been published regarding the application of nanoparticle-doped graphene materials in food sensing by electrochemical methods.[105,106] For example, an immunosensor was developed for sensitive detection of aflatoxin with an enhanced electrochemical performance, which was on the basis of graphene/conducting polymer/gold nanoparticles/ionic liquid composite film on the modified gold electrode with electro-deposition.[106] This biosensor achieved a detection limit of 1 fM and a dynamic range from 3.2 fM to 0.32 pM along with a long-term stability (*i.e.*, over 26 weeks). Because of their wide use in sensing food safety and quality control, graphene and electrochemical sensors have attracted wide attention in food science research and food industrial applications. Graphene is more promising than carbon nanotubes or other materials for electrochemical sensing because it can be easily synthesized (Scotch-tape exfoliation), can achieve direct observation by an optical microscope and has a large area (usually several cm in length) for detection.

10.7 Conclusion

There is a continuous interest in developing and applying graphene and its derivatives (*e.g.*, graphene oxide) as sensing materials because they have outstanding thermal, electrical, mechanical, chemical-adsorption as well as optical features. Graphene-based materials can be readily used to detect and characterize the safety and quality of a variety of food commodities, such as dairy products, meat products, seafood products and fresh produce. In addition, graphene and its derivatives can be integrated as components in the electrochemical sensing platform for agri-food analysis. Future research will be potentially focused on the application of new graphene-based materials for food sensing to achieve a higher sensitivity and selectivity compared to the currently used sensing platforms.

References

1. C. A. Damalas and I. G. Eleftherohorinos, Pesticide exposure, safety issues, and risk assessment indicators, *Int. J. Environ. Res. Public Health*, 2011, **8**, 1402–1419.
2. K. Rebecca, Davidsona *et al.*, Protecting our food: Can standard food safety analysis detect adulteration of food products with selected chemical agents? *TrAC, Trends Anal. Chem.*, 2016, **85**(Part B), 42–46.

3. S. J. Wu, N. Duan, X. Y. Ma, Y. Xia, H. G. Wang, Z. P. Wang and Q. Zhang, Multiplexed fluorescence resonance energy transfer aptasensor between upconversion nanoparticles and graphene oxide for the simultaneous determination of mycotoxins, *Anal. Chem.*, 2012, **84**, 6263–6270.
4. J. C. Vidal, L. Bonel, A. Ezquerra, P. Duato and J. R. Castillo, An electrochemical immunosensor for ochratoxin A determination in wines based on a monoclonal antibody and paramagnetic microbeads, *Anal. Bioanal. Chem.*, 2012, **403**, 1585–1593.
5. L. Bonel, J. C. Vidal, P. Duato and J. R. Castillo, An electrochemical competitive biosensor for ochratoxin Abased on a DNA biotinylated aptamer, *Biosens. Bioelectron.*, 2011, **26**, 3254–3259.
6. C. Yang, Y. Wang, J. L. Marty and X. R. Yang, Aptamer-based colorimetric biosensing of Ochratoxin A using unmodified gold nanoparticles indicator, *Biosens. Bioelectron.*, 2011, **26**, 2724–2727.
7. N. Duran and P. D. Marcato, Nanobiotechnology perspectives. Role of nanotechnology in the food industry: a review, *Int. J. Food Sci. Technol.*, 2013, **48**, 1127–1134.
8. H. F. Wang, L. J. Du, Z. M. Song and X. X. Chen, Progress in the characterization and safety evaluation of engineered inorganic nanomaterials in food, *Nanomedicine*, 2013, **8**, 2007–2025.
9. A. Rae, Real Life Applications of Nanotechnology in Electronics, *OnBoard Technol.*, 2006, **2006**, 28.
10. T. V. Duncan, Applications of nanotechnology in food packaging and food safety: Barrier materials, antimicrobials and sensors, *J. Colloid Interface Sci.*, 2011, **363**, 1–24.
11. M.-C. Daniel and D. Astruc, Gold nanoparticles: Assembly, supramolecular chemistry, quantum-size-related properties, and applications toward biology, catalysis, and nanotechnology, *Chem. Rev.*, 2004, **104**, 293–346.
12. N. Sozer and J. L. Kokini, Nanotechnology and its applications in the food sector, *Trends Biotechnol.*, 2009, **27**, 82–89.
13. S. Lim, O. K. Koo, Y. S. You, Y. E. Lee, M. S. Kim, P. S. Chang, D. H. Kang, J. H. Yu, Y. J. Choi and S. Gunasekaran, Enhancing nanoparticle-based visible detection by controlling the extent of aggregation, *Sci. Rep.*, 2012, **2**, 456.
14. C. Wang and C. X. Yu, Detection of chemical pollutants in water using gold nanoparticles as sensors: a review, *Rev. Anal. Chem.*, 2013, **32**, 1–14.
15. A. K. Singh, D. Senapati, S. Wang, J. Griffin, A. Neely, P. Candice, K. M. Naylor, B. Varisli, J. R. Kalluri and P. C. Ray, Gold nanorod based selective identification of *Escherichia coli* bacteria using two-photon Rayleigh scattering spectroscopy, *ACS Nano*, 2009, **3**, 1906–1912.
16. I.-H. Cho and J. Irudayaraj, In-situ immuno-gold nanoparticle network ELISA biosensors for pathogen detection, *Int. J. Food Microbiol.*, 2013, **164**, 70–75.
17. H. Q. Chen, B. Wang, D. Gao, M. Guan, L. N. Zheng, H. Ouyang, Z. F. Chai, Y. L. Zhao and W. Y. Feng, Broad-spectrum antibacterial

activity of carbon nanotubes to human gut bacteria, *Small*, 2013, **9**, 2735–2746.

18. C. Chellaram, G. Murugaboopathi, A. A. John, R. Sivakumar, S. Ganesan, S. Krithika and G. Priya, Significance of Nanotechnology in Food Industry, *APCBEE Proced.*, 2014, **8**, 109–113.
19. B. Perez-Lopez and A. Merkoci, Nanomaterials based biosensors for food analysis applications, *Trends Food Sci. Technol.*, 2011, **22**, 625–639.
20. M. Reches, K. A. Mirica, R. Dasgupta, M. D. Dickey, M. J. Butte and G. M. Whitesides, Thread as a Matrix for Biomedical Assays, *ACS Appl. Mater. Interfaces*, 2010, **2**, 1722–1728.
21. A. W. Martinez, S. T. Phillips, M. J. Butte and G. M. Whitesides, Patterned paper as a platform for inexpensive, low-volume, portable bioassays, *Angew. Chem., Int. Ed.*, 2007, **46**, 1318–1320.
22. A. W. Martinez, S. T. Phillips, G. M. Whitesides and E. Carrilho, Diagnostics for the developing world: Microfluidic paper-based analytical devices, *Anal. Chem.*, 2009, **82**, 3–10.
23. W. Zhao, M. M. Ali, S. D. Aguirre, M. A. Brook and Y. Li, Paper-Based Bioassays Using Gold Nanoparticle Colorimetric Probes, *Anal. Chem.*, 2008, **80**, 8431–8437.
24. A. Hayat and J. L. Marty, Disposable screen printed electrochemical sensors: Tools for environmental monitoring, *Sensors*, 2014, **14**, 10432–10453.
25. X. X. Chen, B. Cheng, Y. X. Yang, A. N. Cao, J. H. Liu, L. J. Du, Y. F. Liu, Y. L. Zhao and H. F. Wang, Characterization and preliminary toxicity assay of nanotitanium dioxide additive in sugar-coated chewing gum, *Small*, 2013, **9**, 1765–1774.
26. A. Jaeger, D. G. Weiss, L. Jonas and R. Kriehuber, Oxidative stress-induced cytotoxic and genotoxic effects of nano-sized titanium dioxide particles in human HaCaT keratinocytes, *Toxicology*, 2012, **296**, 27–36.
27. K. S. Novoselov, A. K. Geim, S. Morozov, D. Jiang, Y. Zhang, S. V. Dubonos, I. V. Grigorieva and A. A. Firsov, Electric field effect in atomically thin carbon films, *Science*, 2004, **306**, 666–669.
28. Z. H. Wang, H. Cui, J. F. Xia, Q. Han, N. N. Lv, J. G. Gao, X. M. Guo, F. F. Zhang, J. M. Ma and G. K. Su, A novel method for bisphenol A analysis in dairy products using graphene as an adsorbent for solid phase extraction followed by ion chromatography, *Food Anal. Methods*, 2013, **6**, 1537–1543.
29. J. B. Wu, Y. S. Qian, C. L. Zhang, T. L. Zheng, L. Y. Chen, Y. B. Lu and H. Z. Wang, Application of graphene-based solid-phase extraction coupled with ultrahigh-performance liquid chromatography-tandem mass spectrometry for determination of macrolides in fish tissues, *Food Anal. Methods*, 2013, **6**, 1448–1457.
30. L. M. Dai, Functionalization of graphene for efficient energy conversion and storage, *Acc. Chem. Res.*, 2013, **46**, 31–42.
31. P. Avouris and C. Dimitrakopoulos, Graphene: synthesis and applications, *Mater. Today*, 2012, **15**, 86–97.

32. N. O. Weiss, H. L. Zhou, L. Liao, Y. Liu, S. Jiang, Y. Huang and X. F. Duan, Graphene: an emerging electronic material, *Adv. Mater.*, 2012, **24**, 5782–5825.
33. M. Valdés, A. Valdés González, J. GarcíaCalzón and M. Díaz-García, Analytical nanotechnology for food analysis, *Microchim. Acta*, 2009, **166**, 1–19.
34. B. S. Sekhon, Food nanotechnology – an overview, *Nanotechnol. Sci. Appl.*, 2010, **3**, 1–15.
35. Z. Huang and H. K. Lee, Materials-based approaches to minimizing solvent usage in analytical sample preparation, *TrAc, Trends Anal. Chem.*, 2012, **39**, 228–244.
36. J. Hollaender, Rapid assessment of food/package interactions by electrochemical impedance spectroscopy (EIS), *Food Addit. Contam.*, 2009, **14**(6–7), 617–626.
37. M. F. Mabrook and M. C. Petty, Effect of composition on the electrical conductance of milk, *J. Food Eng.*, 2003, **60**, 321–325.
38. J. L. Damez, S. Clerion, S. Abouelkaram and J. Lepetit, Beef meat electrical impedance spectroscopy and anisotropy sensing for non-invasive early assessment of meat ageing, *J. Food Eng.*, 2008, **85**(1), 116–122.
39. C. M. A. Brett, Electrochemical sensors for environmental monitoring. Strategy and examples, *Pure Appl. Chem.*, 2001, **73**, 1969–1977.
40. D. Grieshaber, R. MacKenzie, J. Voros and E. Reimhult, Electrochemical biosensors - Sensor principles and architectures, *Sensors-Basel*, 2008a, **8**, 1400–1458.
41. U. Guth, W. Vonau and J. Zosel, Recent developments in electrochemical sensor application and technology-a review, *Meas Sci. Technol.*, 2009, **20**, 042002.
42. C. B. Liu, K. Wang, S. L. Luo, Y. H. Tang and L. Y. Chen, Direct Electrodeposition of Graphene Enabling the One-Step Synthesis of Graphene-Metal Nanocomposite Films, *Small*, 2011, **7**, 1203–1206.
43. L. Yu, P. W. Li, Q. Zhang, W. Zhang, X. X. Ding and X. P. Wang, Graphene oxide: an adsorbent for the extraction and quantification of aflatoxins in peanuts by high-performance liquid chromatography, *J. Chromatogr. A*, 2013, **1318**, 27–34.
44. M. S. Wilson, Electrochemical immunosensors for the simultaneous detection of two tumor markers, *Anal. Chem.*, 2005, **77**, 1496–1502.
45. M. Eckert, E. Neyts and A. Bogaerts, *CrystEngComm*, 2009, **11**, 1597–1608.
46. M. Valdes, A. Valdes Gonzalez, J. Garcia Calzon and M. Diaz-Garcia, Analytical nanotechnology for food analysis, *Microchim. Acta*, 2009, **166**, 1–19.
47. B. S. Sekhon, Food nanotechnology – an overview, *Nanotechnol. Sci. Appl.*, 2010, **3**, 1–15.
48. N. R. Yaffe and E. W. Blanch, Effects and anomalies that can occur in SERS spectra of biological molecules when using a wide range of aggregating agents for hydroxylamine-reduced and citrate-reduced silver colloids, *Vib. Spectrosc.*, 2008, **48**, 196–201.

49. J. Liu, H. Cai, X. Yu, K. Zhang, X. Li, J. Li, N. Pan, Q. Shi, Y. Luo and X. Wang, Fabrication of graphene nanomesh and improved chemical enhancement for Raman spectroscopy, *J. Phys. Chem. C*, 2012, **116**, 15741–15746.
50. Y. Xie, Y. Li, L. Niu, H. Wang, H. Qian and W. Yao, A novel surface-enhanced Raman scattering sensor to detect prohibited colorants in food by graphene/silver nanocomposite, *Talanta*, 2012, **100**, 32–37.
51. Y. Xie, Y. Li, L. Niu, H. Wang, H. Qian and W. Yao, A novel surface-enhanced Raman scattering sensor to detect prohibited colorants in food by graphene/silver nanocomposite, *Talanta*, 2012, **100**, 32–37.
52. L. K. Massey, H. Roman-Smith and R. A. L. Sutton, Effect of dietary oxalate and calcium on urinary oxalate and risk of formation of calcium oxalate kidney stones, *J. Am. Diet. Assoc.*, 1993, **93**, 901–906.
53. T. A. Ivandini, T. N. Rao, A. Fujishima and Y. Einaga, Electrochemical oxidation of oxalic acid at highly boron-doped diamond electrodes, *Anal. Chem.*, 2006, **78**, 3467–3471.
54. X. Chen, Z. Cai, Z. Huang, M. Oyama, Y. Jiang and X. Chan, Non-enzymatic oxalic acid sensor using platinum nanoparticles modified on graphene nanosheets, *Nanoscale*, 2013, **5**, 5779–5783.
55. T. J. Preston, W. J. Muller and G. Singh, Scavenging of extracellular H2O2 by catalase inhibits the proliferation of HER-2/Neu-transformed rat-1 fibroblasts through the induction of a stress response, *J. Biol. Chem.*, 2001, **276**, 9558–9564.
56. T. Gan, J. Sun, S. Cao, F. Gao, Y. Zhang and Y. Yang, One-step electrochemical approach for the preparation of graphene wrapped-phosphotungstic acid hybrid and its application for simultaneous determination of sunset yellow and tartrazine, *Electrochim. Acta*, 2012, **74**, 151–157.
57. K. S. Rowe and K. J. Rowe, Synthetic food coloring and behavior: a dose response effect in a double-blind, placebo-controlled, repeated-measures study, *J. Pediatr.*, 1994, **125**, 691–698.
58. D. D. Stevenson, R. A. Simon, W. R. Lumry and D. A. Mathison, Adverse reactions to tartrazine, *J. Allergy Clin. Immunol.*, 1986, **78**, 182–191.
59. W. B. Guan, Z. N. Li, H. Y. Zhang, H. J. Hong, N. Rebeyev, Y. Ye and Y. Q. Ma, Amine modified graphene as reversed-dispersive solid phase extraction materials combined with liquid chromatography-tandem mass spectrometry for pesticide multi-residue analysis in oil crops, *J. Chromatogr. A*, 2013, **1286**, 1–8.
60. Z. Li, M. Y. Hou, S. S. Bai, C. Wang and Z. Wang, Extraction of imide fungicides in water and juice samples using magnetic graphene nanoparticles as adsorbent followed by their determination with gas chromatography and electron capture detection, *Anal. Sci.*, 2013, **29**, 325–331.
61. V. K. Ponnusamy and J. F. Jen, A novel graphene nanosheets coated stainless steel fiber for microwave assisted headspace solid phase

microextraction of organochlorine pesticides in aqueous samples followed by gas chromatography with electron capture detection, *J. Chromatogr. A*, 2011, **1218**, 6861–6868.

62. G. Y. Zhao, S. J. Song, C. Wang, Q. H. Wu and Z. Wang, Solid-phase microextraction with a novel graphene-coated fiber coupled with high-performance liquid chromatography for the determination of some carbamates in water samples, *Anal. Methods*, 2011, **3**, 2929–2935.
63. G. Y. Zhao, S. J. Song, C. Wang, Q. H. Wu and Z. Wang, Determination of triazine herbicides in environmental water samples by high-performance liquid chromatography using graphene-coated magnetic nanoparticles as adsorbent, *Anal. Chim. Acta*, 2011, **708**, 155–159.
64. H. Zhang and H. K. Lee, Simultaneous determination of ultraviolet filters in aqueous samples by plunger-in-needle solid-phase microextraction with graphene based sol-gel coating as sorbent coupled with gas chromatography-mass spectrometry, *Anal. Chim. Acta*, 2012, **742**, 67–73.
65. M. M. Tian, W. Feng, J. J. Ye and Q. Jia, Preparation of Fe3O4@TiO2/graphene oxide magnetic microspheres for microchip-based preconcentration of estrogens in milk and milk powder samples, *Anal. Methods*, 2013, **5**, 3984–3991.
66. L. L. Xu, J. J. Feng, J. B. Li, X. Liu and S. X. Jiang, Graphene oxide bonded fused-silica fiber for solid-phase microextraction-gas chromatography of polycyclic aromatic hydrocarbons in water, *J. Sep. Sci.*, 2012, **35**, 93–100.
67. J.-J. Liu, X.-L. Zhang, Z.-X. Cong, Z.-T. Chen, H.-H. Yang and G.-N. Chen, Glutathione functionalized graphene quantum dots as selective fluorescent probes for phosphate-containing metabolites, *Nanoscale*, 2013, **5**, 1810–1815.
68. A.-X. Zheng, Z.-X. Cong, J.-R. Wang, J. Li, H.-H. Yang and G.-N. Chen, Highly-efficient peroxidase-like catalytic activity of graphene dots for biosensing, *Biosens. Bioelectron.*, 2013, **49**, 519–524.
69. Y. Song, K. Qu, C. Zhao, J. Ren and X. Qu, Graphene oxide: intrinsic peroxidase catalytic activity and its application to glucose detection, *Adv. Mater.*, 2010, **22**, 2206–2210.
70. S. Cao, L. Zhang, Y. Chai and R. Yuan, An integrated sensing system for detection of cholesterol based on TiO2-graphene-Pt-Pd hybrid nanocomposites, *Biosens. Bioelectron.*, 2013, **42**, 532–538.
71. M. C. Tonon, D. Lanfray, H. Castel, H. Vaudry and F. Morin, Hypothalamic glucose sensing: role of glia-to-neuron signaling, *Horm. Metab. Res.*, 2013, **45**, 955–959.
72. P. Si, X.-C. Dong, P. Chen and D.-H. Kim, A hierarchically structured composite of Mn3O4/3 D graphene foam for flexible nonenzymatic biosensors, *J. Mater. Chem. B*, 2013, **1**, 110–115.
73. X. Cao, Y. Ye, Y. Li, X. Xu, J. Yu and S. Liu, Self-assembled glucose oxidase/graphene/gold ternary nanocomposites for direct electrochemistry and electrocatalysis, *J. Electroanal. Chem.*, 2013, **697**, 10–14.

74. S. Eissa, C. Tlili, L. L'Hocine and M. Zourob, Electrochemical immunosensor for the milk allergen β-lactoglobulin based on electrografting of organic film on graphene modified screen-printed carbon electrodes, *Biosens. Bioelectron.*, 2012, **38**, 308–313.
75. H. Chen, Z. Gao, Y. Cui, G. Chen and D. Tang, Nanogold-enhanced graphene nanosheets as multienzyme assembly for sensitive detection of low abundance proteins, *Biosens. Bioelectron.*, 2013, **44**, 108–114.
76. C.-H. Lu, J. Li, J.-J. Liu, H.-H. Yang, X. Chen and G.-N. Chen, Increasing the sensitivity and single-base mismatch selectivity of the molecular beacon using graphene oxide as the "nanoquencher", *Chem. – Eur. J*, 2010, **16**, 4889–4894.
77. C.-H. Lu, J. Li, X.-J. Qi, X.-R. Song, H.-H. Yang, X. Chen and G.-N. Chen, Multiplex detection of nucleases by a graphene-based platform, *J. Mater. Chem.*, 2011, **21**, 10915–10919.
78. X. Dominguez-Benetton, S. Sevda, K. Vanbroekhoven and D. Pant, The accurate use of impedance analysis for the study of microbial electrochemical systems, *Chem. Soc. Rev.*, 2012, **41**, 7228–7246.
79. J. Ping, Y. Wang, K. Fan, W. Tang, J. Wu and Y. Ying, High-performance flexible potentiometric sensing devices using free-standing graphene paper, *J. Mater. Chem. B*, 2013, **1**, 4781–4791.
80. P. Bhattacharjee, S. Panigrahi, D. Lin, C. Logue, J. Sherwood, C. Doetkott and M. Marchello, A comparative qualitative study of the profile of volatile organic compounds associated with Salmonella contamination of packaged aged and fresh beef by HS-SPME/GC-MS, *J. Food Sci. Technol.*, 2011, **48**, 1–13.
81. M. Peris and L. Escuder-Gilabert, A 21st century technique for food control: electronic noses, *Anal. Chim. Acta*, 2009, **638**, 1–15.
82. Y. Xu and S. Barringer, Comparison of tomatillo and tomato volatile compounds in the headspace by selected ion flow tube mass spectrometry (SIFT-MS), *J. Food Sci.*, 2010, **75**, C268–C273.
83. J.-F. Feller *et al.*, Graphene quantum resistive sensing skin for the detection of alteration biomarkers, *J. Mater. Chem.*, 2012, **22**, 21754–21766.
84. Q. Shen, L. Gong, J. T. Baibado, W. Dong, Y. Wang, Z. Dai and H.-Y. Cheung, Graphene based pipette tip solid phase extraction of marine toxins in shellfish muscle followed by UPLC-MS/MS analysis, *Talanta*, 2013, **116**, 770–775.
85. J. Li, J. Chen, X.-L. Zhang, C.-H. Lu and H.-H. Yang, A novel sensitive detection platform for antitumor herbal drug aloe-emodin based on the graphene modified electrode, *Talanta*, 2010, **83**, 553–558.
86. X. Lin, Y. Ni and S. Kokot, A novel electrochemical sensor for the analysis of-agonists: the poly(acid chrome blue K)/graphene oxide-nafion/glassy carbon electrode, *J. Hazard. Mater.*, 2013, **260**, 508–517.
87. Q. Liu, Y. Yan, X. Yang, J. Qian, J. Cai and K. Wang, Fe_3O_4-functionalized graphene nanoribbons: preparation, characterization, and improved electrochemical activity, *J. Electroanal. Chem.*, 2013, **704**, 86–89.

88. J. Yang, J. R. Strickler and S. Gunasekaran, Indium tin oxide-coated glass modified with reduced graphene oxide sheets and gold nanoparticles as disposable working electrodes for dopamine sensing in meat samples, *Nanoscale*, 2012, **4**, 4594–4602.
89. P. Lu, S. Liu, G. Dai, Y. Lei and Y. Liang, Enhancement in detection of glucose based on a nickel hexacyanoferrate- reduced graphene oxide-modified glassy carbon electrode, *Aust. J. Chem.*, 2013, **66**, 983–988.
90. J. Yang, J. H. Yu, J. R. Strickler, W. J. Chang and S. Gunasekaran, Nickel nanoparticle chitosan-reduced graphene oxide-modified screen-printed electrodes for enzyme-free glucose sensing in portable microfluidic devices, *Biosens. Bioelectron.*, 2013, **47**, 530–538.
91. J. Tang, D. Tang, Q. Li, B. Su, B. Qiu and G. Chen, Sensitive electrochemical immunoassay of carcinoembryonic antigen with signal dual-amplification using glucose oxidase and an artificial catalase, *Anal. Chim. Acta*, 2011, **697**, 16–22.
92. Q. Wei, Y. Zhao, B. Du, D. Wu, H. Li and M. Yang, Ultrasensitive detection of kanamycin in animal derived foods by label-free electrochemical immunosensor, *Food Chem.*, 2012, **134**, 1601–1606.
93. C. Wu, D. Sun, Q. Li and K. Wu, Electrochemical sensor for toxic ractopamine and clenbuterol based on the enhancement effect of graphene oxide, *Sens. Actuators, B*, 2012, **168**, 178–184.
94. J. Yang, S. Deng, J. Lei, H. Ju and S. Gunasekaran, Electrochemical synthesis of reduced graphene sheet-AuPd alloy nanoparticle composites for enzymatic biosensing, *Biosens. Bioelectron.*, 2011, **29**, 159–166.
95. B. Zhang, D. Tang, B. Liu, Y. Cui, H. Chen and G. Chen, Nanogold-functionalized magnetic beads with redox activity for sensitive electrochemical immunoassay of thyroid-stimulating hormone, *Anal. Chim. Acta*, 2012, **711**, 17–23.
96. B. Zhang, D. Tang, B. Liu, H. Chen, Y. Cui and G. Chen, GoldMag nanocomposite functionalized graphene sensing platform for one-step electrochemical immunoassay of alpha-fetoprotein, *Biosens. Bioelectron.*, 2011, **28**, 174–180.
97. B. Y. Zhao, Q. Wei, C. Xu, H. Li, D. Wu, Y. Cai, K. Mao, Z. Cui and B. Du, Label-free electrochemical immunosensor for sensitive detection of kanamycin, *Sens. Actuators, B*, 2011, **155**, 618–625.
98. J. Zhou, J. Zhuang, M. Miro, Z. Gao, G. Chen and D. Tang, Carbon nanospheres promoted electrochemical immunoassay coupled with hollow platinum nanolabels for sensitivity enhancement, *Biosens. Bioelectron.*, 2012, **35**, 394–400.
99. J. Zhou, D. Tang, L. Hou, Y. Cui, H. Chen and G. Chen, Nanoplatinum-enclosed gold nanocores as catalytically promoted nanolabels for sensitive electrochemical immunoassay, *Anal. Chim. Acta*, 2012, **751**, 52–58.
100. S. Wu, N. Duan, X. Ma, Y. Xia, H. Wang, Z. Wang and Q. Zhang, Multiplexed fluorescence resonance energy transfer aptasensor

between upconversion nanoparticles and graphene oxide for the simultaneous determination of mycotoxins, *Anal. Chem.*, 2012, **84**, 6263–6270.

101. L. Wang, X. Zang, Q. Chang, G. Zhang, C. Wang and Z. Wang, Determination of triazole fungicides in vegetable samples by magnetic solid-phase extraction with graphene-coated magnetic nanocomposite as adsorbent followed by gas chromatography–mass spectrometry detection, *Food Anal. Methods*, 2013, 1–8.
102. W. B. Guan, Z. N. Li, H. Y. Zhang, H. J. Hong, N. Rebeyev, Y. Ye and Y. Q. Ma, Amine modified graphene as reversed-dispersive solid phase extraction materials combined with liquid chromatography-tandem mass spectrometry for pesticide multi-residue analysis in oil crops, *J. Chromatogr. A*, 2013, **1286**, 1–8.
103. J. Yang. Electrochemical sensing for food quality and safety, Thesis (2013).
104. "Electrochemical sensing for food quality and safety", Thesis, By Jiang Yang, University of Wisconsin-Madison, 2013.
105. J. Wu, Q. Liu, K. Wang and J. Cai, *Enhanced Direct Electrochemistry of Glucose Oxidase and Glucose Biosensing Based on TiO2 – Decorated Graphene Nanohybrids*, Trans Tech Publications, Zhengzhou, 2012, pp. 507–510.
106. L. T. Zhou, R. Y. Li and Z. J. Li, An immunosensor for ultrasensitive detection of aflatoxin B1 with an enhanced electrochemical performance based on graphene/conducting polymer/gold nanoparticles/the ionic liquid composite film on modified gold electrode with electrodeposition [J], *Sens. Actuators, B*, 2012, **174**, 359–365.
107. M. S. Wilson and W. Nie, Multiplex measurement of seven tumor markers using an electrochemical protein chip, *Anal. Chem.*, 2006, **78**, 6476–6483.

CHAPTER 11

Smartphone-based Sensing in Food Safety and Quality Analysis

JANE RU CHOI

Department of Biomedical Engineering, Faculty of Engineering, University of Malaya, Kuala Lumpur 50603, Malaysia
Email: janeruchoi@gmail.com

11.1 Introduction

Food safety issues have recently gained significant global attention. In the United States, approximately 33 million people are diagnosed with foodborne illnesses every year, of which more than 9000 people die from the diseases.[1–3] These foodborne illnesses account for an annual medical cost of $10–20 billion in the US.[4] In 2010, it was reported that the most common foodborne pathogens, namely *Escherichia coli* O157:H7 and *Salmonella*, have caused about 1.4 million cases of foodborne illnesses in the US with an estimated medical cost of $2.7 billion.[5] In the developing countries, especially in Africa, incidences of food poisonings have been reported over the past few years, which were mainly caused by food chemicals (*e.g.*, cyanide and aflatoxin) and food pathogens (*e.g.*, *Salmonella*, *Staphylococcus aureus* or pathogenic *E. coli*).[6] In fact, the complexity of food matrices and the presence of a variety of food contaminants in low concentrations make food safety analysis highly challenging. Conventional laboratory-based food analysis techniques, such as high performance liquid chromatography (HPLC) and gas chromatography (GC), are time-consuming, equipment-dependent,

Food Chemistry, Function and Analysis No. 2
Sensing Techniques for Food Safety and Quality Control
Edited by Xiaonan Lu

Published by the Royal Society of Chemistry, www.rsc.org

labour-intensive and require highly trained workers,[7,8] making them less applicable especially in the developing countries with limited access to laboratory infrastructure.[9] To address these challenges, there is a strong need to develop portable and cost-effective quantitative analytical devices for food samples that can be used in remote settings.

With advances in point-of-care (POC) testing, miniaturized sensors, such as microfluidic chip and paper-based sensors, have been introduced to overcome the shortcomings of the conventional analytical devices in food safety analysis and food quality control.[10–17] Even though these sensors offer many advantages, such as their rapidity (*e.g.*, <30 min), low cost ($0.10–3.00 per test) and user-friendliness, they possess several limitations, particularly the lack of quantification.[18–20] To this end, recent efforts have been devoted to develop smartphone-based sensors for quantitative analysis of food contaminants especially in remote and rural areas. For example, lateral flow test strips have been coupled with a smartphone reader for quantification of food analytes, such as bacteria.[21] Several smartphone-based microfluidics paper analytical sensors and chip-based sensors have also been developed to detect food pathogens,[22,23] heavy metals[24] and pesticides.[25] In addition, smartphone-based microscopy has been demonstrated to detect food and water contaminants, including parasites[26] and bacteria.[27] These smartphone-based sensing approaches generally involve the integration of a smartphone into an optical-based biosensor or chemosensor through the use of lighting control in a handheld manner, providing emerging and ubiquitous platforms for POC food safety monitoring.

This chapter focuses on the recent advances in smartphone-based sensing, including paper-based sensing, chip-based sensing, smartphone microscopy and other types of sensing for food safety analysis. We first discuss the historical development of smartphone, microfluidic devices and their integration. Subsequently, we emphasize the beneficial use of smartphone-based sensors for the detection of food analytes, such as chemicals, toxins and pathogens, including proteins and nucleic acids as compared to the conventional benchtop detection technologies. In addition, we summarize the smartphone-based sensing approaches based upon colorimetric, fluorescence, luminescence and electrochemical detection and highlight their progress towards commercialization. We also provide a glimpse of the existing challenges and future perspectives for the application of smartphone-based sensors in food safety and quality control.

11.2 Smartphone-based Sensing

11.2.1 Overview

In 1992, the first smartphone, which contained personal digital assistant applications, was developed by International Business Machines Corporation (Figure 11.1). The early generation had limited performances of processors and sensors with short unplugged power operation time.[28]

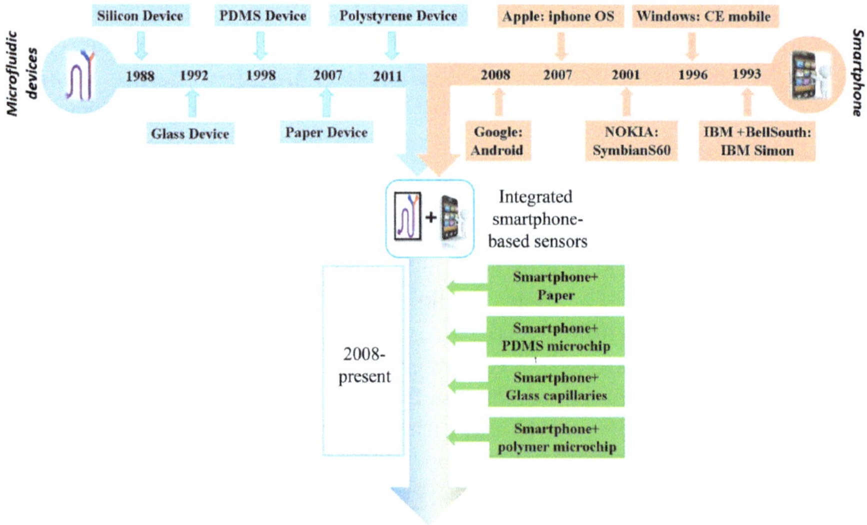

Figure 11.1 The historical development of smartphone-based sensors.[28] Adapted from ref. 28 with permission from the Royal Society of Chemistry.

These were significantly improved by the new generation of smartphones that consist of high-performance processors and back-illuminated sensors with access to on-board power. These features enable the implementation of biosensors or chemosensors into smartphones. The rapid development of smartphones with various sensors enables their application in health monitoring. For instance, smartphone cameras have been used to obtain auto signals and images for heart rate and blood oxygen saturation.[29] Several built-in sensors and external sensors have also been widely used for ultrasound imaging and body temperature measurement.[30] Thereafter, smartphone-based sensors have aimed to allow much easier, rapid and cost-effective POC testing in resource-poor settings. One of the earliest developments in this field was the smartphone-based paper-based sensor, which was developed by Martinez and coworkers in 2008 to detect proteins and glucose concentration in urine.[31] In the past few years, the new generation of smartphones has opened up opportunities for the development of portable and quantitative analytical tools integrated with microfluidic sensors.

In fact, the number of smartphone users worldwide has been growing rapidly, as witnessed by the increase from $1.86 billion in 2015 to $2.1 billion in 2016, especially driven from the developing countries (*e.g.*, India and China).[32,33] It has been reported that China leads the smartphone industry today. The number of smartphone users in China is estimated to rise from 480 million in 2013 to 690 million in 2019. By 2018, it is predicted that at least a quarter of smartphone users worldwide will be located in China. Figure 11.2 shows the total number of smartphone users in the world from

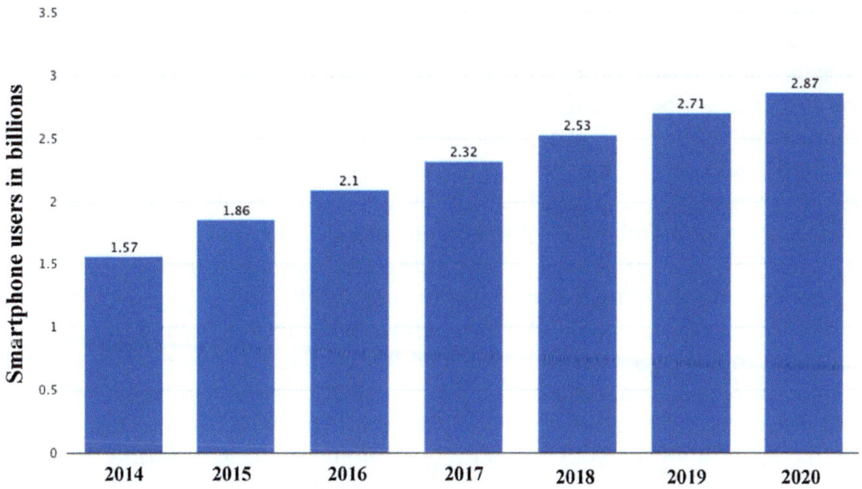

Figure 11.2 Number of smartphone users worldwide from 2014 to 2020.[34] Data reproduced with permission from ref. 34. Copyright 2016 Statista.

2014 to 2020.[34] By 2020, the number of smartphone users worldwide is expected to increase to 2.87 billion, indicating that smartphones have become more accessible even in resource-poor settings with limited healthcare facilities. Therefore, integrating sensors into the smartphone is particularly useful for the detection of biological and chemical targets in remote areas.[35] The data transmission capabilities of smartphones, such as Bluetooth, universal serial bus (USB), wireless fidelity (Wi-Fi) and Global System for Mobile Communication (GSM), allow long-distance communication between the remote testing site and central laboratory for result confirmation and professional guidance.[36]

As aforementioned, recent advances in smartphone technologies have made it possible to integrate the smartphone with chemical and biological sensors to detect various target analytes for medical diagnosis, food safety inspection and environmental monitoring applications. The smartphones are typically integrated through the use of lightening control that fits the sensors (*e.g.*, strip, cuvette/tube, chip, *etc.*)[37] and are normally equipped with processors and memories for data storage and analysis, and a high-resolution screen for test result display.[35] In this case, the integrated platform can function as a light source (through LED flash), a signal detector (through the camera) and a signal processor (through smartphone applications (apps)). Various filters, diffraction gratings, lenses and light sources can also be incorporated into the integrated device for signal enhancement in multiple types of assays, including colorimetric, fluorescence, luminescence and electrochemical assays.[38]

Colorimetric-based sensing has attracted much attention because of its simplicity and cost-effectiveness.[39–42] Unlike fluorescence-based sensors, colorimetric sensors do not require external excitation light sources and collimating beam paths, which immensely improve the simplicity and portability. Generally, the colorimetric assays detect changes in the optical density or intensity of analyte-reagent complexes over a broad range of wavelengths. These changes are normally present due to the plasmon resonance phenomena or structural shift of the complexes. As conventional smartphones rely upon a complementary metal oxide semiconductor (CMOS) array with stringent filters to assign red, green, blue (RGB) values, a smartphone coupled with simple colorimetric sensor without requiring stringent filters is preferred.[43] The integrated smartphone device only requires lighting and image processor for target detection with low-cost CMOS arrays that respond to RGB light, which produces a visible result.[37] In addition, mobile apps that have been widely available for data collection and colour analysis are particularly useful for POC testing. Smartphone-based colorimetric sensing has been widely used for the detection of organic (*e.g.*, toxins, formaldehyde, *etc.*) and inorganic molecules (*e.g.*, fluoride, chlorine, *etc.*), heavy metals (*e.g.*, Cu, Pb, *etc.*) and biomolecules (*e.g.*, nucleic acids and antigens) from buffers to complex samples, such as blood, food and beverages.[44,45]

Apart from colorimetric-based assays, smartphones have also been integrated with fluorescence-based assays. The fluorescence assays detect light emitted upon decay of the target-fluorophore complex and radiative excitation. The fluorescence signal is directly proportional to the intensity of the excitation source, which can be amplified to lower the limit of detection.[46] The spectrum of the sensor can also be adjusted to further enhance the detection sensitivity. By placing long pass filters between the smartphone lens and sensor, the specific emitted lights are allowed to reach the CMOS. When the specific targets are fluorescently tagged, the resulting specific emission peaks allow multiplexed detection.[47,48] Recent smartphone-based fluorescence sensing assays have been demonstrated for the detection of chemicals and pathogens (*e.g.*, viruses and bacteria), including nucleic acids and proteins.[49,50]

Smartphones have also been coupled with luminescence-based assays (*i.e.*, chemiluminescence and bioluminescence).[37] These assays detect narrow-based light emission from chemical and biological reactions between reagent and targets. Unlike fluorescence-based assays, the spontaneous light emission of luminescence-based assay eliminates the need for an excitation light source.[37] This characteristic enables highly sensitive detection of target analytes. However, the reaction intensity is time-dependent, where the intensity may reduce over time and the quantification may not be consistent.[51] To date, smartphones have been integrated into luminescence assays to detect various targets, including various kinds of bacteria[52] and pesticides.[53]

Electrochemical-based assays have also been integrated into smartphones for food safety analysis. This integrated device is insensitive to ambient illumination conditions and insoluble compounds, which demonstrates a higher sensitivity over colorimetric assays.[54,55] Generally, regular electrochemical detection involves a working electrode, a counter electrode and a reference electrode. The principle of electrochemical sensors is based upon the production of electrochemical signal, which can be measured by an electrochemical detector.[56] The platform is able to detect the signal through a measurable potential and charge accumulation (potentiometry), resistance and reactance (impedimetry) or current (amperometry).[57] The smartphones have been linked to the electrical platform to control, record and display the electrochemical detection.[58,59] The fully integrated platform would increase the sensor portability and provide a promising option for real-time detection. Smartphone-based electrochemical sensing has been demonstrated to detect target analytes such as *E. coli*[60] and 2,2′,4,4′-tetrabromodiphenyl ether, a persistent and organic pollutant.[61]

Smartphone-based sensing on the basis of colorimetric, fluorescence, luminescence or electrochemical detection integrated with papers, chips and tubes, microwells or disks will be further discussed in Section 11.3. The advantages of smartphone-based sensing in food safety and quality control applications will be discussed in detail in the following section.

11.2.2 The Advantages in Food Safety Applications

As aforementioned, foodborne diseases have become one of the major causes of morbidity and mortality in the world.[62–64] The conventional and emerging food analysis methods are summarized in Table 11.1. Conventional methods for the detection of food contaminants are gas chromatography (GC) and high performance liquid chromatography (HPLC), whereas for foodborne pathogen detection, the existing methods are culturing and plating assay, enzyme-linked immunosorbent assay (ELISA) and

Table 11.1 Comparison of food testing technologies.[a]

	Pathogen detection			Chemical detection		Pathogen or chemical detection
Diagnostic test	Bacteria culture	ELISA	qPCR	HPLC	GC	Smartphone-based RDT
Time	1–2 days	6 hours	4 hours	20–30 min	20 min	15–20 min
Costs ($)	3–6	10	20	20–30	20–30	2
Specificity (%)	100	70–90	100	95	95	100
Sensitivity (%)	100	61–99	80–100	99	99	80–100
Expertise required	Yes	Yes	Yes	Yes	Yes	No
Special equipment/ facilities required	Yes	Yes	Yes	Yes	Yes	No
Ref.	65	66	67, 69	7	8	78

[a]ELISA: enzyme-linked immunosorbent assay, qPCR: quantitative polymerase chain reaction, HPLC: high-performance liquid chromatography, GC: gas chromatography, RDT: rapid diagnostic test

quantitative polymerase chain reaction (qPCR).[10,65–67] GC and HPLC require high-end equipment and extensive sample preparation, and are time consuming.[7,8] Culturing and plating assay represents the gold standard for the detection of foodborne pathogens. However, it takes hours to days to obtain the results. ELISA is able to detect the pathogens faster but it requires multiple operational steps (*e.g.*, reagent addition and rinsing steps) and its specificity is inferior to the other methods.[68] Similar to ELISA, qPCR requires numerous processing steps (*e.g.*, extraction, amplification and detection of nucleic acids), which takes hours to complete the assays.[67,69] Taken together, the existing detection methods are relatively expensive and time-consuming, and require highly trained personnel,[70] and hence limit their use in resource-poor settings, where access to healthcare facilities is limited and a high incidence of foodborne illnesses is frequently reported.[71]

To address these limitations, smartphone-based sensors have been developed to rapidly, sensitively and specifically detect food contaminants for food safety inspection and control. As aforementioned, the smartphone with miniaturized operating systems, high-quality lenses and internal memory resembles a portable computer, which is more affordable and accessible than analytical laboratory equipment.[36] In most cases, smartphones are integrated with multiple types of accessories for various applications. The integration of smartphones with sensors, such as paper-based and chip-based sensors, enables routine food inspection without the need for external equipment, offering tremendous potential for improving food safety issues, especially in the developing countries.[72] In short, the development of smartphone-based sensors offer several advantages, including (i) the requirement of a low volume of reagents and samples, (ii) cost-effective, (iii) short analysis time, (iv) high portability for on-site analysis, (v) simple operational steps and (vi) high-throughput. The applications of smartphone-based sensing in food safety and quality control are summarized in Table 11.2, which will be briefly discussed in the next section.

11.3 Application of Smartphone-based Sensing in Food Safety and Quality Control

11.3.1 The Integration of Smartphones with Paper-based Assays

Paper-based assays represent one of the most widely used methods for food safety and quality control.[12,73–77] Integrating a smartphone into paper-based assays would significantly improve the assay performance (*e.g.*, enables signal analysis and target quantification).[57] The test results can be imaged using a smartphone built-in-camera. The amounts of targets can then be determined by image post-processing mainly based upon colorimetric, fluorescence and luminescence signals.

Table 11.2 Emerging smartphone-based sensing for food safety and quality control.

Smartphone-based devices	Test principle	Detection target	Notes (limit of detection (LOD), image resolution (IR) or field of view (FOV))	Ref.
Paper-based sensors	Colorimetric	*Listeria monocytogenes*	LOD: 0.5 pg μL^{-1}	21
	Colorimetric	Alkaline phosphatase	LOD: 0.1 U L^{-1}	78
	Colorimetric	Salmonella	LOD: 100 CFU mL^{-1}	80
	Colorimetric	Nitrite	LOD: 0.52 mg L^{-1}	22
	Colorimetric	Mercury ion	LOD: 50 nM	24
	Colorimetric	Malathion and paraoxon	LOD: 10 nM	25
	Fluorescence	*Salmonella Typhimurium*	LOD: 100 CFU mL^{-1}	49
	Luminescence	Pesticide thiram	LOD: 0.1 μM	53
Chip-based sensors	Fluorescence	*Escherichia coli*	LOD: 1 copies reaction $well^{-1}$	23
	Electrochemical	*Plasmodium sp.*	LOD: 16 ng mL^{-1}	83
	Electrochemical	*Escherichia coli*	LOD: 100 nM	60
	Electrochemical	2,2′,4,4′-tetrabromodiphenyl ether	LOD: 10^3–10^4 μg L^{-1}	61
Others (tubes, microwells and disks)	Colorimetric	Peanuts	LOD: 1 ppm	84
	Colorimetric	Mercury ions	LOD: 3.5 ppb	85
	Colorimetric	Fluoride	LOD: 0.1 ppm	86
	Colorimetric	Fluoride	LOD: 0.0256 mg/L^{-1}	87
	Bioluminescence	Bacteria	LOD: 10^6–10^8 CFU mL^{-1}	52
	Colorimetric	Mercury ions	LOD: 0.28 ng mL^{-1}	88
Microscope	Fluorescence	*Giardia lamblia* cysts	IR: ~1.12 μm FOV: ~80 mm^2	26
	Fluorescence	Anti-recombinant bovine somatotropin (rbST) Ab	IR: ~10 μm FOV: ~80 mm^2	96
	Fluorescence	rbST and insulin-like growth factor-1	IR: ~10 μm FOV: ~80 mm^2	97
	Fluorescence	*E. coli* O157:H7	IR: 10 μm FOV: ~81 mm^2	27
	Fluorescence	*Giardia lamblia*	IR: 1.5–2 μm FOV: ~24 mm^2	98
	Fluorescence	*Scenedesmus acuminatus*	IR: 500 nm FOV: —	99

Most smartphone-based paper sensing methods, particularly lateral flow assays, detect analytes based upon colorimetric reaction and detection.[57] For instance, a smartphone app installed in an Android smartphone has been used to analyze the RNA of *Listeria monocytogenes* detected by lateral flow test

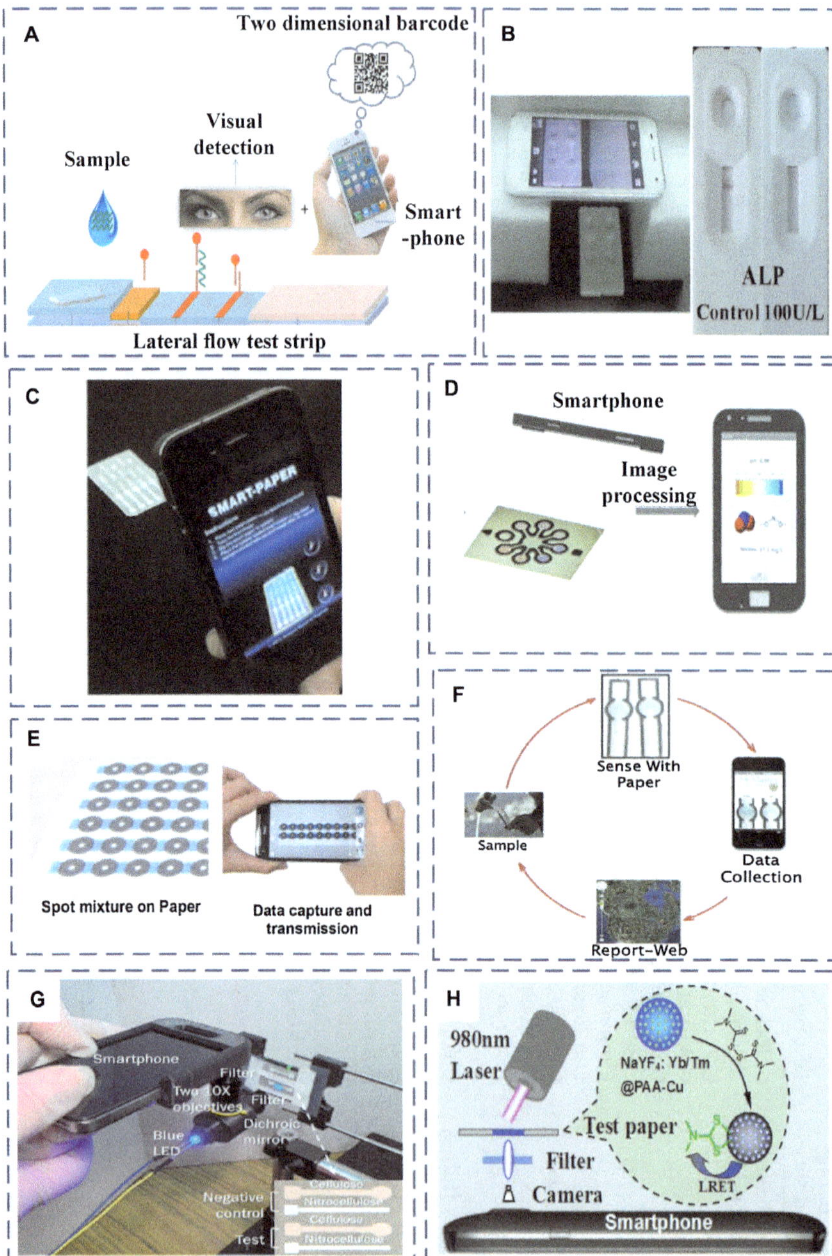

strip (Figure 11.3A).[21] The positive result showed red signals at both control and test lines while the negative result showed red signal only at the control line. The result was captured by an Android-based smartphone and the optical intensities of both lines were recorded by Image J software to determine the area integral and peak height. The result could also be transformed into two-dimensional bar codes for further analysis by distant investigators. Li's group has also quantitatively analyzed alkaline phosphatase activity in raw milk by using smartphone-based lateral flow test strip (Figure 11.3B).[78] Based upon gold nanoparticles-induced colour changes, the colorimetric signals were imaged by a smartphone integrated with digital camera and were further analyzed by a home-programmed MATLAB® code. This technology showed a great promise for food safety analysis in remote settings.

Apart from lateral flow test strips, smartphones have also been integrated into microfluidic paper-based sensors or paper-based analytical devices (μPADs) for food analysis.[79] For example, one study has demonstrated a

Figure 11.3 Integration of smartphone with paper-based assays for food safety and quality control. Smartphone-based lateral flow test strips for the colorimetric detection of (A) *Listeria monocytogenes* RNA[21] and (B) alkaline phosphatase activity[78] in raw milk. Smartphone-based paper analytical device (μPADs) for colorimetric detection of (C) Salmonella,[80] (D) pH and nitrite simultaneously,[22] (E) mercury[24] and (F) organophosphate pesticides.[25] Smartphone-based μPADs for (G) fluorescent detection of *Salmonella Typhimurium*[49] and (H) luminescence detection of pesticide thiram[53] for food safety control.
(A) Adapted from *Biosensors and Bioelectronics*, Volume 62, H. Liu, F. Zhan, F. Liu, M. Zhu, X. Zhou and D. Xing, Visual and sensitive detection of viable pathogenic bacteria by sensing of RNA markers in gold nanoparticles based paper platform, 38–46. Copyright 2014, with permission from Elsevier. (B) Adapted from *Biosensors and Bioelectronics*, Volume 69, L. Yu, Z. Shi, C. Fang, Y. Zhang, Y. Liu and C. Li, Disposable lateral flow-through strip for smartphone-camera to quantitatively detect alkaline phosphatase activity in milk, 307–315. Copyright 2015, with permission from Elsevier. (C) Adapted from ref. 80 with permission from the Royal Society of Chemistry. (D) Adapted from N. Lopez-Ruiz, V. F. Curto, M. M. Erenas, F. Benito-Lopez, D. Diamond, A. J. Palma and L. F. Capitan-Vallvey, *Anal. Chem.*, 2014, 86, 9554–9562. Copyright 2014 American Chemical Society. (E) Adapted from G.-H. Chen, W.-Y. Chen, Y.-C. Yen, C.-W. Wang, H.-T. Chang and C.-F. Chen, *Anal. Chem.*, 2014, 86, 6843–6849. Copyright 2014 American Chemical Society. (F) Adapted from *Water Research*, Volume 70, C. Sicard, C. Glen, B. Aubie, D. Wallace, S. Jahanshahi-Anbuhi, K. Pennings, G. T. Daigger, R. Pelton, J. D. Brennan and C. D. Filipe, Tools for water quality monitoring and mapping using paper-based sensors and cell phones, 360–369. Copyright 2015, with permission from Elsevier. (G) Adapted from ref. 49 with permission from the Royal Society of Chemistry. (H) Adapted from *Biosensors and Bioelectronics*, Volume 75, Q. Mei, H. Jing, Y. Li, W. Yisibashaer, J. Chen, B. N. Li and Y. Zhang, Smartphone based visual and quantitative assays on upconversional paper sensor, 427–732. Copyright 2016, with permission from Elsevier.

smartphone app for colorimetric detection of *Salmonella* using a multi-channel μPAD (Figure 11.3C).[80] Target *Salmonella* bacterial cells were mixed with anti-*Salmonella* antibodies-conjugated nanoparticles to allow immune-agglutination before loading into the paper-based sensor. The platform allowed an angle-dependent and less wavelength-dependent detection method *via* Mie scattering strategy. This strategy was described as the amount of light scattering and the changes in scattering intensities according to the scattering angles. The app showed a trapezoid-shaped box on the screen, allowing the user to fit the image of the paper-based sensor into the box. By pressing the button on the screen, auto-focus and auto-exposure were locked for signal capture and analysis. At the optimum Mie scattering angle, the detection region was then imaged and its normalized light scatter intensity resulting from the antigen–antibody interactions was calculated. With a similar principle, another study developed a smartphone-based μPAD for the detection of *E. coli.*[81] Antibody-beads were initially loaded to the paper channel and dried out prior to the assays. The samples that contain bacterial antigens were added into the sensor. The light scattering intensity arising from the antigen–antibody interactions were then analyzed by using the smartphone app.

In addition, a smartphone-based μPAD has been developed to determine both pH and nitrite concentration simultaneously in water, which represent the indicators of bacterial contamination (Figure 11.3D).[22] Seven sensing areas were designed with corresponding immobilized reagents to produce selective colour changes of sample in the sampling area when the sample is added, and the built-in flash of smartphone camera is used as a light source. The colour information was analyzed by using a smartphone with an Android app for image processing. The app was used to extract the colorimetric information based on HSV colour space for the determination of both pH and nitrite concentration. Under controlled light conditions, the image was processed using an optimized algorithm to prevent any potential detection interference, such as different light sources and position of the paper microfluidic sensor. The determination of pH was based upon two pH indicators, namely chlorophenol red and phenol red. The result of pH 4–6 was shown by the chlorophenol red colour changes from yellow to purple whereas pH 6–9 was shown by the phenol red colour changes from yellow to pink. On the other hand, the colorimetric signal produced from nitrite assay was based upon the Griess-colour reaction, which was quantitatively interpreted by the smartphone.

Similarly, a colorimetric detection method has been developed by another group, which was based upon smartphones for on-site detection of mercury ion (Hg^{2+}) using a μPAD without the need for external bulky equipment (Figure 11.3E).[24] The colorimetric result was observed after the addition of Hg^{2+} with different concentrations and the data were recorded and transmitted for cloud computing through a free app using a smartphone. The RGB values were then analyzed using the Image J software. This platform enables the on-site mercury monitoring of multiple samples in resource-limited settings.

Filipe and coworkers have demonstrated a μPAD in combination with a smartphone app for colorimetric detection of organophosphate pesticides (*i.e.*, malathion and paraoxon) (Figure 11.3F).[25] Acetylcholinesterase (AchE) and indoxyl acetate (IDA) were immobilized on the paper sensor. In the presence of pesticides, AchE reduced the hydrolysis of colourless IDA. In contrast, in the absence of pesticides, AchE hydrolyzed the IDA substrate and converted it into a blue coloured product. Therefore, the colour intensity was reduced with increasing concentration of pesticides. The colour intensity was imaged and analyzed with a smartphone, and the results were displayed in real time on a map through a newly developed website, WaterMap.ca™. This app offers the potential of μPAD for low-cost and real-time monitoring of water quality at a relatively large scale.

Apart from colorimetric signal analysis, several paper-based studies have also reported fluorescence signal analysis by using a smartphone. For example, one study has demonstrated the use of a smartphone-based paper sensor for fluorescence detection of *Salmonella* enterica serovar Typhimurium from poultry packaging solution (Figure 11.3G).[49] The fluorescence signals were captured using an iPhone attached to a miniaturized fluorescent detector and the pixel intensity was analyzed using ImageJ software. All signals were normalized to the negative control. It was found that the data collected using the smartphone app were consistent to those using the benchtop reader system.

Additionally, smartphone-based paper sensors have also been demonstrated by acquiring luminescence signals on the sensors. For instance, a set of optical accessories (*e.g.*, optical filter, low-powered 908 nm laser and mini-cavity) was attached to the smartphone to image the luminescence signal on $NaYF_4$: Yb/Tm upconversion nanoparticles modified test paper and quantitatively analyzed pesticide thiram for food and environmental control (Figure 11.3H).[53] Upon the addition of pesticide thiram, the thiram-Cu^{2+} coordination complex was formed on the test paper, resulting in luminescence quenching through the luminescence resonance energy transfer (LRET) mechanism. The luminescence intensities were detected by using the smartphone for pesticide analysis. The combination of smartphone and upconversion paper sensor offers new opportunities with a rapid and low-cost strategy for POC testing.

11.3.2 The Integration of Smartphones with Chip-based Assays

Apart from paper-based sensors, researchers have also sought to integrate smartphones with chip-based sensors. Microfluidic chips enable precise control and manipulation of low volumes of fluids to achieve automation and high-throughput screening.[82] This has eradicated the need for external components for operations, such as computers, power supplies and benchtop detectors, increasing the overall cost and bulkiness of the sensors.

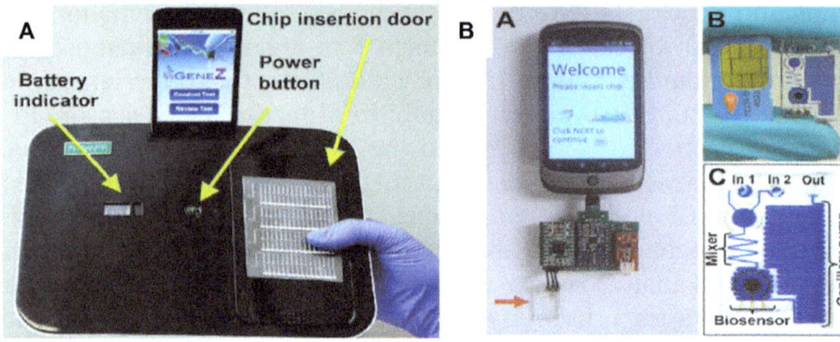

Figure 11.4 Integration of smartphone with chip-based assays for food safety and quality control. Smartphone-based chip sensor for (A) fluorescent detection of *Escherichia coli* and *Staphylococcus aureus* DNA[23] and (B) electrochemical detection of target analytes.[83]
(A) Adapted from ref. 23 with permission from the Royal Society of Chemistry. (B) Adapted from ref. 83 with permission from the Royal Society of Chemistry.

For example, a user-friendly, inexpensive, compact, battery-operated microfluidic chip, namely "Gene-Z", has been introduced for rapid and quantitative detection of genetic markers (*i.e.*, DNA) from *E. coli* and *S. aureus* (Figure 11.4A).[23] The chip consisted of a fluorescence detector and four arrays of fifteen reaction wells, which allowed simultaneous fluorescence analysis of four samples, along with a wireless smartphone device (*i.e.*, iPhone) for automated data collection, fluorescent analysis and reporting *via* a Wi-Fi interface. This platform represents a significant step towards compact smartphone-based tools for POC testing outside the laboratory.

Besides fluorescent detection, most chip-based sensors were designed for electrochemical detection, and are insensitive to ambient illumination conditions.[54] For example, a compact smartphone platform with an integrated chip has been developed for rapid and quantitative electrochemical detection of biomarkers for healthcare, food safety and environmental monitoring (Figure 11.4B).[83] The integrated sensor consisted of a disposable microfluidic chip and an Android smartphone app for signal processing and data analysis. The operating instructions were displayed on the phone to assist the operator through the testing process. The results were then presented on the phone screen for decision-making and were automatically saved for future use. Even though *Plasmodium sp.* was demonstrated as a model analyte for disease diagnosis in this study, the sensor was suggested for future food and water quality monitoring. Unlike the conventional sensor, the proposed smartphone-based sensor allows on-site accurate, quantitative measurement of biomarker concentration. In addition, unlike conventional laboratory techniques (*e.g.*, ELISA) that require 2–3 h for analysis, the proposed assays could be completed in 15 min, which offers immense potential as a POC diagnostic platform, especially in rural areas. Moreover, another study has demonstrated an integrated microfluidic DNA chip platform coupled with a smartphone interface to rapidly detect pathogens (*i.e.*, *E. coli*) with high specificity through electrochemical sensing for water monitoring.[60] The DNA chip was combined with an Android smartphone with smartphone app. The app processed the data and presented the data to the user with a water safety guideline and a graph curve. The app could also keep the previous results by showing a map with blue spots representing "safe water zones" and red spots representing "unsafe water zones". The smartphone-based electrochemical sensor allowed data analysis and interpretation for on-site decision-making based upon water safety guidelines, which could be performed by untrained users, suggesting its potential use for POC water monitoring.

As one of the most common technologies in POC testing, ELISA has been combined with smartphone-based chip sensing for food safety analysis based upon electrochemical analysis. For instance, Pan and coworkers have presented a low cost USB-interfaced smartphone platform for microfluidic chip-based ELISA in detecting food contaminants, such as BDE-47 (2,2′,4,4′-tetrabromodiphenyl ether).[61] The integrated sensor consisted of a microfluidic ELISA chip, micropumps, an Arduino-compatible microcontroller integrated with a printed circuit board and a smartphone. Through the carbon black-polydimethylsiloxane (b-PDMS) composite electrodes, the chip converted the electric current input into a fluidic pump upon the addition of samples. The micropump acted as a liquid flowing controller, which transported the analytes through the microfluidic chip for competitive ELISA. The proposed chip was sensitive for the detection of BDE-47 with a limit of detection down to 10^3–10^4 $\mu g\, l^{-1}$, which was comparable to that of the conventional ELISA. Notably, this sensing design could receive sensing information and control the test procedures, which could significantly improve the automation of ELISA. These advantages allow quantitative POC food and environmental safety monitoring in resource-poor settings.

11.3.3 The Integration of Smartphones with Tube, Microwell or Disk-based Assays

Besides paper-based and chip-based assays, the integration of smartphones into other assays using tubes, microwells and disks has also been reported. For instance, a study has demonstrated the development of a tube-based platform coupled with a smartphone-based digital reader, namely "i-Tube platform", for colorimetric detection of peanut allergens in different food samples (Figure 11.5A).[84] The control and test tubes were placed from the side and illuminated by two light emitting diodes (LEDs). Once the tube was activated, the illumination light was absorbed by the sample, resulting in the change of intensity that was imaged by the smartphone camera. The images were digitally processed using the smartphone app for quantification of food contaminants. Such a cost-effective tool running on a smartphone was useful for food and environmental monitoring in various settings. In another study, a smartphone-based platform has been introduced to quantify Hg^{2+} in a test tube containing water samples with a detection limit of parts per billion (ppb) level.[85] A battery-powered opto-mechanic reader was integrated into the built-in camera module of a smartphone for quantitative colorimetric assays based upon aptamer and plasmonic gold nanoparticles.

A two-colour ratiometric method was employed with LEDs at 523 nm and 625 nm and an Android app was used for image processing. With its portable and cost-effective design as well as wireless data connectivity, this platform could be useful for detecting water contaminants and sharing the contamination information worldwide.

In addition, a smartphone-based chamber has been developed to monitor fluoride in water samples based upon colorimetric detection for food and environmental safety applications (Figure 11.5B).[86] The integrated chamber consisted of a test chamber or tube, filled reagent capsules and a smartphone equipped with digital camera for imaging and an app for recording and analyzing the RGB colour of the image. The fluoride reacted with the commercial zirconium xylenol orange reagent and caused colour changes from pink to yellow as the target concentration increased. Data obtained could be shared *via* an online map, increasing the accessibility of fluoride monitoring in drinking water in remote settings.

Figure 11.5 Integration of smartphones with tube-, microwell- or disk-based assays for food safety and quality control. Smartphones integrated with tubes for colorimetric detection of (A) food allergens, peanuts[84] and (B–C) fluoride in water samples[86,87] and fluorescent detection of (D) bacteria *P. fluorescens M3A.*[52] Smartphone coupled with a microwell for colorimetric detection of (E) mercury ions[88] and (F) food allergens, paralytic shellfish poisons and diarrheic shellfish poisons.[90] (G) A smartphone integrated with a Petri dish for fluorescent detection of microbial spoilage on ground beef.[50] (H) A plastic fibre-based spectrometer platform coupled with a smartphone for dye absorption testing in water for water monitoring.[91]

(A) Adapted from ref. 84 with permission from the Royal Society of Chemistry. (B) Adapted with permission from S. Levin, S. Krishnan, S. Rajkumar, N. Halery and P. Balkunde, *Sci. Total Environ.*, 2016, 551, 101–107 (http://dx.doi.org/10.1016/j.scitotenv.2016.01.156). Originally published in Science of the Total Environment by Elsevier under Creative Commons CC BY License (https://creativecommons.org/licenses/by/4.0/). (C) Reprinted from I. Hussain, K. U. Ahamad and P. Nath, *Anal. Chem.*, 2016, 551, 101–107. Copyright 2016 American Chemical Society. (D) Reprinted with permission from H. Kim, Y. Jung, I.-J. Doh, R. A. Lozano-Mahecha, B. Applegate and E. Bae, *Sci. Rep.*, 2017, 7, 40203. Originally published in *Scientific Reports*, Nature Publishing Group under Creative Commons CC BY License. © The Author(s) 2017. (E) Reprinted with permission from W. Xiao, M. Xiao, Q. Fu, S. Yu, H. Shen, H. Bian and Y. Tang, *Sensors*, 2016, 16, 1871. Originally published in *Sensors*, MDPI under Creative Commons CC BY License (https://creativecommons.org/licenses/by/4.0/). © 2016 by the authors; licensee MDPI, Basel, Switzerland. (F) Reprinted from *Sensors and Actuators B: Chemical*, Volume 238, K. Su, X. Qiu, J. Fang, Q. Zou and P. Wang, An improved efficient biochemical detection method to marine toxins with a smartphone-based portable system—Bionic e-Eye, 1165–1172. Copyright 2017, with permission from Elsevier. (G) Reprinted by permission from Macmillan Publishers Ltd: *Scientific Reports*, (P.-S. Liang, T. San Park and J.-Y. Yoon, Sci. Rep., 2014, 4.), copyright 2014. (H) Reproduced from ref. 91 with permission from the Royal Society of Chemistry.

Another study has also demonstrated robust, portable and low-cost smartphone-based tube-based assays for colorimetric detection of fluoride in drinking water (Figure 11.5C).[87] The LED flashlight and ambient light sensor of the smartphone were utilized as a light source and a light intensity detector, respectively. An Android app was developed to detect and analyze the concentration of fluoride in water samples. Similar to the previously described platform, this app could also be used for data sharing from any remote areas. This platform could be an alternative to the conventional spectrophotometer for accurate and reliable measurement of fluoride level in water samples.

Besides that, smartphones have been coupled with sample tubes for bioluminescence application. For instance, a recent study has demonstrated a smartphone-based bioluminescence detector for quantification of targeted bacteria (Figure 11.5D).[52] The integrated sensing platform consisted of a collection lens, a sample tube and a smartphone to enhance photon-capture efficiency through optimization of hardware (sample chamber that holds the tubes) and software (noise reduction by ensemble averaging (NREA) algorithm). The NREA could reduce the background signal while enhancing the emitted photon signal. In response to the target bacteria *P. fluorescens* M3A (as a model analyte), the platform allowed quantification of the luminescence signal from the sensing element, which emitted photons for detection. This platform has been validated to provide on-site analysis of target analytes for food and environmental monitoring.

Smartphones have also been integrated with microwells for food safety analysis. For example, a study has demonstrated a simple and portable microwell-based colorimetric Hg^{2+} detection system based upon a smartphone equipped with an optical microwell reader app (Figure 11.5E).[88] The platform consisted of a simple light source and a smartphone app, which was used to detect, record and analyze colorimetric signal from the microwell reader. The colorimetric read-out was based upon a specific interaction between Hg^{2+} and aptamer, resulting in a colour change in the solution due to the aggregation of gold nanoparticles. The proposed assay was simple and sensitive, providing new strategies for rapid and on-site assessment of mercury and other heavy metal contaminations in drinking water and/or groundwater. In another study, Xu and coworkers have demonstrated the development of a sensitive and low-cost immunoassay microarray based upon a smartphone to detect various harmful substances in milk.[89] Tetracycline and quinolones, as model analytes, were immobilized in the microwell, into which all the specific antibodies were added. The colorimetric signal of the target was further improved by the silver enhancement technique and the resulting microarray was detected by the smartphone in a mini-cartridge.

Wang and coworkers have demonstrated the development of a smartphone-microwell-based portable system termed bionic electronic eye (Bionic e-Eye) for real time colorimetric analysis of environmental toxins (*e.g.*, paralytic shellfish poisons and diarrheic shellfish poisons), which could cause gastrointestinal food allergy (Figure 11.5F).[90] The Bionic e-Eye integrated image acquisition and data processing by installing the software called i-Plate. With the integration of an ELISA microwell, Bionic e-Eye used HSV and RGB to evaluate its analytical performance in saxitoxin assay and akadaic acid assay for the detection of both toxins, and the result could be shared *via* mobile internet if necessary. It was identified that the precision, sensitivity and dynamic range of this platform were comparable to those of the conventional microplate reader. Therefore, the integrated platform opens up a new opportunity of field-testing of environmental toxins to complement the conventional detection methods.

In addition, a smartphone app coupled with a Petri dish has been developed to detect microbial spoilage on ground beef based upon fluorescence signal analysis (Figure 11.5G).[50] The platform consisted of an 880-nm near infrared LED and a smartphone set containing a software app, a digital camera and a gyro sensor. It was revealed that *E. coli* could change the morphology and size of fat particles in beef, which produced Mie scattering. As a screening tool for food contaminants, the phone could be used to detect and record such changes at different imaging angles. This platform does not require any capturing biomarkers or chemicals, and is inexpensive, handheld and user-friendly, providing a great promise to monitor food contamination.

Other components, such as plastic fibres, have also been integrated into the smartphone-based platform for food safety monitoring. For instance, a compact and ultra-low-cost smartphone spectrometer system coupled with a fluorescent lamp has been developed by Solmaz and coworkers to study dye adsorption in water for water safety management (Figure 11.5H).[91] The plastic fibre-based spectrometer platform was equipped with a smartphone camera, built-in flash and cuvette to study methylene blue absorbance in water using electrospun polyacrylonitrile (PAN)/Zeolite composite nanofibre. The nanofibres have superior mechanical integrity and a higher surface-area-to-volume ratio as compared to other materials. The hybrid nanofibre showed a higher catalytic activity than that of PAN alone because of the presence of zeolite nanoparticles, which enhance surface area and possess high ion exchange capacity, making them highly suitable for adsorption applications. The proposed smartphone-based system is capable of substituting benchtop spectrometers for environmental monitoring.

11.3.4 Smartphone-based Microscopy

A microscope enables the investigation of microorganisms and is broadly used to identify and detect bacteria, cells, viruses and parasites, which cannot be directly visualized by the naked eye.[92,93] A conventional laboratory microscope is expensive and requires highly trained workers, limiting its application for on-site food safety testing, especially in resource-poor settings. To this end, several studies have developed affordable and portable microscopes in combination with smartphones with various lens arrangements for bright field and fluorescence imaging for food safety

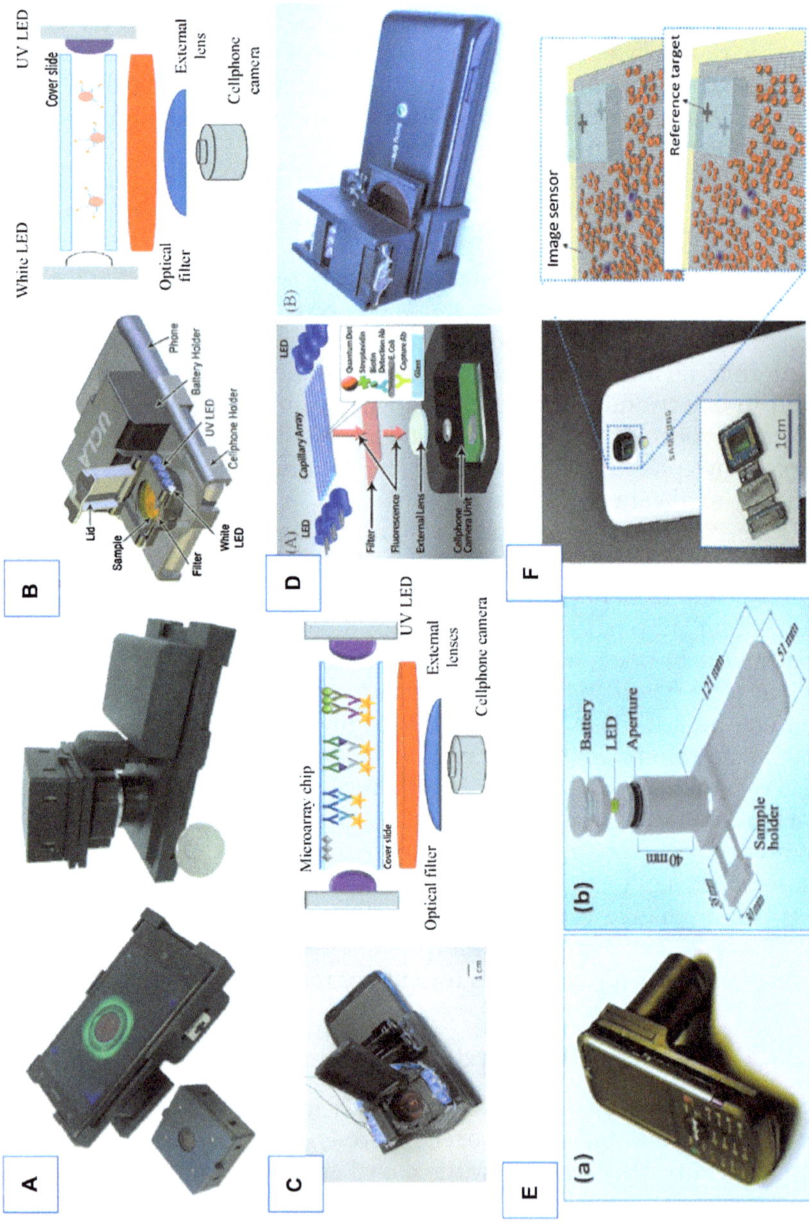
A
B
Lid
Sample
Filter
White LED
UV LED
Cellphone Holder
Battery Holder
Phone
White LED
Cover slide
UV LED
Optical filter
External lens
Cellphone camera
C
Microarray chip
Cover slide
Optical filter
External lenses
Cellphone camera
UV LED
D
(A)
LED
Capillary Array
Filter
Fluorescence
External Lens
Cellphone Camera Unit
(B)
E
(a)
(b)
Battery
LED
Aperture
40 mm
121 mm
51 mm
Sample holder
F
1cm
Image sensor
Reference target

monitoring.[94] Most smartphone-based microscopes are optical microscopes, which are composed of a light source and a lenses system to magnify small objects.[95] In fact, field of view (FOV) and image resolution are two major specifications to evaluate microscopic performance.[36]

To date, much effort has been focused on developing smartphone-based microscopy for POC food safety and environmental testing. For instance, a study has described the detection of common waterborne parasites, *Giardia lamblia* cysts, with a smartphone-based microscope (Figure 11.6A).[26] The platform was composed of a smartphone integrated with an opto-mechanic attachment and a handheld fluorescence microscope with a camera unit to image the water samples in cassettes. The sample cassette consisted of mechanical filter membranes with fixed pore size ~5 μm to avoid the backflow of particles and to capture the fluorescently labelled Giardia cysts. This platform showed a resolution of ~1.12 μm and a large FOV of ~80 mm^2. The images of cysts were captured and transferred to the server using the smartphone through wireless network for target detection and counting, and the results were displayed *via* the custom-developed Windows phone app within 2 min. This field-portable platform allowed analysis of large water volumes (~10–20 mL), which could be beneficial for rapid quantification of various microorganisms for water quality monitoring.

With a similar strategy, a smartphone-based microscope has been developed for detecting food contaminants. A compact 3D printed opto-mechanical module was integrated to detect the presence of anti-recombinant bovine somatotropin (rbST) antibodies in milk (Figure 11.6B),[96] which is a type of illegal proteohormone used to increase milk production. The milk

Figure 11.6 Integration of smartphone with microscope for food safety and quality control. Smartphone-based microscopes for detection of (A) *Giardia lamblia* cysts,[26] (B) anti-recombinant bovine somatotropin (rbST) antibodies[96] and (C) multiple biomarkers (*e.g.*, recombinant bovine somatotropin, insulin-like growth factor *etc.*)[97] in milk. Detection of (D) *E. coli*[27] and (E) *Giardia lamblia*[98] using a smartphone integrated with a microscope. (F) A compact lensless microscope coupled with a smartphone for *Scenedesmus acuminatus* detection in water.[99]

extract was added onto the glass slides, and the target was captured by carboxylated magnetic microsphere, which was labelled by fluorophore quantum dot-labelled goat anti-bovine detection antibody. The fluorescence light was then imaged using the smartphone camera. All microspheres were observed in the dark-field image taken with the smartphone. The images were then analyzed by an Android app running on the same smartphone. The microscope displayed an image resolution of ∼10 μm and a large FOV of ∼80 mm^2. With the proposed cost-effective, handheld, integrated platform, the food safety monitoring process would become more economical and efficient. From the same group, smartphone-based multiplex detection has been demonstrated based upon the same principle (Figure 11.6C).[97] The protein microarray-based fluorescence immunoassay was developed for the detection of multiple biomarkers (*i.e.*, rbST and insulin-like growth factor-1) in milk using specific antibodies for potential on-site food safety monitoring.

Another similar strategy being explored was the utilization of smartphone-based microscopy to detect *E. coli* O157:H7 in liquid samples (Figure 11.6D).[27] Anti-*E. coli* antibody was attached to the glass capillaries as solid substrates to develop a quantum dot-based immunoassay. The fluorescent light emission from the quantum dots was imaged using a smartphone camera and the concentration of *E. coli* O157:H7 was determined. The smartphone-based microscope showed a spatial resolution of 10 μm and a large FOV of ∼81 mm^2. The result indicated that this platform is promising for water and food screening in remote settings. In addition, Ozcan and coworkers have reported a lens-free holographic digital microscope on a smartphone to image various sizes of particles, cells and target waterborne parasites (*e.g.*, *Giardia lamblia*) (Figure 11.6E).[98] The hardware installed on the smartphone was composed of a simple light-emitting diode (LED) for illuminating the samples. This smartphone-based platform showed a spatial resolution of 1.5–2 μm and a large FOV of ∼24 mm^2. Without using any lenses or other optical components, the proposed platform may offer a great potential to address health, food and water safety issues. Collectively, the development of smartphone-based microscopy significantly expands the capability of rapid examination of bacteria, viruses and cells for POC detection of food contamination.

Lee and Yang have demonstrated a compact lensless microscope on a smartphone for healthcare and environmental applications (Figure 11.6F).[99] A microscope was built *via* a simple smartphone camera modification. It was shown that direct shadow images could be achieved even if the users tilted the integrated sensor around an external light source, such as a flashlight, a lamp or sunlight. Image acquisition could be performed using an Android app. Unlike other microscopy strategies, this strategy allowed capturing of samples using ambient illumination as a light source without requiring a dedicated light source. The sample could be imaged at various angles of illumination with a high imaging resolution. As the proof of principle, the integrated system was used to detect *Scenedesmus acuminatus* in water samples with a 500-nm image resolution over a large FOV, demonstrating its promising applications for water monitoring.

11.4 Commercial Smartphone-based Sensors for Potential Food Safety Applications

To date, smartphone-based sensing is still under development. Recently, several new companies have developed smartphone software for POC applications. For example, Holomic LLC has developed smartphone-based platforms for health and environmental management, such as Allergen Tester[84] and Mercury Analyzer.[85] Some companies have developed smartphone software that can read and process the data from a lateral flow test strip. For instance, Mobile Assay Inc. has developed a mobile-based reader based upon Instantaneous Analysis™ software.[38] The app was operated based upon Mobile Image Radiometry (MIR) algorithms created by Cooper and coworker.[100] This app was able to convert the visually detectable colorimetric signal from rapid diagnostic test to a digital signal and interpret the signal by comparing pixel colour, and the data can be stored in the mobile assay secure cloud. It was compatible with Windows, Android and iOS software. Similarly, another company named Navarum has also created smartphone apps, which could read the QR code integrated into the lateral flow cassettes.[101] The apps could interpret the signal by comparing pixel colour.

Some other companies are developing smartphone camera lenses that enhance zoom and image quality, which are similar to benchtop microscopes. For example, the company BLIP Lenses offers cost-effective lenses, which are compatible with any mobile phone. The lenses could be attached onto any phone camera. Even though some commercial smartphone-based sensors have been widely used for medical diagnosis[28,102] (some are not mentioned in the text), they are applicable for the detection of various target analytes, hence may offer tremendous potential for food safety monitoring in the near future.

11.5 Conclusion and Future Perspective

Mobile sensing is a very fast developing field, as reflected by the increasing numbers of studies in the past few years. This chapter presents the recent advances in the development of smartphone-based sensors for food safety and quality control, mostly from 2012 to 2016. The smartphones could be integrated into paper-based sensors, chip-based sensors and tube or microwell assays for the detection of food pathogens, chemicals and toxins mainly based upon colorimetric, fluorescent, luminescent and electrochemical measurement. Some smartphone-based sensors have been commercialized for health, food safety and environmental applications. The smartphone app enables the detection and quantification of target analytes with data transmission *via* wireless network, which could be performed by untrained users. With the cost-effective, quantitative and easy-to-use smartphone, the sensors show great promise for possible development into a stand-alone device for the detection of targeted analytes outside the laboratory, especially in the developing world.

Even though significant progress has been made in this emerging field, there exist some challenges in the applications of sensors on smartphones. For example, the introduction of the sample in the current smartphone-based platform requires external apparatus, such as pumps and pipette tips. Future work should integrate a sample collection device, making it an all-in-one device.[71] Besides that, most sensors require off-chip sample preparation steps, which are highly dependent upon skilful workers. Therefore, it would be ideal to develop a sample-to-answer interface to eliminate the complicated operational steps.[40] Self-powered integrated sensors should be developed to allow on-chip assay automation and fluidic control. Additionally, on-chip reagent storage is important to eradicate the need for a laboratory storage unit (*e.g.*, refrigerator) and the ability of multiplexed detection would significantly increase the assay productivity.[103] Having the ability to monitor on-chip environmental factors (*e.g.*, temperature and relative humidity) would also achieve optimum assay performance even in testing areas with extreme temperature and humidity.[19]

In most resource-poor settings, the wireless network normally suffers from poor connectivity. Therefore, to ensure reliable smartphone-based sensor applications, the platform must be able to support asynchronous data transmission. The integration of alternative power sources, such as solar power and battery technology, would also improve the performance of sensors in rural settings with limited power supply.[85] In the aspect of smartphone modules, a better approach for lighting bias is important to provide a normalization algorithm, which would enable optimum image capturing and processing even under rigorous lighting conditions.[99] In addition, evaluations into the consistency among different smartphone platforms are essential to identify their differences in terms of light sources and/or cameras for analyzing the assays to improve their capability for accurate signal measurement.[36] In short, smartphone-based sensing is a promising field with high potential for scientific research and commercialization. We envision that smartphone-based sensors will be very critical to the monitoring and effective management of food contamination outbreaks in the near future.

References

1. D. G. Nyachuba, *Nutr. Rev.*, 2010, **68**, 257–269.
2. E. Scallan, P. M. Griffin, F. J. Angulo, R. V. Tauxe and R. M. Hoekstra, *Emerging Infect. Dis.*, 2011, **17**, 16.
3. E. Scallan, R. M. Hoekstra, F. J. Angulo, R. V. Tauxe, M. A. Widdowson, S. L. Roy, J. L. Jones and P. M. Griffin, *Emerg Infect. Dis.*, 2011, **17**, 7–15.
4. R. L. Scharff, *J. Food Prot.*, 2012, **75**, 123–131.
5. A. A. Sayad, F. Ibrahim, S. M. Uddin, K. X. Pei, M. S. Mohktar, M. Madou and K. L. Thong, *Sens. Actuators, B*, 2016, **227**, 600–609.

6. M. Uyttendaele, E. Franz and O. Schlüter, *Int. J. Environ. Res. Public Health*, 2015, **13**, 67.
7. B. Miranda, N. M. Lawton, S. R. Tachibana, N. A. Swartz and W. P. Hall, *J. Chem. Educ.*, 2016, **93**, 1770–1775.
8. H. Zhang, Z. Wang and O. Liu, *J. Pharm. Anal.*, 2015, **5**, 223–230.
9. G. Li, S. Liu, Z. Sun, L. Xia, G. Chen and J. You, *Food Chem.*, 2015, **170**, 123–130.
10. Y. T. Atalay, S. Vermeir, D. Witters, N. Vergauwe, B. Verbruggen, P. Verboven, B. M. Nicolaï and J. Lammertyn, *Trends Food Sci. Technol.*, 2011, **22**, 386–404.
11. K. Scida, J. C. Cunningham, C. Renault, I. Richards and R. M. Crooks, *Anal. Chem.*, 2014, **86**, 6501–6507.
12. B. B. Dzantiev, N. A. Byzova, A. E. Urusov and A. V. Zherdev, *TrAC, Trends Anal. Chem.*, 2014, **55**, 81–93.
13. W. Dungchai, O. Chailapakul and C. S. Henry, *Anal. Chim. Acta*, 2010, **674**, 227–233.
14. R. F. Carvalhal, M. Simão Kfouri, M. H. de Oliveira Piazetta, A. L. Gobbi and L. T. Kubota, *Anal. Chem.*, 2010, **82**, 1162–1165.
15. A. Apilux, W. Dungchai, W. Siangproh, N. Praphairaksit, C. S. Henry and O. Chailapakul, *Anal. Chem.*, 2010, **82**, 1727–1732.
16. A. W. Martinez, S. T. Phillips and G. M. Whitesides, *Anal. Chem.*, 2010, **82**, 3–10.
17. Y. Lin, D. Gritsenko, S. Feng, Y. C. Teh, X. Lu and J. Xu, *Biosens. Bioelectron.*, 2016, **83**, 256–266.
18. F. Cimaglia, A. Aliverti, M. Chiesa, P. Poltronieri, E. De Lorenzis, A. Santino and L. A. Sechi, *Nanotechnol. Dev.*, 2012, **2**, e5.
19. J. R. Choi, J. Hu, S. Feng, W. A. B. W. Abas, B. Pingguan-Murphy and F. Xu, *Biosens. Bioelectron.*, 2016, **79**, 98–107.
20. R. Tang, H. Yang, J. R. Choi, Y. Gong, J. Hu, S. Feng, B. Pingguan-Murphy, Q. Mei and F. Xu, *Talanta*, 2016, **152**, 269–276.
21. H. Liu, F. Zhan, F. Liu, M. Zhu, X. Zhou and D. Xing, *Biosens. Bioelectron.*, 2014, **62**, 38–46.
22. N. Lopez-Ruiz, V. F. Curto, M. M. Erenas, F. Benito-Lopez, D. Diamond, A. J. Palma and L. F. Capitan-Vallvey, *Anal. Chem.*, 2014, **86**, 9554–9562.
23. R. D. Stedtfeld, D. M. Tourlousse, G. Seyrig, T. M. Stedtfeld, M. Kronlein, S. Price, F. Ahmad, E. Gulari, J. M. Tiedje and S. A. Hashsham, *Lab Chip*, 2012, **12**, 1454–1462.
24. G.-H. Chen, W.-Y. Chen, Y.-C. Yen, C.-W. Wang, H.-T. Chang and C.-F. Chen, *Anal. Chem.*, 2014, **86**, 6843–6849.
25. C. Sicard, C. Glen, B. Aubie, D. Wallace, S. Jahanshahi-Anbuhi, K. Pennings, G. T. Daigger, R. Pelton, J. D. Brennan and C. D. Filipe, *Water Res.*, 2015, **70**, 360–369.
26. H. C. Koydemir, Z. Gorocs, D. Tseng, B. Cortazar, S. Feng, R. Y. L. Chan, J. Burbano, E. McLeod and A. Ozcan, *Lab Chip*, 2015, **15**, 1284–1293.
27. H. Zhu, U. Sikora and A. Ozcan, *Analyst*, 2012, **137**, 2541–2544.

28. K. Yang, H. Peretz-Soroka, Y. Liu and F. Lin, *Lab Chip*, 2016, **16**, 943–958.
29. D. Gallegos, K. D. Long, H. Yu, P. P. Clark, Y. Lin, S. George, P. Nath and B. T. Cunningham, *Lab Chip*, 2013, **13**, 2124–2132.
30. J. Wojtczak and P. Bonadonna, *Am. J. Emerg. Med.*, 2013, **31**, 573–577.
31. A. W. Martinez, S. T. Phillips, E. Carrilho, S. W. Thomas III, H. Sindi and G. M. Whitesides, *Anal. Chem.*, 2008, **80**, 3699–3707.
32. L. Hjorth and K. Gu, *Continuum*, 2012, **26**, 699–713.
33. S. Muralidharan, C. La Ferle and Y. Sung, *Cyberpsychol. Behav., Soc. Networking*, 2015, **18**, 356–360.
34. Emarketer, Number of smartphone users worldwide from 2014 to 2020., https://http://www.statista.com/statistics/330695/number-of-smartphone-users-worldwide/).
35. D. Lupton and A. Jutel, *Soc. Sci. Med.*, 2015, **133**, 128–135.
36. X. Xu, A. Akay, H. Wei, S. Wang, B. Pingguan-Murphy, B.-E. Erlandsson, X. Li, W. Lee, J. Hu and L. Wang, *Proc. IEEE*, 2015, **103**, 236–247.
37. K. E. McCracken and J.-Y. Yoon, *Anal. Methods*, 2016, **8**, 6591–6601.
38. A. Roda, E. Michelini, M. Zangheri, M. Di Fusco, D. Calabria and P. Simoni, *TrAC, Trends Anal. Chem.*, 2016, **79**, 317–325.
39. J. R. Choi, K. W. Yong, R. Tang, Y. Gong, T. Wen, H. Yang, A. Li, Y. C. Chia, B. Pingguan-Murphy and F. Xu, *Adv. Healthcare Mater.*, 2017, **6**, 1600920.
40. R. H. Tang, H. Yang, J. R. Choi, Y. Gong, S. S. Feng, B. Pingguan-Murphy, Q. S. Huang, J. L. Shi, Q. B. Mei and F. Xu, *Crit. Rev. Biotechnol.*, 2016, 1–18.
41. S. Feng, J. R. Choi, T. J. Lu and F. Xu, *Adv. Porous Flow*, 2015, **5**, 16–29.
42. H. Zhang, L. Ma, L. Ma, M. Z. Hua, S. Wang and X. Lu, *Int. J. Food Microbiol.*, 2016, **243**, 64–69.
43. D. Vilela, M. C. González and A. Escarpa, *Anal. Chim. Acta*, 2012, **751**, 24–43.
44. J. R. Choi, Z. Liu, J. Hu, R. Tang, Y. Gong, S. Feng, H. Ren, T. Wen, H. Yang and Z. Qu, *Anal. Chem.*, 2016, **88**, 6254–6264.
45. J. R. Choi, J. Hu, R. Tang, Y. Gong, S. Feng, H. Ren, T. Wen, X. Li, W. A. B. W. Abas and B. Pingguan-Murphy, *Lab Chip*, 2016, **16**, 611–621.
46. H. S. Jung, X. Chen, J. S. Kim and J. Yoon, *Chem. Soc. Rev.*, 2013, **42**, 6019–6031.
47. A. M. Nicolini, C. F. Fronczek and J.-Y. Yoon, *Biosens. Bioelectron.*, 2015, **67**, 560–569.
48. E. Petryayeva and W. R. Algar, *Analyst*, 2015, **140**, 4037–4045.
49. C. F. Fronczek, T. San Park, D. K. Harshman, A. M. Nicolini and J.-Y. Yoon, *RSC Adv.*, 2014, **4**, 11103–11110.
50. P.-S. Liang, T. San Park and J.-Y. Yoon, *Sci. Rep.*, 2014, **4**, 5953.
51. E. Lebiga, R. E. Fernandez and A. Beskok, *Analyst*, 2015, **140**, 5006–5011.
52. H. Kim, Y. Jung, I.-J. Doh, R. A. Lozano-Mahecha, B. Applegate and E. Bae, *Sci. Rep.*, 2017, **7**, 40203.

53. Q. Mei, H. Jing, Y. Li, W. Yisibashaer, J. Chen, B. N. Li and Y. Zhang, *Biosens. Bioelectron.*, 2016, **75**, 427–432.
54. Y. Xia, J. Si and Z. Li, *Biosens. Bioelectron.*, 2016, **77**, 774–789.
55. E. Silva, M. Santhiago, J. Barragan and L. Kubota, *Anal. Methods*, 2014, **6**, 6133–6136.
56. D. Zhang and Q. Liu, *Biosens. Bioelectron.*, 2016, **75**, 273–284.
57. E. Eltzov, S. Guttel, A. Low Yuen Kei, P. D. Sinawang, R. E. Ionescu and R. S. Marks, *Electroanalysis*, 2015, **27**, 2116–2130.
58. A. J. Bandodkar and J. Wang, *Trends Biotechnol.*, 2014, **32**, 363–371.
59. E. W. Nery and L. T. Kubota, *Anal. Bioanal. Chem.*, 2013, **405**, 7573–7595.
60. U. Kim, S. Ghanbari, A. Ravikumar, J. Seubert and S. Figueira, *IEEE J. Transl. Eng. Health Med.*, 2013, **1**, 3700207.
61. A. Chen, R. Wang, C. R. Bever, S. Xing, B. D. Hammock and T. Pan, *Biomicrofluidics*, 2014, **8**, 064101.
62. J. McEntire, *Infect. Dis. Clin. North Am.*, 2013, **27**, 687–693.
63. A. H. Havelaar, M. D. Kirk, P. R. Torgerson, H. J. Gibb, T. Hald, R. J. Lake, N. Praet, D. C. Bellinger, N. R. De Silva and N. Gargouri, *PLoS Med.*, 2015, **12**, e1001923.
64. M. Iturriza-Gomara and S. J. O'Brien, *Curr. Opin. Infect. Dis.*, 2016, **29**, 495–501.
65. N. H. A. Barudin, S. Sreekantan, M. T. Ong and C. W. Lai, *Food Control*, 2014, **46**, 0e487.
66. I.-H. Cho and J. Irudayaraj, *Int. J. Food Microbiol.*, 2013, **164**, 70–75.
67. W. Ahmed, H. Brandes, P. Gyawali, J. Sidhu and S. Toze, *Water Res.*, 2014, **53**, 361–369.
68. J. R. Choi, J. Hu, S. Wang, H. Yang, W. A. B. Wan Abas, B. Pingguan-Murphy and F. Xu, *Crit. Rev. Biotechnol.*, 2016, 1–12.
69. L.-D. Dinu and S. Bach, *Food Control*, 2013, **31**, 268–273.
70. J. RuáChoi and W. A. B. Abas, *Analyst*, 2016, **141**, 2930–2939.
71. J. R. Choi, R. Tang, S. Wang, W. A. B. W. Abas, B. Pingguan-Murphy and F. Xu, *Biosens. Bioelectron.*, 2015, **74**, 427–439.
72. X. Cui, J. Hu, J. R. Choi, Y. Huang, X. Wang, T. J. Lu and F. Xu, *Anal. Chim. Acta*, 2016, **935**, 207–212.
73. G. Bülbül, A. Hayat and S. Andreescu, *Sensors*, 2015, **15**, 30736–30758.
74. B. Bisha, J. A. Adkins, J. C. Jokerst, J. C. Chandler, A. Pérez-Méndez, S. M. Coleman, A. O. Sbodio, T. V. Suslow, M. D. Danyluk and C. S. Henry, *J. Visualized Exp.*, 2014, **88**, e51414.
75. W. Chen, X. Fang, H. Li, H. Cao and J. Kong, *Sci. Rep.*, 2016, **6**, DOI: 10.1038/srep31948.
76. L. S. A. Busa, S. Mohammadi, M. Maeki, A. Ishida, H. Tani and M. Tokeshi, *Micromachines*, 2016, **7**, 86.
77. Y. Jiang, S. Zou and X. Cao, *Anal. Methods*, 2016, **8**, 6668–6681.
78. L. Yu, Z. Shi, C. Fang, Y. Zhang, Y. Liu and C. Li, *Biosens. Bioelectron.*, 2015, **69**, 307–315.

79. W. C. Mak, V. Beni and A. P. Turner, *TrAC, Trends Anal. Chem.*, 2016, **79**, 297–305.
80. T. San Park, W. Li, K. E. McCracken and J.-Y. Yoon, *Lab Chip*, 2013, **13**, 4832–4840.
81. T. San Park and J.-Y. Yoon, *IEEE Sens. J.*, 2015, **15**, 1902–1907.
82. W. Su, X. Gao, L. Jiang and J. Qin, *J. Chromatogr. A*, 2015, **1377**, 13–26.
83. P. B. Lillehoj, M.-C. Huang, N. Truong and C.-M. Ho, *Lab Chip*, 2013, **13**, 2950–2955.
84. A. F. Coskun, J. Wong, D. Khodadadi, R. Nagi, A. Tey and A. Ozcan, *Lab Chip*, 2013, **13**, 636–640.
85. Q. Wei, R. Nagi, K. Sadeghi, S. Feng, E. Yan, S. J. Ki, R. Caire, D. Tseng and A. Ozcan, *ACS Nano*, 2014, **8**, 1121–1129.
86. S. Levin, S. Krishnan, S. Rajkumar, N. Halery and P. Balkunde, *Sci. Total Environ.*, 2016, **551**, 101–107.
87. I. Hussain, K. U. Ahamad and P. Nath, *Anal. Chem.*, 2016, **89**, 767–775.
88. W. Xiao, M. Xiao, Q. Fu, S. Yu, H. Shen, H. Bian and Y. Tang, *Sensors*, 2016, **16**, 1871.
89. Z. Li, Z. Li, D. Zhao, F. Wen, J. Jiang and D. Xu, *Biosens. Bioelectron.*, 2017, **87**, 874–880.
90. K. Su, X. Qiu, J. Fang, Q. Zou and P. Wang, *Sens. Actuators, B*, 2016, **238**, 1165–1172.
91. G. Kocakusak, A. Bayram, V. Kilic, N. Horzum and M. Solmaz, *Anal. Methods*, 2016, **9**, 579–585.
92. S. Choi, *Biotechnol. Adv.*, 2016, **34**, 321–330.
93. X. Liu, T.-Y. Lin and P. B. Lillehoj, *Ann. Biomed. Eng.*, 2014, **42**, 2205–2217.
94. M. Bates and A. Zumla, *Ann. Transl. Med.*, 2015, **3**.
95. A. Romeo, T. Leung and S. Sánchez, *Lab Chip*, 2016, **16**, 1957–1961.
96. S. K. Ludwig, H. Zhu, S. Phillips, A. Shiledar, S. Feng, D. Tseng, L. A. van Ginkel, M. W. Nielen and A. Ozcan, *Anal. Bioanal. Chem.*, 2014, **406**, 6857–6866.
97. S. K. Ludwig, C. Tokarski, S. N. Lang, L. A. van Ginkel, H. Zhu, A. Ozcan and M. W. Nielen, *PLoS One*, 2015, **10**, e0134360.
98. D. Tseng, O. Mudanyali, C. Oztoprak, S. O. Isikman, I. Sencan, O. Yaglidere and A. Ozcan, *Lab Chip*, 2010, **10**, 1787–1792.
99. S. A. Lee and C. Yang, *Lab Chip*, 2014, **14**, 3056–3063.
100. J. J. Li, A. L. Ouellette, L. Giovangrandi, D. E. Cooper, A. J. Ricco and G. T. Kovacs, *IEEE Trans. Biomed. Eng.*, 2008, **55**, 1560–1571.
101. D. Quesada-González and A. Merkoçi, *Biosens. Bioelectron.*, 2016, **92**, 549–562.
102. A. F. Coskun, R. Nagi, K. Sadeghi, S. Phillips and A. Ozcan, *Lab Chip*, 2013, **13**, 4231–4238.
103. A. Govindarajan, S. Ramachandran, G. Vigil, P. Yager and K. Böhringer, *Lab Chip*, 2012, **12**, 174–181.

Subject Index